Control and Protection of 100% Inverter-based Power Systems

Florian Mahr

Control and Protection of 100% Inverter-based Power Systems

Florian Mahr
Berlin, Germany

Als Dissertation genehmigt von der Technischen Fakultät der Friedrich-Alexander-Universität Erlangen-Nürnberg
Tag der mündlichen Prüfung: 26. April 2024
Gutachter: Prof. Dr.-Ing. Johann Jäger
Em. Univ.-Prof. Dipl.-Ing. Dr. techn. Lothar Fickert

ISBN 978-3-658-47216-0   ISBN 978-3-658-47217-7 (eBook)
https://doi.org/10.1007/978-3-658-47217-7

This work was supported by author.

© The Editor(s) (if applicable) and The Author(s) 2025. This book is an open access publication.

Open Access This book is licensed under the terms of the Creative Commons Attribution 4.0 International License (http://creativecommons.org/licenses/by/4.0/), which permits use, sharing, adaptation, distribution and reproduction in any medium or format, as long as you give appropriate credit to the original author(s) and the source, provide a link to the Creative Commons license and indicate if changes were made.

The images or other third party material in this book are included in the book's Creative Commons license, unless indicated otherwise in a credit line to the material. If material is not included in the book's Creative Commons license and your intended use is not permitted by statutory regulation or exceeds the permitted use, you will need to obtain permission directly from the copyright holder.

The use of general descriptive names, registered names, trademarks, service marks, etc. in this publication does not imply, even in the absence of a specific statement, that such names are exempt from the relevant protective laws and regulations and therefore free for general use.

The publisher, the authors and the editors are safe to assume that the advice and information in this book are believed to be true and accurate at the date of publication. Neither the publisher nor the authors or the editors give a warranty, expressed or implied, with respect to the material contained herein or for any errors or omissions that may have been made. The publisher remains neutral with regard to jurisdictional claims in published maps and institutional affiliations.

This Springer Vieweg imprint is published by the registered company Springer Fachmedien Wiesbaden GmbH, part of Springer Nature.
The registered company address is: Abraham-Lincoln-Str. 46, 65189 Wiesbaden, Germany

If disposing of this product, please recycle the paper.

Abstract

Energy systems are in transition. Sustainable and reliable energy systems are cornerstones of human societies. Renewable energy sources (e.g. wind power plants and photovoltaic power plants) and innovative consumer technologies (e.g. electric vehicles and heat pumps) as well as necessary energy storage devices are connected via power electronic components to the electrical grid. The penetration of the grid by voltage source converters (VSCs) as one type of power electronic components will increase. 100% VSC-based power systems will arise. Protective systems against grid faults are a substantial part of electrical grids. They prevent danger to living beings and damage to technical equipment.

VSCs and protection must collaborate in future grids. Otherwise, the reliability of energy supply is at risk. The characteristics of VSCs differ from the characteristics of synchronous machines used in conventional power plants. Today, protection algorithms are often not adapted to the characteristics of VSCs. At the same time, the flexibility options of VSCs to increase the reliability of the energy supply are not fully utilized today. VSCs and protection algorithms are typically regarded separately from each other. State-of-the-art approaches are often unilateral. One approach is to control VSCs in a way to adopt the characteristics of synchronous machines. Here, the possibilities of VSCs to increase the system resilience against grid faults are restricted. Another approach is to use adaptive distance protection algorithms. Here, the settings depend on the situation and represent only a selective solution.

In this thesis, an alternative approach is presented. VSCs and protection algorithms are not regarded independently of each other. A holistic principle is developed. The key aspects of this thesis to realize sustainable and reliable energy systems are:

- Post-fault characteristics of VSCs
- Neutral point treatment and resonant grounding via VSCs
- Model-based protection algorithm

All control and protection algorithms in this thesis are developed analytically. They are validated in software simulations. In addition, the model-based protection algorithm is validated in real hardware laboratory tests.

The developed control algorithms of VSCs and protection algorithms for 100% VSC-based power systems show significant advantages over existing approaches. Realizing desired properties of frequency and active power in the positive sequence after a grid fault leads to a stable grid behaviour with defined power flows. The realized characteristics in the zero sequence reduce fault currents and allow continuing the grid operation after single-phase-to-ground faults. The model-based protection algorithm is suitable for VSC characteristics and does not restrict flexibility options of VSCs to increase the system resilience against grid faults. Overall, this thesis contributes to the realization of sustainable and reliable energy systems.

Zusammenfassung

Unsere Energiesysteme befinden sich im Wandel. Dabei sind nachhaltige und zuverlässige Energiesysteme Grundpfeiler unserer Gesellschaft. Regenerative Energiesysteme (z. B. Windkraftanlagen und Photovoltaikanlagen), innovative Verbrauchertechnologien (z. B. Elektrofahrzeuge und Wärmepumpen) sowie Energiespeicher sind über leistungselektronische Komponenten als Bindeglieder an das Stromnetz angeschlossen. Stromrichter mit Gleichspannungszwischenkreis stellen eine Klasse leistungselektronischer Komponenten dar. Ihre Anzahl im Stromnetz wird zunehmen und 100%-stromrichterbasierte Energiesysteme werden entstehen. Ferner sind Schutzsysteme ein wesentlicher Bestandteil elektrischer Netze. Sie reduzieren die Gefahr für Lebewesen und Schäden an technischen Anlagen infolge von Netzfehlern.

Sicherlich müssen Stromrichter und Schutzsysteme in zukünftigen Stromnetzen miteinander funktionieren – andernfalls kann die Zuverlässigkeit der Energieversorgung nicht aufrechterhalten werden. Die Eigenschaften von Stromrichtern und Synchronmaschinen unterscheiden sich signifikant voneinander, wobei letztere in konventionellen Kraftwerken eingesetzt werden. Heutzutage sind Schutzalgorithmen an Synchronmaschinen und oft nicht an Stromrichter angepasst. Gleichzeitig warden deren Flexibilitätsmöglichkeiten zur Erhöhung der Zuverlässigkeit der Energieversorgung bislang nicht vollständig ausgeschöpft. Stromrichter und Schutzsysteme werden oft getrennt voneinander betrachtet und Lösungsansätze entstehen häufig aus einer einseitigen Perspektive. Ein Ansatz beinhaltet die Nachbildung der Eigenschaften von Synchronmaschinen mittels Stromrichtern. Gleichwohl sind hierbei die Möglichkeiten von Stromrichtern, die die Resilienz des Systems gegenüber Netzfehlern erhöhen, eingeschränkt.

Der Einsatz adaptiver Distanzschutzalgorithmen stellt einen anderen bekannten Ansatz dar. Aufgrund der situationsabhängigen Schutzeinstellungen werden lediglich punktuelle Lösungen gefunden.

In dieser Arbeit wird ein alternativer Lösungsansatz vorgestellt, wobei Stromrichter und Schutzalgorithmen nicht getrennt voneinander betrachtet werden. Dadurch wird ein ganzheitliches Prinzip etabliert. Die Schlüsselaspekte dieser Arbeit zur Realisierung nachhaltiger und zuverlässiger Energiesysteme sind:

- Charakteristiken von Stromrichtern nach Netzfehlern
- Sternpunktbehandlung und Resonanzsternpunkterdung mittels Stromrichtern
- Modellbasierter Schutzalgorithmus

Sowohl die entwickelten Regelungsalgorithmen von Stromrichtern als auch die Schutzalgorithmen in dieser Arbeit wurden analytisch entwickelt sowie in Softwaresimulationen getestet. Darüber hinaus wurde der modellbasierte Schutzalgorithmus in realen Hardware-Labortests validiert.

Das Zusammenwirken der entwickelten Regelungs- und Schutzalgorithmen führt zu erheblichen Vorteilen gegenüber bestehenden Lösungsansätzen in 100%-strom-richterbasierten Energiesystemen. Die Realisierung gewünschter Charakteristika von Frequenz und Wirkleistung im Mitsystem nach einem Netzfehler führt zu einem stabilen Netzverhalten mit definierten Leistungsflüssen. Die Eigenschaften des Nullsystems führen zu reduzierten Fehlerströmen und ermöglichen daher die Fortführung des Netzbetriebs nach einphasigen Leiter-Erde-Fehlern. Der modellbasierte Schutzalgorithmus ist für den Einsatz in 100%-stromrichterbasierten Energiesystemen ausgezeichnet geeignet und schränkt die Flexibilitätsmöglichkeiten von Stromrichtern zur Erhöhung der Resilienz des Systems gegenüber Netzfehlern nicht ein. Insgesamt trägt diese Arbeit zur Realisierung nachhaltiger und zuverlässiger Energiesysteme bei.

Reading guide

Depending on available time, this thesis opens various doors to gain information. Away from reading the complete thesis, three different ways for time-saving information gathering are provided.

The abstract gives a brief summary of the background, challenge, approach and results of this thesis. The research context reveals.

Combining Sect. 1.2 and Sect. 11.1 provides an extended summary. Research questions are formulated. An overview of the state of the art and the research objectives as well as the discussion of results as the fulfilment of these objectives is given.

Key aspects of this thesis and technical interconnections are provided by combining the introduction in Chap. 1, the conclusion and the research perspectives in Chap. 11 and the summaries of all chapters starting with Sect. 1.4. The background, motivation and discussion of the interaction of voltage source converters and grid protection are continued along the central theme of this thesis.

A deep dive into the control and protection of 100% inverter-based power systems and understanding details is possible by reading the complete thesis. The table of contents and the outline in Fig. 1.2 help to navigate along this way.

The visualization of this reading guide is provided in Fig. 1. In this figure, all chapters and sections are linked, providing the possibility for quick navigation in the digital document.

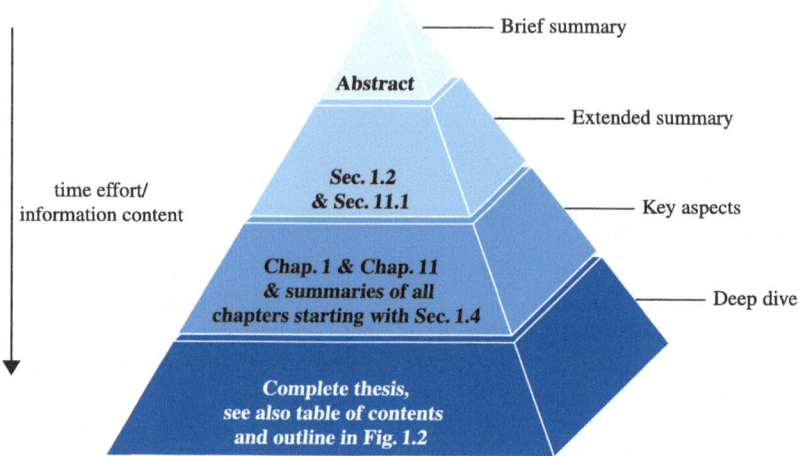

Fig. 1 Visualization of the reading guide of this thesis with the possibility for quick navigation in the digital document

Contents

1	**Introduction**		1
	1.1 Background and Motivation		1
		1.1.1 Energy Systems in Transition	1
		1.1.2 Power Electronics as Key Technology	2
		1.1.3 Grid Architectures for the Integration of RES	4
		1.1.4 Future Grid Operation Strategies	5
	1.2 Interaction of VSCs and Grid Protection		6
		1.2.1 Research Questions	6
		1.2.2 State of the Art	8
		1.2.3 Research Objectives	10
	1.3 Outline of this Thesis		11
	1.4 Summary		13
2	**Electrical Energy Supply in 100% IBPS**		15
	2.1 Fundamentals of VSCs		15
		2.1.1 Half-bridge Topology	16
		2.1.2 Three-phase Bridge Topology	19
	2.2 Frequency Characteristics of 100% IBPS		20
		2.2.1 Interpretation of Frequency	20
		2.2.2 Requirements on the Properties of VSCs	22
		2.2.3 Grid Control	24
	2.3 Requirements on VSC Control and Grid Protection		28
	2.4 Research Approach		29
	2.5 Summary		30

3	Electrical quantities, state-space modelling and system properties		33
	3.1	Description of Electrical Quantities	33
		3.1.1 Classification of Types of Description	33
		3.1.2 Description as Complex Numbers	35
		3.1.3 General Representation of a Three-phase Four-wire System	43
	3.2	State-space Models	45
		3.2.1 State Variables and State-Space Representation	45
		3.2.2 State-space Models of Selected Circuit Structures	47
		3.2.3 Discretization of State-space Models	52
	3.3	System Properties in the State-space	54
		3.3.1 Solution of the System of State Differential Equations	54
		3.3.2 Asymptotic Stability	55
		3.3.3 Controllability	56
		3.3.4 Observability	59
		3.3.5 Duality of Controllability and Observability	62
	3.4	Summary	62
4	Synchronization of VSCs to the Grid		65
	4.1	Grid Connection of a Voltage Source Converter	65
		4.1.1 Grid Equivalent Circuit	65
		4.1.2 Energy Transfer between the Grid and the VSC	68
	4.2	Synchronization Principles	75
	4.3	Voltage-based Synchronization	77
		4.3.1 PLL-controller	77
		4.3.2 SRF-PLL	78
		4.3.3 DSRF-PLL	79
		4.3.4 MSHDC-PLL	80
	4.4	Power-based Synchronization	84
		4.4.1 Active Power Synchronization	85
		4.4.2 Reactive Power Synchronization	90
		4.4.3 Small-signal Power Synchronization Stability	91
		4.4.4 Scenario: Different CND-control Parameters	92
		4.4.5 Scenario: Two GFM-VSCs and their Interactions	93
	4.5	Hybrid Synchronization	97
		4.5.1 Principle of the Hybrid Synchronization	97

		4.5.2	Scenario: Grid Behaviour of a GFM-VSC and a H-VSC	98
	4.6	Summary		100
5	**Grid-side Characteristics and V/I-Control of VSCs**			103
	5.1	Hierarchical Control Structure and Functional Chain		103
	5.2	Calculation of Reference Variables		106
		5.2.1	Symmetrical Components Equivalent Circuit and Overview of Grid-Side Characteristics	106
		5.2.2	Design of Voltage Sources	110
		5.2.3	Design of Current Sources	114
		5.2.4	Realization of V/I-Characteristics	123
	5.3	Design of the State-Feedback V/I-Controller		127
		5.3.1	Derivation of the Control Law and Analysis of the Closed Control Loop	128
		5.3.2	Feedforward Control of Reference Variables	130
		5.3.3	Feedforward Control of Disturbance Variables	131
		5.3.4	State-Feedback Controller Based on the LQR-Algorithm	132
		5.3.5	Implementation of Integral Characteristics	135
	5.4	Summary		139
6	**Enhanced Grid Fault Characteristics of VSCs**			141
	6.1	Resilience of 100% IBPS Against Grid Faults		141
	6.2	Resilience of the Grid Operation		143
		6.2.1	Hardware Configuration	144
		6.2.2	Software Configuration in the Zero Sequence	150
		6.2.3	Resonant Grounding System Mode	155
	6.3	Resilience of the VSC		169
		6.3.1	Transient State Current Limiting	169
		6.3.2	Steady State Current Limiting	180
	6.4	Summary		185
7	**Grid Fault Modelling**			189
	7.1	Model-Based Fault Analysis		189
	7.2	State-Space Models of Selected Grid Topologies		192
		7.2.1	Single Π-Section Line	192
		7.2.2	Combination of Multiple Π-Section Lines	195
	7.3	Modelling of Faults		197
		7.3.1	Transverse Faults	198

		7.3.2 Longitudinal Faults	199
	7.4	State-Space Models of Lines	200
		7.4.1 Model of a Healthy Line	201
		7.4.2 Universal Representation of Faulty Lines	202
		7.4.3 Three-Phase-to-Ground Fault	204
		7.4.4 Three-Phase Break	205
	7.5	Summary	206
8	**Model-Based Protection Algorithm**		**209**
	8.1	Model-Based Fault Analysis Based on Estimated Values	209
	8.2	State Estimation	210
		8.2.1 State Observer	210
		8.2.2 Kalman Filter	213
	8.3	Protection Criterion Based on Hypothesis Testing	218
	8.4	Model-Based Protection Scheme	222
		8.4.1 Basic Structure and Properties	222
		8.4.2 Fault Identification	225
		8.4.3 Fault Characterization	227
		8.4.4 Fault Localization	229
		8.4.5 Distinction From Alternative Protection Principles	231
	8.5	Summary	233
9	**Validation of the MBP by Laboratory Tests**		**235**
	9.1	Laboratory Environment and Workflow	235
	9.2	Scenario 1: Double-sided Infeed	239
	9.3	Scenario 2: Double-sided Infeed with Fault Localization	245
	9.4	Scenario 3: Parallel Line Topology	248
	9.5	Scenario 4: Intermediate Infeed Topology	252
	9.6	Summary	257
10	**Model-based Protection Scenarios in 100% IBPS**		**261**
	10.1	VSC-specific Fault Characteristics and Simulation Scenarios	261
	10.2	Scenario 5: 3p-fault and Voltage Reduction	263
	10.3	Scenario 6: 2p-fault and Reactive Current Injection	269
	10.4	Scenario 7: 1p-fault and RGS-mode	273
	10.5	Scenario 8: 1p-fault and Transient Current Limiting	278
	10.6	Summary	281

11	**Conclusion and Research Perspectives**	285
	11.1 Conclusion	285
	11.2 Research Perspectives	289

References 291

Abbreviations

AC	alternating current
AI	artificial intelligence
AT	actuator tracking
BESS	battery energy storage system
BIBO	bounded-input bounded-output
CDF	cumulative distribution function
CIG	converter interfaced generation
CL	current limiting
CM	common mode
CND	configurable natural droop
CT	current transformer
DAB	dual active bridge
DC	direct current
DER	distributed energy resource
DM	differential mode
DSE	dynamic state estimation
DSRF	double synchronous reference frame
EKF	extended Kalman filter
ES	energy storage
ESS	energy storage system
EV	electric vehicle
FC	fault characterization
FCS	finite control set
FI	fault identification

FL	fault localization
GC	grid-connected
GFL	grid-following
GFM	grid-forming
GPS	Global Positioning System
GSC	grid-side converter
GaN	gallium nitride
HESS	hybrid energy storage system
HIL	hardware-in-the-loop
HP	heat pump
HPF	high-pass filter
HVDC	high voltage direct current
IBPS	inverter-based power system
IBR	inverter-based resource
IGBT	insulated-gate bipolar transistor
IIDG	inverter-interfaced distributed generator
KCL	Kirchhoff's current law
KF	Kalman filter
KVL	Kirchhoff's voltage law
LOHC	liquid organic hydrogen carrier
LPF	low-pass filter
LQR	linear-quadratic regulator
MBP	model-based protection
MIMO	multiple-input multiple-output
MMC	modular multilevel converter
MOSFET	metal-oxide semiconductor field-effect transistor
MP	model prediction
MPC	model predictive control
MSHDC	multi-sequence harmonic decoupling cell
NPC	neutral point clamped
NPT	neutral point treatment
OBP	observer-based protection
PC	primary control
PDF	probability density function
PI	proportional-integral
PLL	phase-locked loop
PV	photovoltaics
PWM	pulse-width modulation
RES	renewable energy source

RGS	resonant grounding system
RMS	root-mean-square
RoCoF	rate of change of frequency
SC	secondary control
SCR	short-circuit ratio
SG	synchronous generator
SGS	solid grounding system
SLP	settingless protection
SM	synchronous machine
SO	second order
SOC	state of charge
SRF	synchronous reference frame
SST	solid-state transformer
SVM	space vector modulation
SiC	silicon carbide
TC	tertiary control
UGS	ungrounded system
UKF	unscented Kalman filter
V/I	voltage/current
VA	virtual admittance
VI	virtual impedance
VPP	virtual power plant
VSC	voltage source converter
VT	voltage transformer
WBG	wide-bandgap
WPP	wind power plant
WT	wind turbine
ZCS	zero current switching
ZOH	zero-order hold
ZVS	zero voltage switching

Introduction 1

This chapter opens the thesis. In Sect. 1.1, the background and the motivation of this thesis regarding energy systems in transition and the role of power electronics as the key technology is presented. Grid architectures for the integration of renewable energy sources (RES) and future grid operation strategies are explained. Sect. 1.2 introduces the main topic of this thesis: The interaction between voltage source converters (VSCs) and grid protection. Open research questions concerning this topic are formulated as the technical challenge. State-of-the-art solutions and their drawbacks are summarized. Based on this perspective, the research objectives of this thesis are formulated. In Sect. 1.3, a visual outline of this thesis is given.

1.1 Background and Motivation

1.1.1 Energy Systems in Transition

The imperative of the reduction of CO_2-emissions requires the use of renewable energy sources (RES) and innovative consumer technologies.

Fundamental properties of RES are summarized in [1]. RES are inexhaustible on a human timescale and represent distributed energy resources (DERs). The available power is location-dependent and volatile over time. In this context, the availability of wind power plants (WPPs) and photovoltaic (PV) power plants can complement each other in time. For the time-independent power consumption, the use of energy storage systems (ESS) is essential.

There are different energy storage technologies [2]. ESS can decouple the power flow between the electrical power grid and the RES or the consumer. The use of distributed ESS near the RES and consumers is preferable to keep their power volatility and power peaks out of the grid.

Electric vehicles (EVs) and heat pumps (HPs) are levers to reduce the primary energy consumption at the consumer side. The efficiency of EVs regarding the primary energy consumption is increased by a factor of approximately 3 to 4 compared to conventional vehicles based on combustion technologies [3]. The efficiency of HPs regarding the primary energy consumption indicated by the coefficient of performance is also increased by a factor of approximately 3 to 4 compared to conventional heating systems [4].

In the past, the energy transport of the mobility sector and the heating sector was realized via chemical energy carriers. In the future, the energy transport of these sectors will be realized via the electrical power grid. The coupling of the electricity sector, the mobility sector and the heating sector will be a reality.

1.1.2 Power Electronics as Key Technology

Power electronic components are converters of electrical power without energy storage capability based on power semiconductor devices [143]. Power electronic converters, which connect an alternating current (AC) system and a direct current (DC) system, are categorized into self-commutated and line-commutated converters [5]. Subsequently, these converters are called AC/DC-converters.

Voltage source converters (VSCs) are self-commutated converters based on an approximately constant voltage DC-link. They allow defined power flows from the AC-system to the DC-system and vice versa. Consequently, VSCs are operated in the rectifier mode or the inverter mode [5]. The V/I-characteristics depend on their control algorithm and thus they represent software-defined components. In this thesis, VSCs are regarded exclusively, because of their technical advantages over alternative converter technologies and their wide application range.

There is a wide application range of VSCs connected to the AC power grid, see Fig. 1.1. These grid-connected (GC) VSCs integrate RES to the grid [6]. They are used as inverters in PV power plants [7]. Furthermore, DC/DC-converters are applied to realize maximum power point (MPP) tracking [8]. GC VSCs are also used in WPPs (see [9]) and hydro power plants (see [10, 11]) to allow an optimal operation by decoupling the rotation speed of electrical machines from the electrical grid frequency.

Beyond that, different ESS are integrated by GC VSCs into the grid. The application range of VSCs extends from battery energy storage systems (BESS) (see [6]) to pumped hydroelectric energy storage (see [12]), fuel cell applications (see [13]) and hydrogen production via water electrolysis (see [14]).

1.1 Background and Motivation

On the consumer side, VSCs allow bidirectional charging of EVs (see [3, 15]). Moreover, VSCs are used in inverter HPs to adjust power via rotation speed control.

Future VSC technologies will profit from material developments. Wide-bandgap (WBG) semiconductors, such as silicon carbide (SiC) and gallium nitride (GaN), allow power semiconductor devices with higher voltage ratings and higher switching frequency. The beneficial properties of SiC materials and the applications of SiC devices are presented in [16]. Zero voltage switching (ZVS) and zero current switching (ZCS) methods can reduce or even avoid switching losses [5, 17].

Further trends involve hardware topologies of VSCs. Switch-based multilevel topologies (see [18, 19]) and modular multilevel topologies enable the application of VSCs in medium and high voltage grids. Different categories of modular multilevel AC/DC topologies are presented in [20, 21]. Energy balancing in the modular multilevel converter (MMC) under unbalanced grid conditions is presented in [144]. The multidirectional energy flow control via a MMC with integrated battery energy storage systems (BESS) is presented in [22] and applied to a multiport device in [23]. The model-based predictive control of a MMC is shown in [145]. A modular multilevel matrix AC/AC-converter (see [24]) and a modular multilevel DC/DC-converter (see [25]) are presented in the literature. A hybrid energy storage system (HESS) using a modular VSC approach is presented in [146]. The application of interleaved VSC topologies increases the current ratings of the system and reduces the current ripple [26].

On the system level, solid-state transformers (SSTs, see [5]) and smart transformers (see [27]) based on VSC technology arise. They often use the dual active bridge (DAB) as an isolated DC/DC-converter. The DAB (see [28, 29]) can be applied in a modular concept, e.g. in series on the high voltage side and in parallel on the low voltage side.

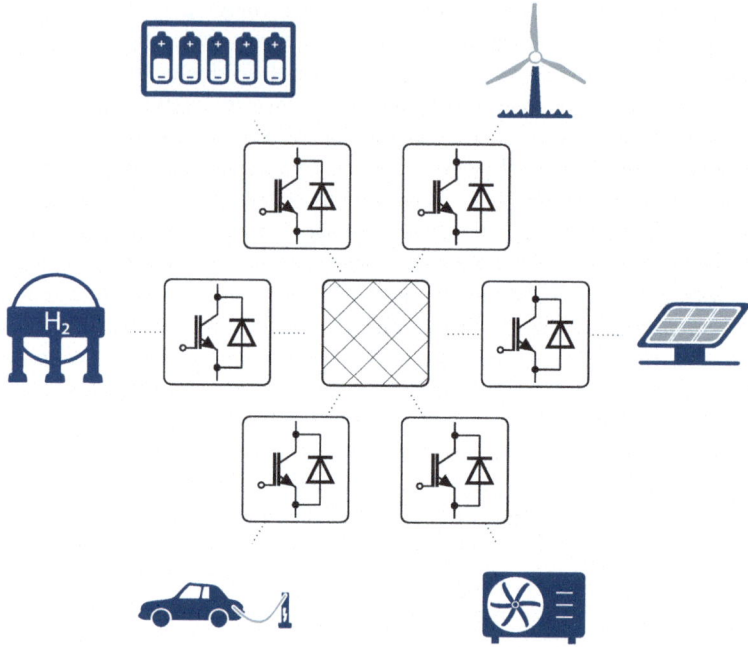

Fig. 1.1 Application range of voltage source converters (VSCs) including renewable energy sources (RES), battery energy storage systems (BESS), hydrogen systems and innovative consumer technologies such as electric vehicles (EVs) and heat pumps (HPs)

1.1.3 Grid Architectures for the Integration of RES

From the considerations in Sect. 1.1.1 and Sect. 1.1.2 it becomes clear, that the penetration of VSCs in the electrical power grid will increase significantly in the future. At the same time, the share of electrical machines as the connecting link to the grid will decrease. Pure and nearly pure power electronic grids will arise in the future. In this thesis, the term "100 % inverter-based power systems (IBPS)" is used. A future with inverter-based resources (IBRs) (see [30]) and the operation paradigm of an all-converter interfaced generation (CIG) power system (see [31]) are discussed in the literature.

1.1 Background and Motivation

In contrast to conventional thermal power plants, RES are often integrated into medium or low voltage grids due to lower nominal power of the plant. Different grid architectures for the integration of RES are possible.

Legal requirements and time pressure often lead to the operational integration of RES into an existing grid [32]. The complexity of the grid increases and its structure becomes unclear. Bidirectional power flows are caused by RES and voltage problems arise. Overload situations can occur and the reliability of electrical energy supply is threatened [32].

To overcome these problems, collector grids can be used. Often, collector grids are based on ESS. The concept of aggregating the capacity of RES to make them more accessible and manageable to create a single operating profile is called "virtual power plant (VPP)" [33, 34]. VPPs can have one or more connection points to the grid [32]. Using only one connection point carries the risk, that a fault event at the connection point can lead to the disconnection of the entire collector grid. Since power plants based on RES are distributed locally, the use of collectors grids with only one connection point unnecessarily threatens the reliability of electrical energy supply.

1.1.4 Future Grid Operation Strategies

There are different grid operation strategies for strategically planned grids with a clear structure. These grids can represent a combination of a utility grid and VPPs, see Sect. 1.1.3.

The strategy of the cellular approach is to balance energy production and consumption at the lowest possible level [35–37]. This grid operation strategy includes the consideration of consumers by definition. The objective of the cellular approach is to reduce the transport distance of energy and ideally the direct consumption at the place of production. The transmission grid serves as the connection between different energy cells to balance energy among these cells. Long distances between non-autonomous cells or semi-autonomous cells can be overcome by an overlay grid [38]. Here, the advantages of DC transmission are substantial, especially if cables instead of overhead lines are used.

The load-following concept represents the conventional grid operation philosophy. Here, the electrical power production is adjusted to the power consumption to allow maximum flexibility of use. Since power flows inside the grid are not adjusted and not actively regulated, the limits of the transport capacity of lines can be exceeded.

The strategy of the energy management concept using energy packages is to decouple the power profiles of production and consumption. One BESS at the production side and one BESS at the consumption side enable adjusted and actively regulated power flows inside the grid realized by power electronic energy routers. Consequently, effects caused by the volatility of RES and time-dependant power demand do not affect the grid and thus overload situations are avoided. This grid operation strategy is presented in [147–149]. The use of energy packages during a grid blackout is discussed in [150].

Using the energy packages concept, the entire available energy transport capacity of lines is used [149]. The grid frequency and the grid voltage as grid operation parameters are decoupled from power production and power consumption profiles. Besides, instantaneous power and reactive power of loads are decoupled from the grid. Furthermore, alternative energy carriers (e.g. liquid organic hydrogen carriers (LOHC)) with complementary physical properties to the electrical energy transport path can be integrated in this holistic energy management concept [148]. Additional energy transport paths increase the reliability of energy supply and the energy transportation flexibility. Moreover, the prioritisation of energy supply of critical infrastructures is possible.

Different grid operation strategies are reported in the literature. In [39], the "Intergrid" as a hybrid mix of AC and DC architectures comprising subgrids of different sizes that are dynamically decoupled and hierarchically interconnected. The "Energy Internet" is an internet-type network of all the components of a power system, which closely interacts with others by sharing both energy and information [40]. The Energy Internet extends the smart grid concept. In contrast to the smart grid concept, all agents, including prosumers and consumers, are able to sell/generate and buy/consume energy [40]. Furthermore, a plug-and-play interface is a fundamental requirement of the Energy Internet.

1.2 Interaction of VSCs and Grid Protection

1.2.1 Research Questions

Sustainability and reliability are two essential aspects of electrical energy systems. They must be ensured when designing future energy systems.

Sustainable solutions for energy production, storage and consumption are based on power electronic technology, see Sect. 1.1.2. The penetration of the electrical power grid by voltage source converters (VSCs) will increase significantly in the future and 100 % inverter-based power systems (IBPS) will arise. The characteristics

1.2 Interaction of VSCs and Grid Protection

of grid-connected VSCs depend on their control algorithm. Especially, frequency characteristics can be designed.

> **Research questions 1**
> What are the requirements on the frequency characteristics of VSCs in 100 % IBPS and how do VSCs synchronize in 100 % IBPS? How does the resulting dynamic behaviour look like and what are the impacts on the grid stability—especially after a grid fault?

Grid faults lead to changes in the topology of overhead lines and cables. After the fault is cleared, the grid topology is changed. The fault current depends on the current grid topology, the current line topology and the fault current sources. VSCs are fault current sources and open new flexibility options, especially in the negative and the zero sequence during grid faults. The reliability of energy supply is closely linked to the resilience of the system against grid faults.

> **Research questions 2**
> What are the flexibility options of VSCs during grid faults? How can selected flexibility options of VSCs increase the resilience of the grid and VSCs against grid faults?

In addition, an appropriate protection principle must be used to prevent danger to living beings and damage to technical components.

> **Research questions 3**
> How must the protection principle be designed in order to not restrict flexibility options of VSCs during grid faults and at the same time to reliably detect grid faults fed by VSCs?

1.2.2 State of the Art

Distance protection algorithms are frequently used in medium voltage grids, see [41, 143]. These algorithms are also adapted to the characteristics of synchronous machines (see [143]), which represent sources of fault currents. The power flows during normal operating conditions inside the grid are often assumed to be unidirectional from higher voltage levels to lower voltage levels.

The physics of VSCs differ fundamentally from the physics of synchronous machines, see Sect. 2.1 and [42]. In the past, positive sequence current controllers were used to define the grid characteristics of VSCs [43, 44]. Today, the share of voltage controllers (see [45]) increases. Often, proportional-integral (PI) controllers are used [46]. The injection of reactive current in the negative sequence of medium voltage grids is standardized in [47].

An obvious approach to preserve existing distance protection algorithms in future grids is to control VSCs in a way to adopt the characteristics of synchronous machines. For this imitating approach on different detail levels, the terms "synchronverter" (see [48]) or "virtual synchronous machine" (see [49]) are used for example. Further variants are found in [50, 51].

But, the physics of VSCs differ fundamentally from the physics of synchronous machines and cannot be manipulated. The approaches in the literature do not exploit all possibilities of VSC, since they are restricted by the imitating approach. For example, the possibility of VSCs to define the characteristics of the zero sequence is not used. Neutral point treatment is realized via the VSC transformer today. On the other side, four-leg converters draw more and more attention [52].

Numerous problems regarding the interaction between VSCs and distance protection algorithms are reported in the literature.

The results of hardware-in-the-loop (HIL) tests in [53] reveal, that the impedance measuring error caused by the fault resistance can be enlarged and unpredictable with the presence of VSCs. Furthermore, the measuring error is affected by the current limit of the VSC [53]. The injection of reactive power in the negative sequence also leads to problems. The associated indeterminacy aggravates, since the measuring errors are also affected by the relative relationship between positive- and negative-sequence reactive power and pre-fault power flow conditions [53].

In [54] it is reported, that the reliability and the speed of distance protection can be adversely affected due to a lower short circuit current level and unconventional short circuit current characteristics. The authors propose to avoid the use of constant reactive power control together with a distance relay because it can cause the relay to be unable to calculate the impedance accurately [54].

1.2 Interaction of VSCs and Grid Protection

The authors of [55] report, that the limitation in fault current magnitude along with phase angle modulation in the presence of VSCs affects the performance of a conventional distance relay. Especially in case of faults with significant fault resistance, conventional distance relaying algorithms fail to identify the faults in the prescribed zones [55].

The effects of a VSC-based high voltage direct current (HVDC) system on the distance protection of transmission lines are investigated in [56]. It is concluded, that the limited reactive positive sequence and suppressed negative sequence fault current from the VSC impact the distance relay heavily [56]. Wrong tripping actions due to relay under-reach are observed, if reactive current injection to support the network voltage along with the active current blocking limitation strategy is activated [56]. The saturation limits of the converter affect the distance relay in a way, that the smaller the saturation limit is, the higher the under-reach is [56]. It is determined, that "an effective solution to update the protection system is the need of the hour" to accommodate VSC-based HVDC connection into the network [56].

The common assumption regarding the phase difference between the line-end currents during a zone-one fault is not valid for a line that emanates from a VSC-interfaced RES plant is reported in [57].

Problems of state-of-the-art distance protection in electricity grids fed by VSCs are emphasized in [58]. It is summarized, that three fault current injection strategies stated in current grid codes made distance protection on the VSC side ineffective [58]. The authors demand, that the control system needs to be optimized to bring the fault currents quicker into the steady state [58].

Distance protection of AC grids with HVDC-connected offshore wind generators are investigated in [59]. The impact of the reactive power control of VSCs on distance relays has become an important subject [59]. The authors summarize, that the issue of miscoordination of zone 2 protection caused by the control of the HVDC systems needs to be addressed [59].

Further issues with HVDC systems are addressed in [60]. It is stated, that the VSC-based HVDC system has a significant influence on the zone 2 and zone 3 relay performance due to the active and reactive power source connected in-between the relay location and the fault point [60]. It is obtained, that the malfunction of the zone 2 relay and the zone 3 relay can be mitigated by altering the reactive power injected into the system using AC voltage reference input variations [60].

Some solutions to solve the problems regarding the interaction between VSCs and distance protection algorithms are found in the literature.

A control-based solution is provided in [61]. The prime objective of this method is to imitate certain features of the symmetrical components of the fault currents of synchronous machines that affect distance relays [61]. The fault current angle of the

VSC is controlled in a way to imitate the characteristics of a synchronous machine [61].

In [62] an adaptive distance protection algorithm for lines connected to VSC-interfaced renewable power plants is proposed. Following structural and operational changes in the power system, an adaptive zone 1 setting is presented in [63].

The interaction between VSCs and distance protection algorithms causes different problems. Solutions from the literature can be organized into two categories. Solutions of the first category are based on the obvious approach of controlling VSCs in a way to adopt the characteristics of synchronous machines. Here, the possibilities of VSCs to increase the system resilience against grid faults are restricted. Solutions of the second category are based on the use of adaptive distance protection algorithms. Here, the settings depend on the situation and represent a selective solution, but not a holistic solution.

The interaction between VSCs and distance protection algorithms is a crucial research gap and sophisticated solutions are missing up to now.

1.2.3 Research Objectives

The superordinate research objective of this thesis is to point out a perspective of future energy systems based on renewable energy sources (RES). Following on the research questions in Sect. 1.2.1, a holistic control and protection concept for 100 % inverter-based power systems (IBPS) based on voltage source converters (VSCs) is developed. The aim is to contribute closing the research gap described in Sect. 1.2.2 by meeting the following research objectives.

> **Research objective 1**
> Evaluating the synchronization principles of VSCs as a fundamental requirement for the operation of 100 % IBPS and determining the connection to grid dynamics and grid stability—especially after a grid fault.

> **Research objective 2**
> Identifying the flexibility options of VSCs during grid faults to increase the system resilience and realizing them by using a suitable control algorithm.

> **Research objective 3**
> Developing an appropriate protection algorithm for 100 % IBPS without restricting the flexibility options of VSCs during grid faults.

To define an approach to meet these research objectives, the physical constraints of VSCs and the frequency characteristics of 100 % IBPS described in Chap. 2 have to be considered.

1.3 Outline of this Thesis

In Chap. 1, the current situation of electrical energy supply is reflected in particular based on ecological aspects. Research questions are identified and the state of the art is presented. The research objectives of this thesis are formulated. The fundamentals of voltage source converters (VSCs) and frequency characteristics of 100 % inverter-based power system (IBPS) are presented in Chap. 2. Furthermore, constraints meeting the research objectives are derived and the research approach is described.

Chap. 3 contains the description of electrical quantities, state-space modelling and the analysis of system properties. This chapter forms the common basis for the control algorithms and the protection algorithms in this thesis. Chap. 4 describes the synchronization of VSCs to the grid and the frequency and active power characteristics after a grid fault. In Chap. 5, the characteristics and the V/I-control of VSCs are presented. This chapter gives an overview of flexibility options in the positive sequence and the negative sequence. The enhanced grid fault characteristics of VSCs and the options in the zero sequence are the content of Chap. 6.

The grid fault modelling is introduced in Chap. 7. These models are the basic prerequisite of the model-based protection principle. In Chap. 8, the model-based protection (MBP) principle is presented. The MBP is validated by laboratory tests in Chap. 9. In Chap. 10, the MBP is tested based on the enhanced grid fault characteristics of VSCs in a 100 % IBPS. Chap. 11 presents the results of this thesis and gives an overall conclusion. The results are evaluated regarding the research subjects and the research objectives. Besides, future research perspectives are pointed out.

Figure 1.2 illustrates the relations between all the chapters of this thesis and helps to define an individual chronology of reading. It also opens the possibility for quick navigation in the digital document.

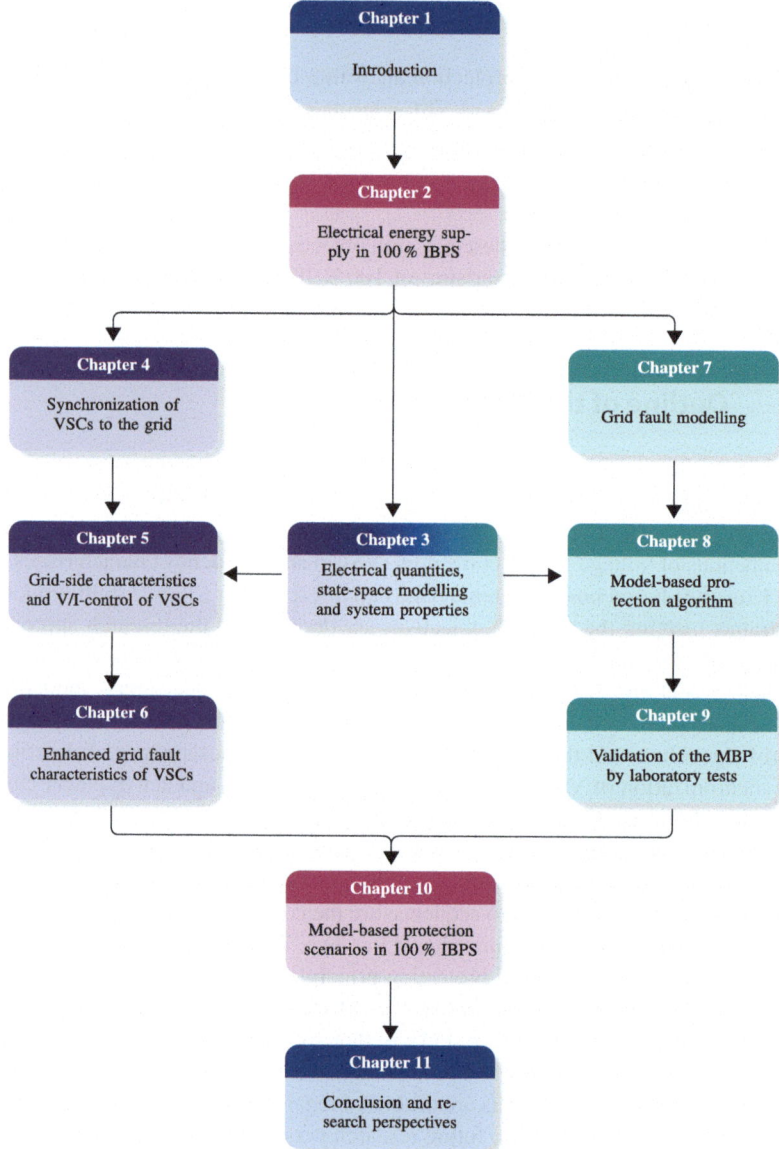

Fig. 1.2 Outline of this thesis and relations between the chapters with the possibility for quick navigation in the digital document

1.4 Summary

This chapter opens this thesis entitled "Control and Protection of 100 % Inverter-based Power Systems". The background and the motivation of this thesis are presented and regarded in a broader context. Research questions are identified as the challenges to be addressed in this thesis. Approaches from the literature to fill the gap between power electronics and grid protection are summarized. Their drawbacks are also pointed out. The research objectives are defined and the outline of this thesis is visualized.

Energy systems are in transition due to the imperative of the reduction of CO_2-emissions. Today, the momentum of power electronics is considerable. Different energy sectors converge. Power electronic components are the link between renewable energy sources (RES), energy storage systems (ESS), innovative consumer technologies and the electrical power grid. Power electronics are a key technology with various flexibility options. Sustainable 100 % inverter-based power systems (IBPS) will arise in the future. Because of the properties of RES and the increasing transport of electrical energy due to energy sector-coupling, new grid architectures and grid operation strategies have to be considered.

The interaction between voltage source converters (VSCs) and grid protection is crucial for the operation of 100 % IBPS. Existing approaches in the literature to fill the gap between VSCs and grid protection show different drawbacks. Adopting the characteristics of synchronous machines to VSCs restricts the possibilities of VSCs to increase the system resilience against grid faults. Adaptive distance protection algorithms represent a selective but not a holistic solution. Sophisticated solutions are missing up to now. New solutions regarding the interaction between VSCs and grid protection have to be developed. A holistic control and protection concept for 100 % IBPS is indispensable.

To define the approach to meet the research objectives, the physical constraints of VSCs and frequency characteristics of 100 % IBPS have to be considered. Therefore, the physical and technical fundamentals of VSCs as well as the frequency characteristics of 100 % IBPS are presented in Chap. 2. To follow the key aspects of the central theme of this thesis on a shortcut: The conclusion of the next chapter is provided in Sect. 2.5.

Open Access This chapter is licensed under the terms of the Creative Commons Attribution 4.0 International License (http://creativecommons.org/licenses/by/4.0/), which permits use, sharing, adaptation, distribution and reproduction in any medium or format, as long as you give appropriate credit to the original author(s) and the source, provide a link to the Creative Commons license and indicate if changes were made.

The images or other third party material in this chapter are included in the chapter's Creative Commons license, unless indicated otherwise in a credit line to the material. If material is not included in the chapter's Creative Commons license and your intended use is not permitted by statutory regulation or exceeds the permitted use, you will need to obtain permission directly from the copyright holder.

Electrical Energy Supply in 100% IBPS 2

This chapter presents the electrical energy supply in 100% inverter-based power systems (IBPS). In Sect. 2.1, the physical and technical fundamentals of voltage source converters (VSCs) are presented. The frequency characteristics of 100% IBPS are described in Sect. 2.2. It contains the remastered interpretation of frequency in 100% IBPS and the requirements on the frequency characteristics of VSCs. Furthermore, the frequency characteristics in the steady state are explained based on an illustrative example in the context of grid control after a contingency. Section 2.3 points out the requirements on VSC control and grid protection of a 100% IBPS. Based on the identified physical and technical constraints, the research approach of this thesis is presented in Sect. 2.4.

2.1 Fundamentals of VSCs

The fundamentals of voltage source converters (VSCs) are explained based on a three-level neutral point clamped (NPC) VSC [64]. A three-level NPC VSC offers the possibility to be connected to the medium voltage grid without a transformer (see Sect. 6.2.1.2).

The main advantages of the NPC VSC compared to a two-level VSC using the identical carrier frequency are lower switching losses per semiconductor power device [19]. Furthermore, the power quality and the electromagnetic compatibility are superior [19]. The drawbacks of the NPC VSC are higher overall conduction losses and the additional need for balancing of the direct current (DC) link voltages. Moreover, more power semiconductor devices are required.

2.1.1 Half-bridge Topology

The half-bridge topology as a part of a VSC is an electrical switching circuit, that connects the DC-link to another circuit. The half-bridge topology consists of a number of active switching semiconductor power devices. For example, insulated-gate bipolar transistors (IGBTs) based on silicon or metal-oxide-semiconductor field-effect transistor (MOSFETs) based on silicon carbide (SiC) are used for grid-connected (GC) VSCs. The allowed blocking voltage and the current of a semiconductor power device are limited. For example, the current limit must not exceed its nominal value approximately by the factor 1.2.

The three-level half-bridge consists of four switching cells (see [143]). Each switching cell is built from one semiconductor power device and one antiparallel diode. The midpoint of the DC-link is connected via two additional diodes. The equivalent circuit of a three-level half-bridge using four IGBT power devices (S_{11} to S_{14}) is shown in Fig. 2.1. The first index is used to indicate the number of the phase, to which the half-bridge is connected. The DC-link voltage is named v_{DC} and the converter (index: c) output voltage of phase 1 is named v_{c1}.

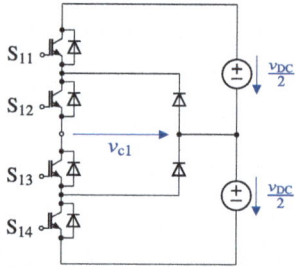

Fig. 2.1 Equivalent circuit of a three-level NPC VSC half-bridge used in phase 1

To evoke current flows, one of three electrical potentials of the DC-link is connected to the alternating current (AC) side by selecting an appropriate combination of switching states of S_{11} to S_{14}. The output voltages v_{c1} take values from a finite set. In this way, currents with positive and negative sign from and to each DC-link potential can flow. Electrical energy is transferred from the DC-side to the AC-side or vice versa while being converted from one form to another.

During normal operating conditions, three different switching states S_{HB1} of the half-bridge are used. Each switching state of the half-bridge is defined by the combination of switching states of S_{11} to S_{14} and leads to a certain value of v_{c1} as

2.1 Fundamentals of VSCs

a fraction of the DC-link voltage v_{DC}. Table 2.1 summarizes the combinations of S_{HB1}, S_{11} to S_{14} and v_{c1}.

Table 2.1 Switching states S_{HB1} of a NPC VSC half-bridge in phase 1 during normal operating conditions neglecting the blocking state.

S_{HB1}	S_{11}	S_{12}	S_{13}	S_{14}	v_{c1}
1	on	on	off	off	$\frac{v_{DC}}{2}$
2	off	on	on	off	0 V
3	off	off	on	on	$-\frac{v_{DC}}{2}$

Applying Kirchhoff's voltage law (KVL) in Fig. 2.1 and considering Table 2.1, it becomes obvious, that v_{DC} should be balanced regarding the midpoint of the DC-link. In addition, the current value of v_{DC} must be greater than twice the amplitude of the grid voltage on the AC-side. Since the current value of v_{DC} represents the current energy inside the DC-link, it is calculated from the difference of the instantaneous power on the DC-side and on the AC-side. Therefore, v_{DC} is controlled via the active power on the AC-side by the NPC VSC or via a DC/DC-converter connecting for example a battery energy storage system (BESS) to the DC-link.

By an appropriate temporal distribution of two switching states S_{HB1} (see Table 2.1) of a half-bridge within a carrier period T_{car}, the pulse width of v_{c1} is adjusted. Using this concept, the short-term mean value of v_{c1} named $\overline{v}_{c1}(t)$ (see Eq. (2.1)) is adjusted as desired within T_{car} representing a value from a continuous set. In this way, an arbitrary reference signal $v_{c1,ref}$ is approximately synthesized on the AC-side of the NPC VSC half-bridge via v_{c1}. The carrier frequency of silicon-based IGBTs is in the range of a few kHz, whereas the carrier frequency of SiC-based MOSFETS is in the range of a few ten or hundred kHz. This concept is called pulse-width modulation (PWM) [65]. The current value of $\overline{v}_{c1}(t)$ is calculated according to Eq. (2.1).

$$\overline{v}_{c1}(t) = \frac{1}{T_{car}} \int_{t-\frac{T_{car}}{2}}^{t+\frac{T_{car}}{2}} v_{c1}(\tau) d\tau \qquad (2.1)$$

Often, a carrier-based PWM is applied. By comparing the reference signal $v_{c1,ref}$ with two carrier signals ($v_{c1,car1}$ and $v_{c1,car2}$) showing the period duration T_{car}, the current switching state S_{HB1} of the half-bridge is determined according to a defined switching logic, see Table 2.2.

Table 2.2 Switching logic of a NPC VSC half-bridge in phase 1 during normal operating conditions neglecting the blocking state

Comparison of signals	S_{HB1}
$v_{c1,ref} > v_{c1,car1}$	1
$v_{c1,car1} > v_{c1,ref} > v_{c1,car2}$	2
$v_{c1,car2} > v_{c1,ref}$	3

Fig. 2.2 exemplary shows the synthesis of a sinusoidal 50 Hz-reference signal $v_{c1,ref}$ via the three-level output voltage v_{c1}. Here, two triangular carrier signals ($v_{c1,car1}$ and $v_{c1,car2}$) in phase opposition with a carrier frequency of 800 Hz are used. From Fig. 2.2 it becomes obvious, that the amplitude of $v_{c1,ref}$ is limited by the maximum and minimum value of v_{c1}. These values are dependant on the current value of v_{DC} (see Table 2.1).

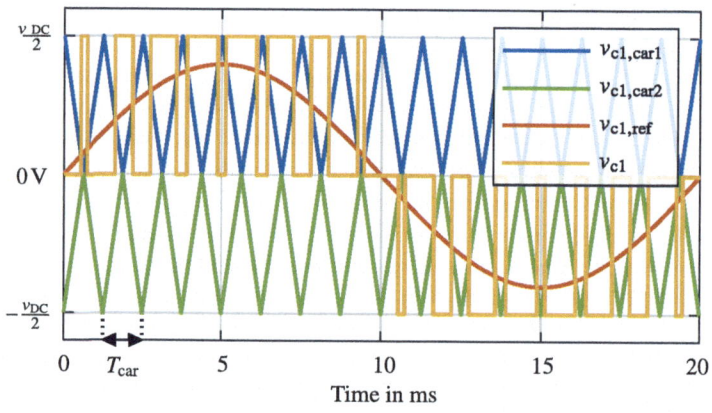

Fig. 2.2 Synthesis of a sinusoidal 50 Hz-reference signal $v_{c1,ref}$ via the three-level output voltage v_{c1} using the PWM concept with a carrier frequency of 800 Hz

The switching characteristics of VSCs cause harmonic frequency components of voltages and currents regarding the switching frequency. These undesired frequency components reduce the power quality in the AC-grid. To attenuate these frequency components, a filter between the VSC and the grid is commonly used, see [143] and Sect. 3.2.2.

2.1 Fundamentals of VSCs

2.1.2 Three-phase Bridge Topology

The three-phase bridge topology represents the fundamental circuit of a VSC. It consists of the parallel connection of three half-bridges (see Fig. 2.1) at the same DC-link. The equivalent circuit of a three-level NPC VSC connected via LCL-filter (see Sect. 3.2.2.3) to the grid is shown in Fig. 2.3.

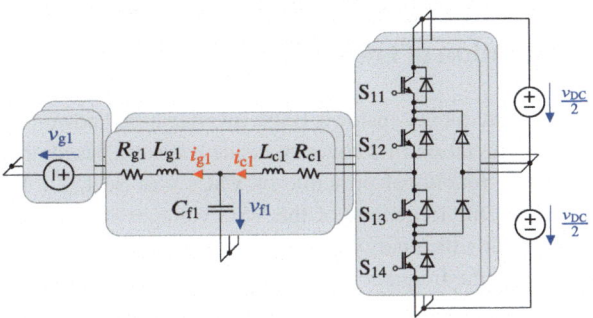

Fig. 2.3 Equivalent circuit of a three-level NPC VSC connected to the grid via a LCL-filter

According to the operating principle of a half-bridge connecting different DC-link potentials to the AC-side (see Table 2.1 and Fig. 2.2), each half-bridge is modelled in a simplified manner by a voltage source. Consequently, the three-phase bridge is modelled by three voltage sources connected in parallel, see Fig. 2.4.

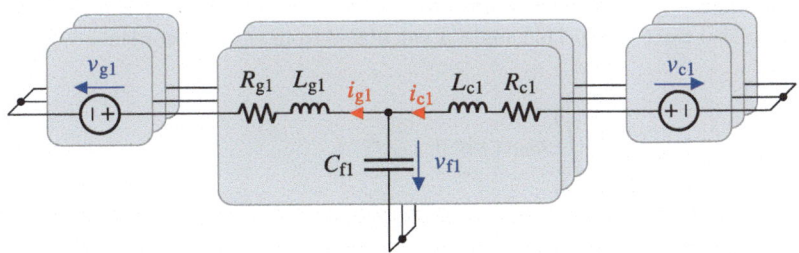

Fig. 2.4 Simplified model of a three-level NPC VSC connected to the grid via a LCL-filter

The model from Fig. 2.4 is valid for different VSC topologies (see [143]) and is used for the design process of the controller of the VSC, see Sect. 5.1. It has to be considered, that the voltage sources of each phase x with $x \in \{1, 2, 3\}$ represent values v_{cx} from a finite set, see Sect. 2.1.1. Furthermore, the currents flowing through the semiconductor power devices i_{cx} are limited, see Sect. 2.1.1. The three half-bridges of the VSC can synthesize output voltages v_{cx} independently of each other.

2.2 Frequency Characteristics of 100% IBPS

2.2.1 Interpretation of Frequency

The interpretation of frequency inside a 100% inverter-based power system (IBPS) differs from the interpretation of frequency inside a system based on synchronous machines. Due to the functional principle of VSCs (see Sect. 2.1.1), there is no physical link between the frequency of the VSC voltages v_{cx} with $x \in \{1, 2, 3\}$ and a variable describing the energy of an energy storage component. Regarding a synchronous machine, the frequency of its output voltages is linked to the rotation speed [42]. In contrast, VSCs can operate at any frequency (depending on the carrier frequency, see Sect. 2.1.1) and a jump discontinuity of the frequency of v_{cx} is possible. The decoupling of frequency and energy allows defining minimum and maximum values of the frequency deviation from its nominal value (see $\Delta\omega_m$ in Eq. (2.2c)) to limit the rate of change of power.

To enable effective methods for the limitation of VSC currents (see also Sect. 6.3), the filter capacitor voltages v_{fx} (see Fig. 2.3 and Fig. 2.4) are controlled and their angular frequency ω_m is adjusted, see Sect. 4.4.1 and Sect. 5.2.2. The filter capacitor voltages v_{fx} are described by continuous functions.

The angle γ_m^+ equals the positive sequence angle of the filter capacitor voltages v_{fx} (see also Eq. (3.16)). This angle is calculated from ω_m in accordance with Eq. (2.2).

$$\gamma_m^+(t) = \int_0^t \omega_m(\tau)d\tau + \varphi_m(t) \tag{2.2a}$$

$$= \int_0^t \big(\omega_0 + \Delta\omega_m(\tau)\big)d\tau + \big(\varphi_{(1),0} + \Delta\varphi_m(t)\big) \tag{2.2b}$$

$$= \omega_0 t + \varphi_{(1),0} + \int_0^t \Delta\omega_m(\tau)d\tau + \Delta\varphi_m(t) \tag{2.2c}$$

$$= \gamma_{m,0}^+ + \Delta\gamma_m^+(t) \tag{2.2d}$$

2.2 Frequency Characteristics of 100% IBPS

In Eq. (2.2), the index 0 represents the nominal operating point and the Δ-variables represent the deviations from the nominal value. The variable ω_0 represents the nominal operating angular frequency and $\varphi_{(1),0}$ represents the nominal positive sequence angle given by the load flow calculation as a part of the tertiary control, see Sect. 2.2.3.

Regarding Eq. (2.2c) as a calculation rule for the reference value of γ_m^+, the third addend and the fourth addend reveal as degrees of freedom to be adjusted by the VSC. Here, $\Delta\varphi_m$ is often set to zero to decelerate the system response and to avoid a jump discontinuity. In this way, the active power of the VSC is adjusted indirectly via the voltage angle difference, see Sect. 4.1.2.2. More details about the power-based synchronization is presented in Sect. 4.4.

There is a predefined correlation between ω_m and the active power according to the primary grid control principle (see Sect. 2.2.3 and Eq. (4.58)). Regarding this correlation and neglecting the secondary and tertiary grid control, the grid frequency in a 100% IBPS in the steady state is an indicator of the deviations of the active powers of VSCs from their nominal active powers inside the 100% IBPS, see also Sect. 4.4.5.

In contrast to a grid based on synchronous machines, returning the grid frequency of a 100% IBPS to the nominal frequency does not require any additional positive or negative energy flow into or from the grid. The reason is, that no rotating masses have to be accelerated or decelerated during the secondary control, see Sect. 2.2.3.

The frequency deviation from its nominal value is a possibility to share active power according to a predefined ratio between different VSCs after changing load conditions or grid faults by using different droop constants to adjust the voltage angle difference, see Sect. 4.4.5.

Since a VSC is a component without energy storage capability, the power and energy on the DC-side have always to be considered. If the DC-link voltage falls below a certain value, the necessary amplitude of the AC-voltage cannot be reached, see Sect. 2.1.1.

A variable describing the energy on the DC-side (e.g. the state of charge (SOC) of a BESS) can be reflected to the frequency of v_{fx} via an additional droop addend in $\Delta\omega_m$ [66]. If each VSC connected to the 100% IBPS uses this principle, the grid frequency becomes an indicator of the total energy connected to the grid. This concept is similar to the total energy of rotating masses connected to a grid based on synchronous machines. It can be used to balance energy between different components connected to the 100% IBPS.

2.2.2 Requirements on the Properties of VSCs

Basically, a VSC is operated in the grid-forming (GFM) mode or in the grid-following (GFL) mode. Both modes refer to the type of synchronization of the VSC to the grid regarding the active power transfer in the positive sequence, see Sect. 4.2. They distinguish from each other regarding the calculation of the reference angle γ^+ to which the VSC-controller is adapted, see Sect. 4.4 and Sect. 4.3.

GFM-VSCs and GFL-VSCs show different properties. In contrast to GFL-VSCs, GFM-VSCs show the capability to form an island grid and define the grid frequency (see Sect. 2.2.1) without any other reference despite the active power. The grid restoration process after a blackout is linked to this context. Detailed information including also renewable energy sources (RES) connected via VSCs and high voltage direct current (HVDC) systems into the grid restoration process is given in [67]. Further investigations of GFM HVDC systems are presented in [151, 152]. Since 100% IBPS are operated without synchronous machines and the previously describes properties are crucial, the GFM-mode in the positive sequence is selected and investigated subsequently.

Beyond the capability of island operation, GFM-VSCs must provide further properties regarding power and frequency for a satisfactory grid operation. These properties are summarized in Fig. 2.5 and discussed below.

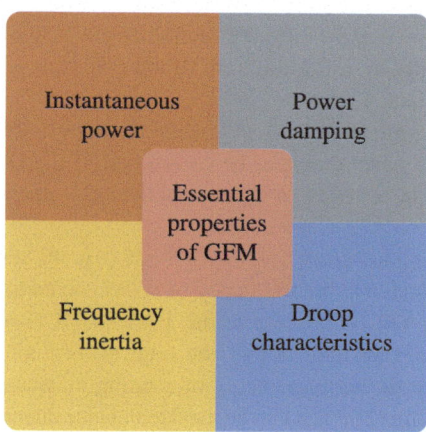

Fig. 2.5 Essential properties of grid-forming (GFM) VSCs inside a 100% inverter-based power system (IBPS)

2.2 Frequency Characteristics of 100% IBPS

The provision of instantaneous power is essential after changing load conditions and grid faults. Power balance after contingencies is always ensured due to thermodynamic laws. Power imbalances should not be balanced by parasitic inductive and capacitive elements of overhead lines and cables due to their low-energy capacity. Instead, GFM-VSCs ensure power balance through the energy from the DC-link and energy storage systems (ESS) with higher energy capacity. Therefore, instantaneous power must be provided. In contrast to GFL-VSCs depending on the measurement of the frequency at the point of interconnection, GFM-VSCs inherently provide instantaneous power.

Frequency inertia provides time immediately after changing load conditions and grid faults to balance power in the desired way in the first instants, see Sect. 4.4.4. It slows down the natural reaction of the system and buys the controllers and the operator time to take actions [68]. Frequency inertia can be quantified by the rate of change of frequency (RoCoF) [68]. GFM-VSCs do not realize frequency inertia inherently, but using appropriate controllers, such as the configurable natural droop (CND)-controller (see Sect. 4.4.1.2), allows its realization. In contrast to synchronous machines, instantaneous power and frequency inertia are not necessarily linked to each other [143].

After the time interval dominated by instantaneous power and frequency inertia after a contingency, the power imbalance is distributed among different VSCs inside the grid representing an oscillatory system. Due to stability aspects, it is favourable to damp power oscillations, see Sect. 4.4.5. Power damping is not realized by GFM-VSCs inherently. Appropriate controllers, such as the CND-controller (see Sect. 4.4.1.2), allow its realization.

Droop characteristics allow the distribution of active power after a contingency among different VSCs in the desired way, see Sect. 4.4.5. The shares of active power are often calculated from the nominal active power of the VSCs. Creating a correlation (see Eq. (4.58)) between the measured active power and the controlled frequency of the filter capacitor voltages $v_{f,x}$ (see Fig. 2.3 and Fig. 2.4) adjusts the voltage angle (see Eq. (2.2c) and Eq. (4.28b)) and consequently the active power share appropriately, see Eq. (4.24). GFM-VSCs do not realize droop characteristics inherently. CND-controllers allow its realization, see Sect. 4.4.1.2.

It is presented in Sect. 4.4.1.2, that frequency inertia, power damping and droop characteristics are set independently of each other using three individual parameters of the CND-controller. These parameters are adjusted via software and realized to grid via the VSC as an actuator, see Sect. 5.1. For practical applications, the voltage and current limits of semiconductor power devices have to be considered (see Sect. 2.1.1) and power and energy have to be provided by the DC-side on a sufficient level.

2.2.3 Grid Control

The grid control strategy defines the frequency and voltage characteristics after a grid fault and after the fault clearance. It influences the dynamics and the stability of the grid. The basic objective is to maintain the grid operation. The grid control of 100% IBPS is based on the grid control of grids based on synchronous machines, but the interpretation of frequency of 100% IBPS from Sect. 2.2.1 has to be considered. The grid control strategy is based on the physical relations between active power and angle/frequency (see Eq. (2.2c), Eq. (4.28b) and Eq. (4.24)) and reactive power and voltage (see Eq. (4.25)) assuming dominant inductive grid characteristics.

Subsequently, the steady state after the provision of instantaneous power is observed. The power inside the positive sequence is regarded. Active power is transformed into other forms of energy (e.g. mechanical energy or chemical energy). Reactive power is needed to increase and decrease the strength of electrical and magnetic fields in overhead lines and cables periodically.

Two system parameters are relevant regarding grid control. The grid frequency is a system-wide parameter, whereas the grid voltage is a local system parameter according to Kirchhoff's voltage law (KVL). The grid frequency is identified as an indicator of the deviations of the active powers of VSCs from their nominal active powers inside the 100% IBPS, see Sect. 2.2.1. In general, VSCs can operate at any frequency (depending on the carrier frequency, see Sect. 2.1.1). The grid voltage has to be limited so that the maximum field strength of components is not exceeded. The following considerations are related to the grid frequency.

The fundamental principle of the grid control regarding frequency and active power represents a hierarchical structure based on three layers with individual objectives: Primary control, secondary control and tertiary control. These layers are separated in time after a contingency, see Fig. 2.6.

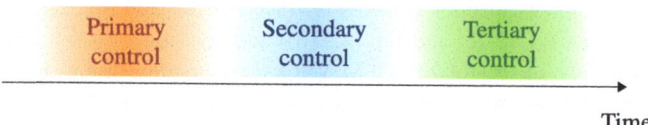

Fig. 2.6 Fundamental principle of the grid control represented as a hierarchical structure based on three layers separated in time with individual objectives

The three layers are activated and deactivated at different points in time after a contingency. Since they are not yet standardised for 100% IBPS, no time values are

2.2 Frequency Characteristics of 100% IBPS

given in Fig. 2.6. This standardisation is crucial and has to be put into practice in the future.

Subsequently, the interactions of the primary control, the secondary control and the tertiary control are explained in Fig. 2.7. A GFM-VSC connected to a BESS on the DC-side is regarded using the active sign convention. Here, the angular frequency ω_m (see also Sect. 2.2.1, Eq. (2.2c) and Sect. 4.4.1) represents the reference value of the angular frequency of the filter capacitor voltages v_{fx} (see Fig. 2.3 and Fig. 2.4). The reference value has to be realized by the controller as the actual value. In contrast to the actual value of ω_m, its reference value can show a jump discontinuity. The actual value and the reference value can differ from each other. In the steady state, it is assumed that they equal each other. For reasons of clarity, the index ref is omitted.

The nominal operating point (OP 1) is defined by the nominal angular frequency ω_0 and the nominal active power P_0 of the VSC, see Fig. 2.7. The active power balance between power production and power consumption inside the 100% IBPS is assumed.

A contingency (e.g. grid fault or changing load conditions) is assumed in a way, that the power consumption inside the 100% IBPS exceeds the power production. The instantaneous powers of all GFM-VSCs (see Sect. 2.2.2) ensure the power balance inside the 100% IBPS. The active power of the regarded VSC increases.

The objective of the primary control (PC) is to distribute active power among different VSCs in the desired way (see also Sect. 4.4.5). Doing without communication between different VSCs, it is expedient to establish an artificial linear relation between the measured active power deviation from P_0 and the reference value of ω_m. Using these droop characteristics as a proportional controller, the active power is adjusted using the integral of $\Delta\omega_{m,PC}$ in Eq. (2.2c), see also Sect. 2.2.1 and Sect. 2.2.2. The active power contribution of each VSC to balance the active power depends on its droop constant $k_{P,PC}$ as the slope of the linear function, see Eq. (2.3) [69]. In the steady state, a new grid operating point is reached. Here, the angular frequencies of each VSC are identical and deviate from the nominal angular frequency. The active power of each VSC contributing to the PC deviates from its nominal value in general.

$$\Delta\omega_{m,PC}(t) = \frac{1}{k_{P,PC}}(P_0 - P(t)) \tag{2.3}$$

The angular frequency ω_m of the regarded VSC contributing to the PC is reduced and P increases according to an increased voltage angle difference. The OP 2 is reached, see Fig. 2.7.

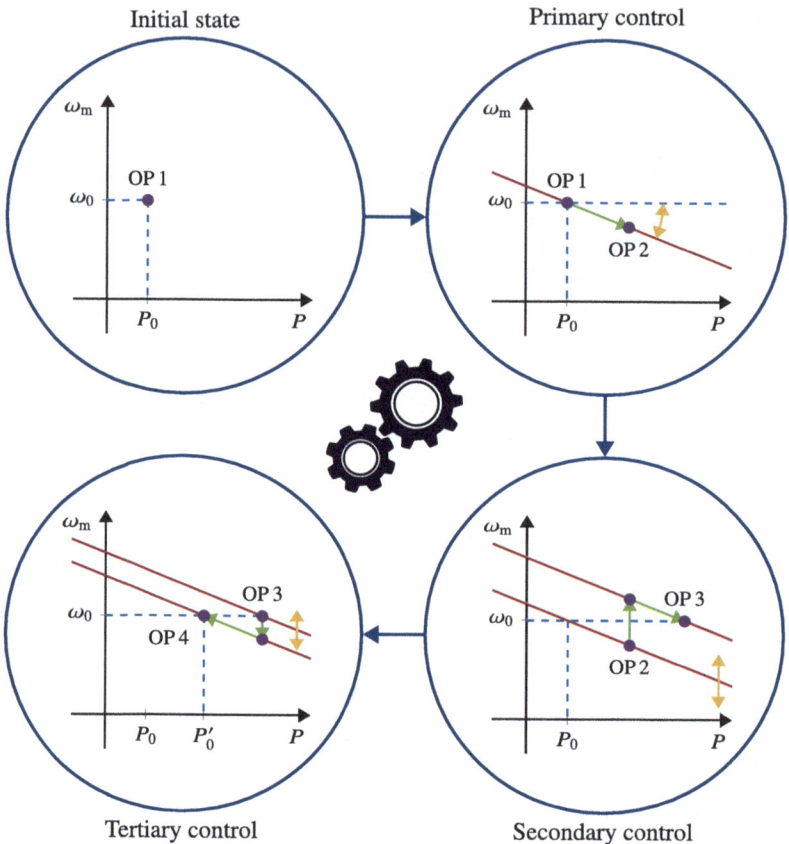

Fig. 2.7 Primary control, secondary control and tertiary control and their interactions explained based on an exemplary contingency grid scenario

The secondary control (SC) is activated after the PC. The objective of the SC is to return the grid frequency to its nominal value. A proportional-integral (PI) controller is used to calculate $\Delta\omega_{m,SC}$ (see Eq. (2.4)) from the frequency deviation in Eq. (2.2c) [70]. In this way, the droop is shifted vertically. In the steady state, the angular frequencies of each VSC are constant and equal their nominal value.

$$\Delta\omega_{m,SC}(t) = \frac{1}{k_{P,SC}}\big(\omega_0 - \omega_m(t)\big) + \frac{1}{k_{I,SC}}\int\big(\omega_0 - \omega_m(t)\big)dt \qquad (2.4)$$

2.2 Frequency Characteristics of 100% IBPS

It is assumed, that not each VSC contributes to the SC in the investigated scenario. That means that the overall additional active power shared during the PC is shared among the VSCs contributing to the SC. The VSCs, that do not contribute to the SC, return to their initial state. The angular frequency ω_m of the regarded VSC contributing to the SC is returned to the nominal frequency after a step-up (due to the proportional gain, see Eq. (2.4)) and P increases. The OP 3 is reached, see Fig. 2.7.

If all VSCs contribute to the SC using identical angular frequency deviations $\Delta\omega_{m,SC}$, their active power does not change and the nominal frequency is reached in the steady state. These characteristics are exclusive for 100% IBPS, since no rotating masses have to be accelerated or decelerated. The frequency is not linked to a variable describing the energy of a component. The angular frequency deviation $\Delta\omega_m$ is solely used to adjust the active power.

The tertiary control (TC) is activated after the SC. The objective of the TC is to define the active power P_0' of new operating points at the nominal frequency depending on the current grid situation based on power flow calculations [71]. A PI-controller is used to calculate $\Delta\omega_{m,TC}$ (see Eq. (2.5)) from the active power deviation from P_0' in Eq. (2.2c) [70]. In this way, the droop is also shifted vertically. In the steady state, the angular frequencies of each VSC are constant and equal their nominal value.

$$\Delta\omega_{m,TC}(t) = \frac{1}{k_{P,TC}}(P_0' - P(t)) + \frac{1}{k_{I,TC}} \int (P_0' - P(t))dt \qquad (2.5)$$

Assuming in the investigated scenario, that additional power production is realized, the angular frequency ω_m of the regarded VSC is returned to the nominal frequency after a step-down (due to the proportional gain, see Eq. (2.5)) and P is reduced. A new active power balance situation is realized and the OP 4 is reached, see Fig. 2.7.

The design of $\Delta\omega_m$ in Eq. (2.2c) applying PC, SC and TC add as follows, see Eq. (2.6).

$$\Delta\omega_m(t) = \Delta\omega_{m,PC}(t) + \Delta\omega_{m,SC}(t) + \Delta\omega_{m,TC}(t) \qquad (2.6)$$

In this context, the energy management concept using energy packages (see Sect. 1.1.4 and [147–149]) is interpreted as an alternative way to define new operating points during TC. Designing energy packages and their distribution are a particular form of power flow calculation using P/Q-nodes (see also [71]) to define energy sender and receiver. Furthermore, additional energy transport paths using complementary energy carriers are opened. Since the current SOCs of BESS and load predictions are considered, this concept is predestined for future grids, see Sect. 1.1.1.

2.3 Requirements on VSC Control and Grid Protection

Analysing the frequency characteristics of 100% IBPS (see Sect. 2.2), different requirements on the VSC control appear. Several physical and technical constraints (see Sect. 2.1) realizing these requirements have to be considered.

The capability of island operation and the properties discussed in Sect. 2.2.2 (instantaneous power, frequency inertia, power damping and droop characteristics) are essential for electrical energy supply in 100% IBPS. Furthermore, VSCs must operate in strong grids and especially in weak grids (see Sect. 4.1.1.2). Renewable energy sources (RES) are often located in remote sites and the line impedance tends to be high. The control accuracy of currents or voltages has to be ensured to reach operation points exactly regarding grid control, see Sect. 2.2.3. Intuitive tuning of parameters of the VSC controller is advantageous for the usability by technical staff. The tuning of parameters of commonly used PI-controller is not a straightforward process [143].

During grid faults, flexibility options of the negative and the zero sequence open new ways to increase the resilience of the system during grid faults. State-of-the-art VSCs are connected as a three-phase three-wire system to the grid and represent an open circuit in the zero sequence. Current and voltage limits of semiconductor power devices must not be exceeded in the transient state and the steady state to prevent the destruction of these devices, see Sect. 2.1.1. Today, especially the current limiting in the transient state is insufficient. VSC-controllers provide fast response after contingencies, since the available time interval to provide instantaneous power is limited due to energy reasons. Today, PI-controllers do not exploit the full potential of GFM-VSCs, because of slow cascaded control structures [143].

Grid protection algorithms must ensure the requirements on the properties selectivity, sensitivity and speed. Characteristics of voltages and currents during grid faults induced by VSCs distinguish from the characteristics induced by synchronous machines. Today, some assumptions of distance protection algorithms designed for grids based on synchronous machines no longer apply or only apply to a limited extent. These assumptions refer for example to dynamic phase angles between voltages and currents and a higher voltage drop at the relay location based on the source impedance ratio using VSCs [143].

Protection algorithms must operate independently of signal characteristics. VSCs cause harmonic frequency components of voltages and currents regarding the switching frequency due to their operating principle, see Sect. 2.1.1. Moreover, VSC currents are limited due to the ratings of semiconductor power devices. Deviations from sinusoidal characteristics during transient current limiting occur, see Sect. 6.3.1.5. Beyond that, the half-bridges of VSCs can be controlled independently of each other (see Sect. 2.1.2) and asymmetrical three-phase systems including the positive sequence, the negative sequence and the zero sequence appear.

Protection algorithms must apply for new methods of neutral point treatment (NPT) realized by VSCs causing specific characteristics of voltages and currents, see [153] and Sect. 6.2.2.2. The type of NPT is realized by VSC controllers. Thus, NPT will be more dynamic in the future and changing the type of NPT will be realized at high speed. Today, NPT is based on a static and low speed philosophy.

Furthermore, protection algorithms must apply for new grid architectures (see Sect. 1.1.3) and grid operation strategies (see Sect. 1.1.4). The integration of renewable energy sources (RES) will lead to intermediate infeed (see Sect. 7.2.2.3) due to their local distribution. Parallel lines (see Sect. 7.2.2.1) will be used to increase the transport capacity needed for innovative CO_2-friendly consumer technologies such as electric vehicles (EVs) and heat pumps (HPs), see Sect. 1.1.1. The direction of power flows will not be predetermined any more, e.g. due to the integration of energy storage systems (ESS).

2.4 Research Approach

In this chapter, the approach for the control and protection of 100% inverter-based power systems (IBPS) meeting the research objectives from Sect. 1.2.3 and fulfilling the requirements from Sect. 2.3 is defined. The overarching objective is to contribute to a sustainable and reliable energy system.

This thesis focuses on three-phase AC-systems. However, many principles presented in this thesis can also be adopted for DC-systems in future research activities.

A holistic approach for the control and protection of 100% IBPS is chosen. The control of voltage source converters (VSCs) and the grid protection algorithm are not regarded separately from each other. These research areas are connected to each other and dependencies from each other exist. The developed VSC control algorithms as well as the developed grid protection algorithm are based on state-space models. Dual system theory properties are used for their design.

A grid-forming (GFM) controller enabling island operation, instantaneous power, frequency inertia, power damping and frequency droop characteristics in the positive sequence is designed. To realize constant active power or the control of the

DC-link voltage from the AC-side for selected VSCs, this concept is combined with an advanced phase-locked loop (PLL) to a hybrid synchronization mode.

A state-space controller is developed to control the positive sequence, the negative sequence and the zero sequence of an extended three-phase four-wire VSC. The flexibility options of VSCs are used to increase the resilience of 100% IBPS. State-space controllers enable improved dynamic performance of GFM-VSCs compared to proportional-integral (PI) controllers. Additional system properties (asymptotic stability, controllability and observability) are identified in the state-space. The parameters of the presented state-space controllers are adjusted intuitively using weighting matrices.

A model-based protection algorithm is designed. This algorithm meets the requirements on the properties selectivity, sensitivity and speed. The flexibility options of VSCs are not restricted by the protection algorithm. Due to the model-based approach, a universal protection principle regarding voltage and current characteristics induced by VSCs, dynamic neutral point treatment, grid architectures and grid operation is developed.

2.5 Summary

In this chapter, the electrical energy supply in 100% inverter-based power systems (IBPS) is presented. The fundamentals of voltage source converters (VSCs) and the grid control define the physical and technical constraints in this thesis.

This chapter describes the technical correlations between the research areas discussed in Chap. 1. A research approach to meet the research objectives formulated in Sect. 1.2.3 is defined.

VSCs are power electronic converters of electrical power without energy storage capability. Reference voltage signals are synthesized by realizing different switching states and connecting different electrical potentials of the DC-link to the AC-side. Therefore, the value of the DC-link voltage must not fall below a limit given by the AC-voltage. The current limit of semiconductor power devices must not exceed a certain value, which is indicated by its nominal value. The voltage/current characteristics of VSCs depend on their control algorithm. The half-bridges representing basic elements of VSCs can be operated independently of each other. Multiple flexibility options of VSCs appear.

Due to the functional principle of VSCs, there is no physical link between the frequency of their output voltages and a variable describing the energy of an energy storage component. The frequency is a measure to adjust the active power via the voltage angle. Instantaneous power, frequency inertia, power damping and fre-

2.5 Summary

quency droop characteristics are essential properties of grid-connected VSCs inside a 100% IBPS. These properties and the capability of island operation are realized applying the grid-forming (GFM) control mode of VSCs in the positive sequence.

The grid control of 100% IBPS after changing load conditions or grid faults is similar to grids based on synchronous machines. Frequency droop characteristics, which are based on three layers (primary control, secondary control and tertiary control) separated in time, are used. Regarding secondary control, no additional positive or negative energy flow into or from the grid is required. In contrast to grids based on synchronous machines, no rotating masses have to be accelerated or decelerated.

The physical and technical constraints imposed by the hardware of the VSC have to be considered during the control design. In addition, the grid control defines requirements for the VSC controller.

The grid protection algorithm has to be universal regarding the characteristics of the VSC. Furthermore, the grid protection algorithm has to match suitable grid architectures and grid operation strategies for the integration of renewable energy sources.

A holistic research approach for the control and protection of 100% IBPS is chosen. This research approach is pursued consequently throughout this thesis. In the following chapters, the constraints and requirements of this chapter are considered. To follow the key aspects of the central theme of this thesis on a shortcut: The conclusion of the next chapter is provided in Sect. 3.4.

Open Access This chapter is licensed under the terms of the Creative Commons Attribution 4.0 International License (http://creativecommons.org/licenses/by/4.0/), which permits use, sharing, adaptation, distribution and reproduction in any medium or format, as long as you give appropriate credit to the original author(s) and the source, provide a link to the Creative Commons license and indicate if changes were made.

The images or other third party material in this chapter are included in the chapter's Creative Commons license, unless indicated otherwise in a credit line to the material. If material is not included in the chapter's Creative Commons license and your intended use is not permitted by statutory regulation or exceeds the permitted use, you will need to obtain permission directly from the copyright holder.

3 Electrical quantities, state-space modelling and system properties

This chapter presents the mathematical description of electrical quantities, the state-space modelling of electrical systems and the analysis of system properties. After a fundamental classification of types of description of electrical quantities, phasors and space vectors are presented in Sect. 3.1. Furthermore, a general representation of a three-phase four-wire system is given. State-space models are clear representations of physical laws of comprehensive systems. In Sect. 3.2, the concept of the state-space representation is introduced and models for L-, LC- and LCL-structures are developed. Discretization methods for state-space models are presented to enable control and protection algorithms to be executed on digital computers. In Sect. 3.3, the asymptotic stability, the controllability and the observability are introduced as system properties. Moreover, the duality of controllability and observability is shown.

3.1 Description of Electrical Quantities

3.1.1 Classification of Types of Description

Voltages and currents are time-dependent electrical quantities. Their time curves represent numerical sequences of real instantaneous values. The shape of their time curve is categorized into direct signals, alternating signals and mixed signals.

Physical laws (e.g. Kirchhoff's voltage law (KVL) and Kirchhoff's current law (KCL)) and voltage/current-characteristics (V/I-characteristics) describing electrical properties of components represent relations between voltages and currents [72]. These relations are often formulated as ordinary differential equations. So, the unknown quantities consist of functions and derivatives of one variable [73]. Analysing locally limited electrical circuits, voltages and currents are partly unknown quantities which are only dependent on time.

The analysis of electrical circuits often leads to heterogeneous, linear differential equations. The solution of a heterogeneous, linear differential equation is composed by a homogeneous and a particular solution [73]. The homogeneous solution describes the transient state and disappears after the transient state. The particular solution describes the steady state. The background and the aim of the investigation defines, whether the transient state behaviour or the steady state behaviour or both are of interest. For example, a steady state analysis is expedient for normal grid operation or steady state fault behaviour. In contrast, a transient state analysis is necessary for the investigation of control and protection algorithms. The transient behaviour caused by step changes (e.g. changes of the reference value or the grid topology) is crucial.

Voltages and currents are described by real or complex numbers. Complex numbers allow simpler mathematical handling of equations compared to equations based on real numbers. Here, complex algebraic equations instead of differential equations have to be solved. [72]

The types of description of voltages and currents can further be classified according to their calculation basis. The steady state of periodic signals is characterized by repeating time sequences. Therefore, a time interval is investigated. The length of this time interval is mostly chosen to the period duration. Mean- and root-mean-square- (RMS-) values are integral values of one time interval [72]. In the transient state, single points in time are relevant.

The Tab. 3.1 provides a classification of types of description for voltages and currents. They are distinguished by the number system and the calculation basis. Subsequently, the complex description of voltages and currents is used. These types of description simplify developing control and protection algorithms.

Table 3.1 Classification of types of description of voltages and currents

| | | number system | |
		real	complex
calculation basis	point in time	instantaneous value	space vector
	time interval	mean/RMS-value	RMS/amplitude-phasor

3.1.2 Description as Complex Numbers

3.1.2.1 Phasors

Electrical alternating current (AC) power grids are operated based on sinusoidal voltages and currents. An electromagnetic transformer can be used and the basic time curve is preserved after the time derivation according to the electromagnetic induction law [74]. Furthermore, reactive power can be used to adjust the voltage independent of active power [143]. Moreover, the tripping of circuit breakers is facilitated due to zero crossings of voltages and currents. The following descriptions are based on voltages, but they are identical for currents in general.

Sinusoidal time curves are represented as follows, see Eq. (3.1).

$$v(t) = \hat{V}\cos(\gamma_v(t)) \tag{3.1}$$

It is assumed, that only the steady state of an electrical circuit with monofrequency sources and linear elements is investigated. The argument $\gamma_v(t)$ is a linear function of time and ω_v is assumed to be independent of time, see Eq. (3.2).

$$\gamma_v(t) = \omega_v t + \varphi_v \tag{3.2}$$

To simplify the calculation of the particular solution of the differential equations, a complex algebraic system of equations in the frequency domain based on phasors is formulated. The term phasor is a portmanteau of phase vector. The variable $v(t)$ is transformed from the time domain to the frequency domain, see Eq. (3.3).

$$v(t) \multimap \underline{\hat{V}} \tag{3.3}$$

The variable $\underline{\hat{V}}$ (see Eq. (3.4)) is named the amplitude phasor.

$$\underline{\hat{V}} = \hat{V} e^{j\varphi_v} \tag{3.4}$$

The solution of the system of equations is based on a time-independent consideration. That means that the calculation in the frequency domain is independent of the frequency, since time variance is expressed by the product $\omega_v t$ in Eq. (3.2). As a consequence, the amplitude and the phase angle determine the solution of the system of equations. They are regarded as time-independent in this case.

Phasors can be interpreted as stationary vectors in the complex plane. The formulations of phasors are based on Euler's formula, which is seen in the inverse transformation, see Eq. (3.5) [75].

$$v(t) = \operatorname{Re}\left\{\underline{\hat{V}} e^{j\omega_v t}\right\} \tag{3.5}$$

3.1.2.2 Symmetrical Components

Electrical AC power grids are generally operated as three-phase four-wire systems. One benefit over single-phase systems is, that electrical energy is transferred at constant instantaneous power [76].

The steady state of the according quantities are calculated using phasors (see Sect. 3.1.2.1) of each phase (index 1, 2, 3). Often, phase-to-neutral voltages and phase currents are used. In this way, the active power during normal operation conditions is calculated.

The calculation of three-phase four-wire systems in the steady state is simplified by using symmetrical components to describe voltages and currents by transformed phasors. The transformation rule is shown in Eq. (3.6) [71].

$$\begin{pmatrix} \hat{\underline{V}}_1 \\ \hat{\underline{V}}_2 \\ \hat{\underline{V}}_3 \end{pmatrix} = \mathbf{\underline{T}}_S \begin{pmatrix} \hat{\underline{V}}_{(1)} \\ \hat{\underline{V}}_{(2)} \\ \hat{\underline{V}}_{(0)} \end{pmatrix} \tag{3.6}$$

The symmetrical components are composed of three components. The index (1) represents the positive sequence, the index (2) represents the negative sequence and the index (0) represents the zero sequence. The transformation matrix from Eq. (3.7) and the complex number $\underline{a}$ from Eq. (3.8) is used.

$$\mathbf{\underline{T}}_S = \begin{pmatrix} 1 & 1 & 1 \\ \underline{a}^2 & \underline{a} & 1 \\ \underline{a} & \underline{a}^2 & 1 \end{pmatrix} \tag{3.7}$$

$$\underline{a} = e^{j\frac{2}{3}\pi} \tag{3.8}$$

Symmetrical components are a special case of modal transformations. The basic idea of modal transformations is to diagonalize a symmetric diagonal-cyclic matrix. These matrices often appear describing three-phase systems. In this way, three single-phase electrical circuits appear. They are decoupled from each other during normal operation conditions and are calculated independently. If a grid fault occurs, the three single-phase circuits can be coupled. Symmetrical components are used to decouple complex algebraic equations for each phase and represent transformed phasors. [71]

3.1.2.3 Space Vectors

To describe transient states in three-phase four-wire systems, symmetrical components (see Sect. 3.1.2.2) are not suitable. Due to the integral behaviour of phasors, crucial effects of control and protection, e.g. changes of disturbance variables and transient grid fault characteristics, are not represented in detail.

Applying the principle of the time-variant space vector and the time-dependent zero sequence component, instantaneous values are decoupled from each other. The calculation basis is a point in time. In contrast, symmetrical components decouple time-independent phasors from each other and the calculation basis is a time interval.

The following transformation from the complex space vector $\underline{v}$, the complex conjugate space vector $\underline{v}^*$ and the double real zero sequence component v_h to phase quantities is shown in Eq. (3.9) [71]:

$$\begin{pmatrix} v_1 \\ v_2 \\ v_3 \end{pmatrix} = \frac{1}{2}\underline{\mathbf{T}}_S \begin{pmatrix} \underline{v} \\ \underline{v}^* \\ v_h \end{pmatrix} \qquad (3.9)$$

As well as symmetrical components, this transformation is a special case of modal transformations and the transformation matrix $\underline{\mathbf{T}}_S$ from Eq. (3.7) is used. Apart from the constant factor 0.5, the transformation matrix of space vector/zero sequence component and symmetrical components is identical. This fact leads to the advantage, that the fault conditions and the calculation algorithm using symmetrical components can be adopted to calculate grid fault conditions using space vectors and zero sequence components [71].

The real part of the space vector $\underline{v}$ in a stationary reference frame is called the alpha-component v_α. The imaginary part of the space vector $\underline{v}$ in a stationary reference frame is called the beta-component v_β. The zero sequence component is described as v_0. The relations are given in Eq. (3.10).

$$\begin{pmatrix} \underline{v} \\ \underline{v}^* \\ v_h \end{pmatrix} = \begin{pmatrix} 1 & j & 0 \\ 1 & -j & 0 \\ 0 & 0 & 2 \end{pmatrix} \begin{pmatrix} v_\alpha \\ v_\beta \\ v_0 \end{pmatrix} \qquad (3.10)$$

It is seen, that v_α, v_β and v_0 contain the information about the phase quantities v_1, v_2 and v_3. So, the complex conjugate space vector $\underline{v}^*$ contains no additional information compared to the complex space vector $\underline{v}$. It is only introduced for mathematical reasons.

Using Eq. (3.9) and Eq. (3.10), the following direct transformation from $\alpha\beta 0$-quantities to 123-quantities is gained, see Eq. (3.11).

$$\begin{pmatrix} v_1 \\ v_2 \\ v_3 \end{pmatrix} = \frac{1}{2} \begin{pmatrix} 2 & 0 & 2 \\ -1 & \sqrt{3} & 2 \\ -1 & -\sqrt{3} & 2 \end{pmatrix} \begin{pmatrix} v_\alpha \\ v_\beta \\ v_0 \end{pmatrix} \qquad (3.11)$$

Here, the following transformation matrix according to Eq. (3.12) is used.

$$\mathbf{T}_{\alpha\beta 0 \to 123} = \frac{1}{2} \begin{pmatrix} 2 & 0 & 2 \\ -1 & \sqrt{3} & 2 \\ -1 & -\sqrt{3} & 2 \end{pmatrix} \qquad (3.12)$$

From Eq. (3.11) is seen, that the zero sequence component effects all phase quantities in the same way. So, the zero sequence component of phase-to-neutral voltage can be interpreted as a shift of the neutral point.

Using Eq. (3.10) and Eq. (3.11), the following calculation rules for the space vector and the zero sequence component from the phase quantities are gained, see Eq. (3.13) and Eq. (3.14).

$$\underline{v} = \frac{2}{3}(v_1 + \underline{a}v_2 + \underline{a}^2 v_3) \qquad (3.13)$$

$$v_0 = \frac{1}{3}(v_1 + v_2 + v_3) \qquad (3.14)$$

It is obvious, that the space vector and the zero sequence component are independent of each other.

Applying Eq. (3.14) to phase currents, this equation can be interpreted as a KCL with a constant factor. That means that the zero sequence component of phase currents can be interpreted as a scaled current in the neutral line.

3.1 Description of Electrical Quantities

3.1.2.4 Synchronous Reference Frame

Subsequently, the steady state of a monofrequency three-phase system is investigated. The system is described by a positive sequence phasor, see Eq. (3.15). The negative and the zero sequence phasors are zero.

$$\underline{\hat{V}}_{(1)} = \hat{V}_{(1)} e^{j\varphi_{(1)}} \tag{3.15}$$

This steady state leads to a special case of a space vector (see Sect. 3.1.2.3) and is of great importance for the normal grid operation.

Assuming ω to be independent of time and using Eq. (3.6) and Eq. (3.5), the phase quantities appear according to Eq. (3.16).

$$\begin{pmatrix} v_1 \\ v_2 \\ v_3 \end{pmatrix} = \hat{V}_{(1)} \begin{pmatrix} \cos\left(\omega t + \varphi_{(1)}\right) \\ \cos\left(\omega t + \varphi_{(1)} - \frac{2}{3}\pi\right) \\ \cos\left(\omega t + \varphi_{(1)} + \frac{2}{3}\pi\right) \end{pmatrix} \tag{3.16}$$

Applying Eq. (3.13), the space vector is described as follows, see Eq. (3.17).

$$\underline{v} = \underline{\hat{V}}_{(1)} e^{j\omega t} \tag{3.17}$$

Regarding one period, the space vector draws a circular path in the complex plane. The Eq. (3.17) shows the relation between the positive sequence phasor and the space vector in this special case.

If the space vector $\underline{v}$ is regarded in a reference system, which is rotated by the angle γ_T mathematically positive, the new components v_d and v_q appear. This transformation leads to benefits regarding the properties of the space vector according to Eq. (3.17). The components in the rotated reference frame are calculated by Eq. (3.18) using the transformation matrix in Eq. (3.19).

$$\begin{pmatrix} v_d \\ v_q \end{pmatrix} = \begin{pmatrix} \cos(\gamma_T) & \sin(\gamma_T) \\ -\sin(\gamma_T) & \cos(\gamma_T) \end{pmatrix} \begin{pmatrix} v_\alpha \\ v_\beta \end{pmatrix} \tag{3.18}$$

$$\mathbf{T}(\gamma_T) = \begin{pmatrix} \cos(\gamma_T) & \sin(\gamma_T) \\ -\sin(\gamma_T) & \cos(\gamma_T) \end{pmatrix} \tag{3.19}$$

The Eq. (3.18) is equivalent to the multiplication of $\underline{v}$ from Eq. (3.17) with $e^{-j\gamma_T}$.

In general, the transformation angle γ_T is freely selectable. So, γ_T can be chosen to be linear dependent from time, see Eq. (3.20).

$$\gamma_T(t) := \omega_T t + \varphi_T \qquad (3.20)$$

If the angular frequency ω_T of the transformation angle is adjusted to the angular frequency ω of the space vector, the space vector gets time-independent regarding the rotating reference frame, see also Eq. (3.17). In this case, the transformation angle γ_T is calculated according to Eq. (3.21).

$$\gamma_T(t) \stackrel{!}{=} \omega t + \varphi_T \qquad (3.21)$$

This special reference frame is called the synchronous reference frame (SRF). The components v_d and v_q represent direct current (DC) quantities and are time-independent. It is beneficial to design control algorithms based on models using these components. The reason is, that the operating frequency of the control loop should be much lower than the crossover frequency of the open-loop [143].

The Fig. 3.1 illustrates the principle of a SRF and the components of a stationary reference frame and SRF-components of the space vector $\underline{v}$.

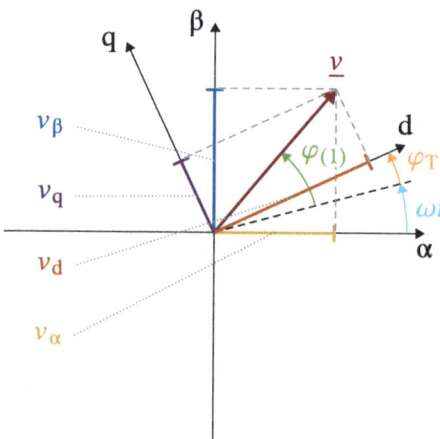

Fig. 3.1 Space vector $\underline{v}$ and its components in a stationary reference frame (v_α and v_β) and a positive rotating synchronous reference frame (v_d and v_q) as projections on the respective coordinate axis

3.1 Description of Electrical Quantities

If additionally φ_T is selected to equal $\varphi_{(1)}$, then the space vector is aligned to the d-axis and v_q equals to zero.

3.1.2.5 Double Synchronous Reference Frame

In this section, a monofrequency three-phase system is investigated in the steady state. The system is described by the sum of a positive sequence phasor (see Eq. (3.15)) and a negative sequence phasor (see Eq. (3.22)). The zero sequence phasor is zero.

$$\underline{\hat{V}}_{(2)} = \hat{V}_{(2)} e^{j\varphi_{(2)}} \tag{3.22}$$

This steady state leads to a special case of a space vector (see Sect. 3.1.2.3) and extends the considerations of Sect. 3.1.2.4. It is of great importance for asymmetrical grid fault conditions.

Assuming ω to be independent of time and using Eq. (3.6) and Eq. (3.5), the phase quantities appear according to Eq. (3.23).

$$\begin{pmatrix} v_1 \\ v_2 \\ v_3 \end{pmatrix} = \hat{V}_{(1)} \begin{pmatrix} \cos(\omega t + \varphi_{(1)}) \\ \cos(\omega t + \varphi_{(1)} - \frac{2}{3}\pi) \\ \cos(\omega t + \varphi_{(1)} + \frac{2}{3}\pi) \end{pmatrix} + \hat{V}_{(2)} \begin{pmatrix} \cos(\omega t + \varphi_{(2)}) \\ \cos(\omega t + \varphi_{(2)} + \frac{2}{3}\pi) \\ \cos(\omega t + \varphi_{(2)} - \frac{2}{3}\pi) \end{pmatrix} \tag{3.23}$$

Applying Eq. (3.13), the space vector is described as Eq. (3.24).

$$\underline{v} = \underline{\hat{V}}_{(1)} e^{j\omega t} + \underline{\hat{V}}^*_{(2)} e^{-j\omega t} \tag{3.24}$$

Regarding one period, the space vector draws an elliptical path in the complex plane. The Eq. (3.24) shows the relation between the two phasors and the space vector in this special case.

According to Eq. (3.24) the space vector $\underline{v}$ is interpreted as the sum of a positive sequence space vector $\underline{v}^+$ and a negative sequence space vector $\underline{v}^-$, see Eq. (3.25).

$$\underline{v} = \underline{v}^+ + \underline{v}^- \tag{3.25}$$

In the following investigations, the space vectors $\underline{v}^+$ and $\underline{v}^-$ are regarded independently of each other. Applying the principle of Sect. 3.1.2.4 individually to the space vectors $\underline{v}^+$ and $\underline{v}^-$ to gain time-independent components, two different rotating reference frames are used.

The positive sequence space vector $\underline{v}^+$ is transformed analogous to Sect. 3.1.2.4 using γ_T^+. If ω_T^+ equals to ω, then $\underline{v}^+$ is time-independent and represents a stationary space vector in the positive synchronous reference frame. In this case, the transformed components v_d^+ and v_q^+ are DC-quantities. If additionally φ_T^+ is selected to equal $\varphi_{(1)}$, then the $\underline{v}^+$ is aligned to the d^+-axis and v_q^+ equals zero. The component v_d^+ equals the amplitude of the positive sequence $\hat{V}_{(1)}$.

The negative sequence space vector $\underline{v}^-$ is transformed using γ_T^- (see also Eq. (3.20)). If ω_T^- equals $-\omega$, then $\underline{v}^-$ is time-independent and represents a stationary space vector in the negative synchronous reference frame. In this case, the transformed components v_d^- and v_q^- are DC-quantities. If additionally φ_T^- is selected to equal $-\varphi_{(2)}$, then the $\underline{v}^-$ is aligned to the d^--axis and v_q^- equals zero. The component v_d^- equals the amplitude of the negative sequence $\hat{V}_{(2)}$.

The reference frame of the positive sequence rotates mathematically positive with the angular frequency of the space vector $\underline{v}^+$. The reference frame of the negative sequence rotates mathematically negative with the angular frequency of the space vector $\underline{v}^-$. This principle with two synchronous rotating reference frames is called the double synchronous reference frame (DSRF) and four DC-quantities are gained.

As explained in Sect. 3.1.2.4, this fact is beneficial for the design of control algorithms based on models using these quantities. In this way, the positive and the negative sequence can be controlled independently of each other.

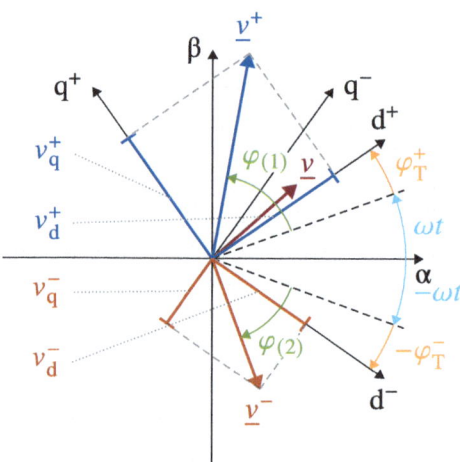

Fig. 3.2 Space vector $\underline{v}$ composed by a positive sequence space vector $\underline{v}^+$ and a negative sequence space vector $\underline{v}^-$ and their components in a positive rotating synchronous reference frame (v_d^+ and v_q^+) and a negative synchronous rotating reference frame (v_d^- and v_q^-), respectively

3.1 Description of Electrical Quantities

The Fig. 3.2 illustrates the space vector $\underline{v}$, the space vector $\underline{v}^+$ of the positive sequence and the space vector $\underline{v}^-$ of the negative sequence. Besides, the principle of the DSRF is shown.

This concept also shows, that regarding the positive sequence space vector $\underline{v}^+$ in the negative sequence reference frame leads to AC-components with the double angular frequency. At the same time, regarding the negative sequence space vector $\underline{v}^-$ in the positive sequence reference frame also leads to AC-components with the double angular frequency. This effect is used to calculate v_d^+, v_q^+, v_d^- and v_q^- from the space vector $\underline{v}$ (see Eq. (3.13)) using low-pass filters, see Sect. 4.3.3.

3.1.3 General Representation of a Three-phase Four-wire System

In this section, the steady state of a polyfrequency three-phase four-wire system is investigated. The steady state is described by arbitrary periodic signals, in general. The Fourier analysis is applied on arbitrary periodic signals to decompose these signals into frequency components [72]. So, a periodic signal is described as a sum of frequency components at multiples of the fundamental frequency, which are independent of each other. The spectral form is a certain representation of the sum of frequency components.

According to this principle, a polyfrequency three-phase four-wire system is described according to Eq. (3.26) [77]. The mean values $\overline{v}_1$, $\overline{v}_2$ and $\overline{v}_3$ are also represented. They are separated by low-pass filters.

$$\begin{pmatrix} v_1 \\ v_2 \\ v_3 \end{pmatrix} = \begin{pmatrix} \overline{v}_1 \\ \overline{v}_2 \\ \overline{v}_3 \end{pmatrix} + \sum_{n=1}^{\infty} \hat{V}_{(1)}^n \begin{pmatrix} \cos\left(n\omega t + \varphi_{(1)}^n\right) \\ \cos\left(n\omega t + \varphi_{(1)}^n - \tfrac{2}{3}\pi\right) \\ \cos\left(n\omega t + \varphi_{(1)}^n + \tfrac{2}{3}\pi\right) \end{pmatrix} \quad (3.26)$$

$$+ \sum_{n=1}^{\infty} \hat{V}_{(2)}^n \begin{pmatrix} \cos\left(n\omega t + \varphi_{(2)}^n\right) \\ \cos\left(n\omega t + \varphi_{(2)}^n + \tfrac{2}{3}\pi\right) \\ \cos\left(n\omega t + \varphi_{(2)}^n - \tfrac{2}{3}\pi\right) \end{pmatrix}$$

$$+ \sum_{n=1}^{\infty} \hat{V}_{(0)}^n \begin{pmatrix} \cos\left(n\omega t + \varphi_{(0)}^n\right) \\ \cos\left(n\omega t + \varphi_{(0)}^n\right) \\ \cos\left(n\omega t + \varphi_{(0)}^n\right) \end{pmatrix}$$

Generalizing and extending the principles of Sect. 3.1.2.5, the positive sequence space vector and the negative sequence space vector are composed by sums of positive or negative sequence space vectors for each frequency component. These sums add to the resulting space vector, see Eq. (3.27). The separation of frequency components of positive and negative sequence vectors is shown in Sect. 4.3.4.1.

$$\underline{v} = \sum_{n=1}^{\infty} \underline{v}^{+n} + \sum_{n=1}^{\infty} \underline{v}^{-n} \tag{3.27}$$

In Eq. (3.26) the zero sequence is explicitly considered. The zero sequence is also composed as a sum of different frequency components, see Eq. (3.28). The separation of the sum of the frequency components of the zero sequence from the phase quantities v_1, v_2 and v_3 is realized by Eq. (3.14). The frequency components are separated from each other by low-pass filters.

$$v_0 = \sum_{n=1}^{\infty} \hat{V}_{(0)}^n \cos\left(n\omega t + \varphi_{(0)}^n\right) \tag{3.28}$$

The representation in Eq. (3.26) and the possibility to separate symmetrical components from each other and their individual frequency components lead to subsystems. These subsystems are handled independent of each other using, see for example the principles of Sect. 3.1.2.4 and Sect. 3.1.2.5. The handling of the zero sequence is shown in Sect. 6.2.2.1. That also means that each subsystem is controlled independently. This fact is also valid for the mean values, see Sect. 6.2.1.2.

3.2 State-space Models

3.2.1 State Variables and State-Space Representation

Differential equations are essential to describe relationships between voltages and currents by Kirchhoff's laws and V/I-characteristics. The order of the highest differential coefficient is called the order of the differential equation. Each n-th order linear differential equation can be reformulated as n first-order linear differential equations [78].

The following assumptions are made for the subsequent investigations: The regarded systems are time-invariant. That means, among other things, that the properties of passive components (e.g. R, L and C) are not time-dependent. Beyond that, linear systems are regarded, since many components of electrical power systems behave approximately linear or are at least linearisable at operating points. Moreover, systems with feedthrough properties are not investigated.

For each first-order differential equation, one state variable is defined. State variables describe voltages or currents, which are time-derived in the differential equations. So, the number of state variables equals the order of the system. In electrical systems, state variables reflect the energy of a component, e.g. the capacitor voltage or the inductor current.

State variables are inner system variables. The entirety of all state variables of a system describes the state of the system in the state-space. If the state of one point in time and the time functions of all input variables are known, the state and the output variables of other points in time can be calculated [78]. The system of differential equations is described according to Eq. (3.29).

$$\frac{d}{dt}\mathbf{x}(t) = \mathbf{A}\mathbf{x}(t) + \mathbf{B}\mathbf{u}(t) + \mathbf{E}\mathbf{z}(t) \qquad (3.29)$$

The system matrix $\mathbf{A}$ characterizes the influence of the state variables $\mathbf{x}$ on their time derivatives. The input matrix $\mathbf{B}$ defines the influence of the actuating variables $\mathbf{u}$ on the time derivative of $\mathbf{x}$. Actuating variables are manipulated actively. The disturbance matrix $\mathbf{E}$ specifies the influence of the disturbance variables $\mathbf{z}$ on the time derivative of $\mathbf{x}$. Disturbance variables cannot be manipulated actively.

A system of algebraic equations describes the relation between the state variables $\mathbf{x}$ and the output variables $\mathbf{y}$. These equations are called the output equations, see Eq. (3.30). In some cases, these equations can be designed in a way, that the output variables are a subset of the state variables.

$$\mathbf{y}(t) = \mathbf{C}\mathbf{x}(t) \tag{3.30}$$

The systems of state differential equations (see Eq. (3.29)) and output equations (see Eq. (3.30)) are called the state equations. A state-space model of a system is based on the state equations. The state-space model is visualized as a block diagram, see Fig. 3.3.

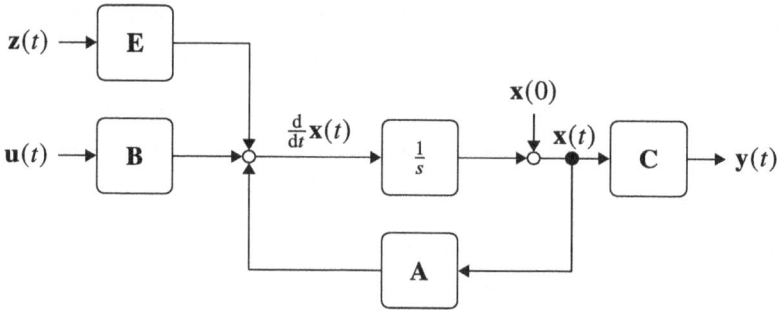

Fig. 3.3 State-space model based on the state equations visualized as a block diagram

State-space models allow a compact and clear representation of a system. Essential variables are explicitly categorized as input variables, state variables and output variables. Input variables can further be categorized as actuating variables and disturbance variables. Moreover, state-space models allow the easy handling of multiple-input multiple-output (MIMO) systems with more than one input and output variables. In addition to the input-output behaviour, the inner system is also regarded. This is crucial for maximum values and stability in the inner system. Moreover, important system properties like controllability (see Sect. 3.3.3) and observability (see Sect. 3.3.4) can be investigated.

The use of state-space models based on space vector components (see Sect. 3.1.2.3, Sect. 3.1.2.4 and Sect. 3.1.2.5) are beneficial, since they allow the description of transient states by differential equations. In general, space vector components of a stationary or rotating reference frames are used. By transforming the input variables, state variables and output variables, the state-space models can be reformulated. Important system properties (e.g. asymptotic stability, controllability and observability) are not affected [78].

Transfer functions can be derived from state-space models. The representation of a system based on transfer functions is for example used in the tuning process

3.2 State-space Models

of proportional-integral (PI) controllers [143]. In this input-output view, the inner system properties are not regarded explicitly. By using the Laplace-variable s, the system of differential equations in Eq. (3.29) in the time domain can be transformed into a system of algebraic equations in the frequency domain, see Eq. (3.31) and Eq. (3.32).

$$s\mathbf{X}(s) = \mathbf{A}\mathbf{X}(s) + \mathbf{B}\mathbf{U}(s) + \mathbf{E}\mathbf{Z}(s) \tag{3.31}$$

$$\mathbf{X}(s) = (s\mathbf{I} - \mathbf{A})^{-1}\mathbf{B}\mathbf{U}(s) + (s\mathbf{I} - \mathbf{A})^{-1}\mathbf{E}\mathbf{Z}(s) \tag{3.32}$$

The system of output equations is described according to Eq. (3.33).

$$\mathbf{Y}(s) = \mathbf{C}(s\mathbf{I} - \mathbf{A})^{-1}\mathbf{B}\mathbf{U}(s) + \mathbf{C}(s\mathbf{I} - \mathbf{A})^{-1}\mathbf{E}\mathbf{Z}(s) \tag{3.33}$$

Based on this representation, the matrix of the transfer functions of the actuating variables are formulated, see Eq. (3.34).

$$\mathbf{G_u}(s) = \left.\frac{\mathbf{Y}(s)}{\mathbf{U}(s)}\right|_{\mathbf{Z}(s)=0} \tag{3.34a}$$

$$= \mathbf{C}(s\mathbf{I} - \mathbf{A})^{-1}\mathbf{B} \tag{3.34b}$$

Moreover, the matrix of the transfer functions of the disturbance variables can formulated according to Eq. (3.35).

$$\mathbf{G_z}(s) = \left.\frac{\mathbf{Y}(s)}{\mathbf{Z}(s)}\right|_{\mathbf{U}(s)=0} \tag{3.35a}$$

$$= \mathbf{C}(s\mathbf{I} - \mathbf{A})^{-1}\mathbf{E} \tag{3.35b}$$

3.2.2 State-space Models of Selected Circuit Structures

In this section, the theoretical considerations from Sect. 3.2.1 are applied to essential circuit structures of electrical power systems. State-space models for monofrequency (angular frequency ω) three-phase four-wire systems are shown. By using components of the positive and the negative sequence space vectors (see Sect. 3.1.2.5) and the zero sequence (see Sect. 3.1.2.3), transient states are described. These models are based on KVL, KCL and V/I-characteristics. The presented models are especially relevant for the control and protection algorithms in this thesis.

3.2.2.1 L-structures

The equivalent circuit of a lossy three-phase four-wire L-structure connecting two voltage sources is shown in Fig. 3.4.

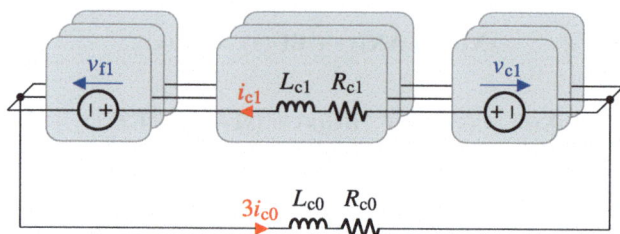

Fig. 3.4 Equivalent circuit of a lossy three-phase four-wire L-structure connecting two voltage sources

This equivalent circuit describes for example a simplified L-filter of a grid-connected voltage source converter (VSC) modelled by the VSC voltages v_{cx}. This filter can be realized by one choke per phase.

It is assumed that the passive components are equal in each phase x: $R_{cx} = R_c$, $L_{cx} = L_c$ with $x \in \{1, 2, 3\}$. The according state-space model using the space vector components with two state variables, two actuating variables and two disturbance variables is described in accordance with Eq. (3.36).

$$\frac{d}{dt}\begin{pmatrix} i_{cd}^{\pm} \\ i_{cq}^{\pm} \end{pmatrix} = \begin{pmatrix} -\frac{R_c}{L_c} & \pm\omega \\ \mp\omega & -\frac{R_c}{L_c} \end{pmatrix} \begin{pmatrix} i_{cd}^{\pm} \\ i_{cq}^{\pm} \end{pmatrix} + \begin{pmatrix} \frac{1}{L_c} & 0 \\ 0 & \frac{1}{L_c} \end{pmatrix} \begin{pmatrix} v_{cd}^{\pm} \\ v_{cq}^{\pm} \end{pmatrix} \qquad (3.36)$$

$$+ \begin{pmatrix} -\frac{1}{L_c} & 0 \\ 0 & -\frac{1}{L_c} \end{pmatrix} \begin{pmatrix} v_{fd}^{\pm} \\ v_{fq}^{\pm} \end{pmatrix}$$

The state-space model using the space vector components in the stationary reference frame ($\alpha\beta$-components) reveals by substituting the d-components by the α-components and the q-components by the β-components in Eq. (3.36). Moreover, the angular frequency ω has to be set to zero, since the equations of α- and β-components are not coupled.

The state-space model of the zero sequence with one state variable, one actuating variable and one disturbance variable is described according to Eq. (3.37).

3.2 State-space Models

$$\frac{\mathrm{d}}{\mathrm{d}t}i_{c0} = -\frac{R_c^*}{L_c^*}i_{c0} + \frac{1}{L_c^*}v_{c0} - \frac{1}{L_c^*}v_{f0} \tag{3.37}$$

The following abbreviations are used:

$$R_c^* = R_c + 3R_{c0} \tag{3.38}$$

$$L_c^* = L_c + 3L_{c0} \tag{3.39}$$

3.2.2.2 LC-structures

The equivalent circuit of a lossy three-phase four-wire LC-structure connecting a voltage source and a current source is shown in Fig. 3.5.

This equivalent circuit describes for example a simplified LC-filter of a grid-connected VSC modelled by the VSC voltages v_{cx}. The grid is represented by the current source in this case. The capacitive part can be realized by one capacitor per phase. Alternatively, this structure can be used as a Γ-equivalent circuit of an overhead line section.

As in Sect. 3.2.2.1, the passive components are assumed to be equal in each phase x. In addition to the assumptions in Sect. 3.2.2.1, the following assumptions are made: $C_{fx} = C_f$ with $x \in \{1, 2, 3\}$. The according state-space model using the space vector components with four state variables, two actuating variables and two disturbance variables is described in accordance with Eq. (3.40).

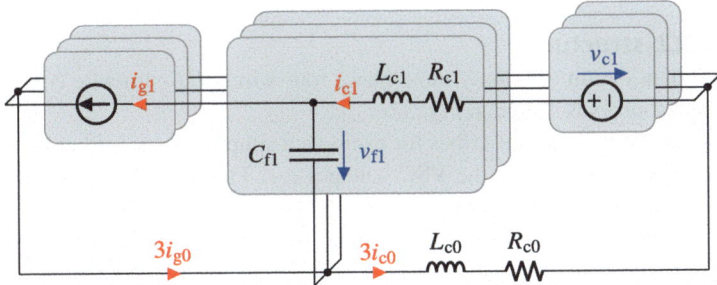

Fig. 3.5 Equivalent circuit of a lossy three-phase four-wire LC-structure connecting a voltage source and a current source

$$\frac{d}{dt}\begin{pmatrix} i_{cd}^{\pm} \\ i_{cq}^{\pm} \\ v_{fd}^{\pm} \\ v_{fq}^{\pm} \end{pmatrix} = \begin{pmatrix} -\frac{R_c}{L_c} & \pm\omega & -\frac{1}{L_c} & 0 \\ \mp\omega & -\frac{R_c}{L_c} & 0 & -\frac{1}{L_c} \\ \frac{1}{C_f} & 0 & 0 & \pm\omega \\ 0 & \frac{1}{C_f} & \mp\omega & 0 \end{pmatrix} \begin{pmatrix} i_{cd}^{\pm} \\ i_{cq}^{\pm} \\ v_{fd}^{\pm} \\ v_{fq}^{\pm} \end{pmatrix} + \begin{pmatrix} \frac{1}{L_c} & 0 \\ 0 & \frac{1}{L_c} \\ 0 & 0 \\ 0 & 0 \end{pmatrix} \begin{pmatrix} v_{cd}^{\pm} \\ v_{cq}^{\pm} \end{pmatrix}$$

$$+ \begin{pmatrix} 0 & 0 \\ 0 & 0 \\ -\frac{1}{C_f} & 0 \\ 0 & -\frac{1}{C_f} \end{pmatrix} \begin{pmatrix} i_{gd}^{\pm} \\ i_{gq}^{\pm} \end{pmatrix} \quad (3.40)$$

Substituting the d-components by the α-components and the q-components by the β-components in Eq. (3.40), the state-space model using the space vector components in the stationary reference frame (αβ-components) appear. Furthermore, the angular frequency ω has to be set to zero.

The state-space model of the zero sequence with two state variables, two actuating variables and two disturbance variables according to Eq. (3.41). The abbreviations from Eq. (3.38) and Eq. (3.39) are applied.

$$\frac{d}{dt}\begin{pmatrix} i_{c0} \\ v_{f0} \end{pmatrix} = \begin{pmatrix} -\frac{R_c^*}{L_c^*} & -\frac{1}{L_c^*} \\ \frac{1}{C_f} & 0 \end{pmatrix} \begin{pmatrix} i_{c0} \\ v_{f0} \end{pmatrix} + \begin{pmatrix} \frac{1}{L_c^*} \\ 0 \end{pmatrix} v_{c0} + \begin{pmatrix} 0 \\ -\frac{1}{C_f} \end{pmatrix} i_{g0} \quad (3.41)$$

3.2.2.3 LCL-structures

The equivalent circuit of a lossy three-phase four-wire LCL-structure connecting two voltage sources is shown in Fig. 3.6.

This equivalent circuit describes for example a simplified LCL-filter of a grid-connected VSC modelled by the VSC voltages v_{cx}. This filter can be realized by two chokes and one capacitor per phase.

It is assumed that the passive components are equal in each phase x. In addition to the assumptions in Sect. 3.2.2.2, the following assumptions are made: $R_{gx} = R_g$, $L_{gx} = L_g$ with $x \in \{1, 2, 3\}$. The according state-space model using the space vector components with six state variables, two actuating variables and two disturbance variables is described in accordance with Eq. (3.42).

3.2 State-space Models

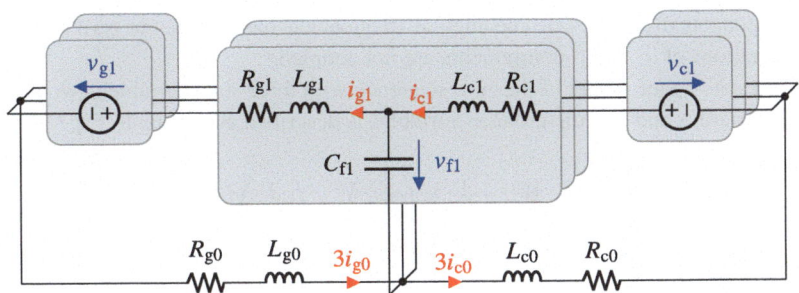

Fig. 3.6 Equivalent circuit of a lossy three-phase four-wire LCL-structure connecting two voltage sources

$$\frac{d}{dt}\begin{pmatrix} i_{cd}^{\pm} \\ i_{cq}^{\pm} \\ i_{gd}^{\pm} \\ i_{gq}^{\pm} \\ v_{fd}^{\pm} \\ v_{fq}^{\pm} \end{pmatrix} = \begin{pmatrix} -\frac{R_c}{L_c} & \pm\omega & 0 & 0 & -\frac{1}{L_c} & 0 \\ \mp\omega & -\frac{R_c}{L_c} & 0 & 0 & 0 & -\frac{1}{L_c} \\ 0 & 0 & -\frac{R_g}{L_g} & \pm\omega & \frac{1}{L_g} & 0 \\ 0 & 0 & \mp\omega & -\frac{R_g}{L_g} & 0 & \frac{1}{L_g} \\ \frac{1}{C_f} & 0 & -\frac{1}{C_f} & 0 & 0 & \pm\omega \\ 0 & \frac{1}{C_f} & 0 & -\frac{1}{C_f} & \mp\omega & 0 \end{pmatrix} \begin{pmatrix} i_{cd}^{\pm} \\ i_{cq}^{\pm} \\ i_{gd}^{\pm} \\ i_{gq}^{\pm} \\ v_{fd}^{\pm} \\ v_{fq}^{\pm} \end{pmatrix} + \begin{pmatrix} \frac{1}{L_c} & 0 \\ 0 & \frac{1}{L_c} \\ 0 & 0 \\ 0 & 0 \\ 0 & 0 \\ 0 & 0 \end{pmatrix} \begin{pmatrix} v_{cd}^{\pm} \\ v_{cq}^{\pm} \end{pmatrix}$$

(3.42)

$$+ \begin{pmatrix} 0 & 0 \\ 0 & 0 \\ -\frac{1}{L_g} & 0 \\ 0 & -\frac{1}{L_g} \\ 0 & 0 \\ 0 & 0 \end{pmatrix} \begin{pmatrix} v_{gd}^{\pm} \\ v_{gq}^{\pm} \end{pmatrix}$$

As in Sect. 3.2.2.1 and Sect. 3.2.2.2, the state-space model using the space vector components in the stationary reference frame ($\alpha\beta$-components) is derived by substituting the d-components by the α-components and the q-components by the β-components in Eq. (3.42). The angular frequency ω has to be set to zero because the equations of α- and β-components are not coupled.

The state-space model of the zero sequence with three state variables, two actuating variables and two disturbance variables is described according to Eq. (3.43).

$$\frac{d}{dt}\begin{pmatrix} i_{c0} \\ i_{g0} \\ v_{f0} \end{pmatrix} = \begin{pmatrix} -\frac{R_c^*}{L_c^*} & 0 & -\frac{1}{L_c^*} \\ 0 & -\frac{R_g^*}{L_g^*} & \frac{1}{L_g^*} \\ \frac{1}{C_f} & -\frac{1}{C_f} & 0 \end{pmatrix} \begin{pmatrix} i_{c0} \\ i_{g0} \\ v_{f0} \end{pmatrix} + \begin{pmatrix} \frac{1}{L_c^*} \\ 0 \\ 0 \end{pmatrix} v_{c0} + \begin{pmatrix} 0 \\ -\frac{1}{L_g^*} \\ 0 \end{pmatrix} v_{g0} \tag{3.43}$$

The abbreviations from Eq. (3.38) and Eq. (3.39) are used for the zero sequence. Additionally, the following variables are used:

$$R_g^* = R_g + 3R_{g0} \tag{3.44}$$

$$L_g^* = L_g + 3L_{g0} \tag{3.45}$$

3.2.3 Discretization of State-space Models

Control and protection algorithms are often executed on digital computers. Consequently, the algorithms in this thesis are developed based on discrete-time state-space models. In this section, it is presented how to derive time-discrete state-space models from continuous-time state-space models, see Sect. 3.2.1.

Sampled systems are characterized by measurement data acquisition at $t = kT_S$. The time interval between two sampling points is called the sampling time T_S. The reciprocal value of T_S equals the sampling frequency f_S.

A discrete-time state-space model (matrix index d) equivalent to Eq. (3.29) and Eq. (3.30) is characterised according to Eq. (3.46) and Eq. (3.47) [79]. The output equations in Eq. (3.47) do not change, since they represent algebraic equations.

$$\mathbf{x}(k+1) = \mathbf{A}_d\mathbf{x}(k) + \mathbf{B}_d\mathbf{u}(k) + \mathbf{E}_d\mathbf{z}(k) \tag{3.46}$$

$$\mathbf{y}(k) = \mathbf{C}\mathbf{x}(k) \tag{3.47}$$

The discrete-time state-space model is visualized as a block diagram, see Fig. 3.7.

3.2 State-space Models

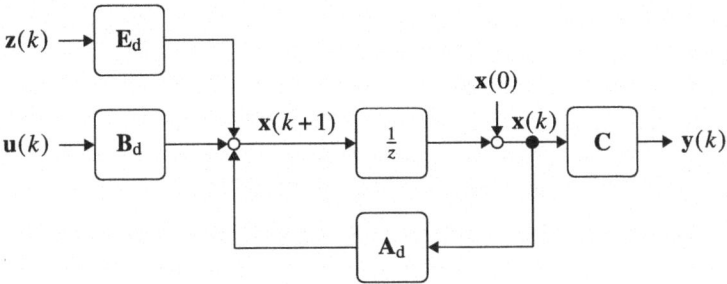

Fig. 3.7 Discrete-time state-space model visualized as a block diagram

One option to discretize the system of state differential equations (see Eq. (3.29)) is to approximate the differential quotient by the difference quotient, see Eq. (3.48).

$$\left.\frac{d}{dt}\mathbf{x}(t)\right|_{t=kT_S} \approx \frac{\mathbf{x}(kT_S) - \mathbf{x}(kT_S - T_S)}{T_S} \quad (3.48)$$

This approximation method is called Forward-Euler method [80]. The integral solution of the system of state differential equations in Eq. (3.29) is approximated by a sum of rectangular areas represented by a width of T_S. The Forward-Euler method is also applicable for the discretization of non-linear state-space models.

Applying the Forward-Euler method, the discretization of Eq. (3.29) equals Eq. (3.49). The unity matrix **I** is used.

$$\mathbf{x}(k+1) = (T_S\mathbf{A} + \mathbf{I})\mathbf{x}(k) + T_S\mathbf{B}\mathbf{u}(k) + T_S\mathbf{E}\mathbf{z}(k) \quad (3.49)$$

Comparing Eq. (3.49) and Eq. (3.46) leads to the calculation rules of the Forward-Euler method in Eq. (3.50), Eq. (3.51) and Eq. (3.52).

$$\mathbf{A}_d = T_S\mathbf{A} + \mathbf{I} \quad (3.50)$$

$$\mathbf{B}_d = T_S\mathbf{B} \quad (3.51)$$

$$\mathbf{E}_d = T_S\mathbf{E} \quad (3.52)$$

3.3 System Properties in the State-space

3.3.1 Solution of the System of State Differential Equations

The state trajectories represented by the vector of the state variables $\mathbf{x}(t)$ provide information about the system behaviour through the influence of the initial state $\mathbf{x}(0)$ and the input variables $\mathbf{u}(t)$ and $\mathbf{z}(t)$. The state trajectories are calculated by solving the heterogeneous system of state differential equations, see Eq. (3.29).

First, the homogeneous solution is calculated. The homogeneous system of the state differential equations equals Eq. (3.53).

$$\frac{d}{dt}\mathbf{x}(t) = \mathbf{A}\mathbf{x}(t) \qquad (3.53)$$

Its solution describes the free motion of the system and is found via an exponential function and a time-independent factor, see Eq. (3.54). [78].

$$\mathbf{x}(t) = \mathbf{\Phi}(t)\mathbf{x}(0) \qquad (3.54)$$

The state-transition matrix $\mathbf{\Phi}(t)$ is a matrix exponential function (see Eq. (3.55a)), which can also be represented as a power series (see Eq. (3.55b)) [75].

$$\mathbf{\Phi}(t) = e^{\mathbf{A}t} \qquad (3.55a)$$

$$= \sum_{k=0}^{\infty} \mathbf{A}^k \frac{t^k}{k!} \qquad (3.55b)$$

Second, the particular solution is calculated. It describes the forced motion of the system and is found via an exponential function and a time-dependent factor [78].

The solution of the system of the state differential equations equals the superposition of the homogeneous and the particular solution, see Eq. (3.56) [79].

$$\mathbf{x}(t) = \mathbf{\Phi}(t)\mathbf{x}(0) + \int_0^t \mathbf{\Phi}(t-\tau)\mathbf{B}\mathbf{u}(\tau)d\tau + \int_0^t \mathbf{\Phi}(t-\tau)\mathbf{E}\mathbf{z}(\tau)d\tau \qquad (3.56)$$

3.3 System Properties in the State-space

The iterative application of the state difference equations (see Eq. (3.46)) leads to the solution of the discrete-time system according to Eq. (3.57).

$$\mathbf{x}(k) = \mathbf{A}_d^k \mathbf{x}(0) + \sum_{j=0}^{k-1} \mathbf{A}_d^{k-1-j} \mathbf{B}_d \mathbf{u}(j) + \sum_{j=0}^{k-1} \mathbf{A}_d^{k-1-j} \mathbf{E}_d \mathbf{z}(j) \tag{3.57}$$

The solution of the system of state differential equations directly reveals an alternative discretization method to the Forward-Euler method from Sect. 3.2.3. By evaluating Eq. (3.56) at $t = kT_S$ and $t = (k+1)T_S$ and comparing these solutions with Eq. (3.46), the relations Eq. (3.58), Eq. (3.59) and Eq. (3.60) appear [79].

$$\mathbf{A}_d = \Phi(T_S) \tag{3.58}$$

$$\mathbf{B}_d = \int_0^{T_S} \Phi(\alpha) d\alpha \mathbf{B} \tag{3.59}$$

$$\mathbf{E}_d = \int_0^{T_S} \Phi(\alpha) d\alpha \mathbf{E} \tag{3.60}$$

The input variables $\mathbf{u}(t)$ and $\mathbf{z}(t)$ are time-independent within one sampling interval T_S due the sample and hold principle of digital computers.

This discretization method is called zero-order hold (ZOH) principle [80]. In contrast to the Forward-Euler method, which is an approximate method, the ZOH method is an exact method.

3.3.2 Asymptotic Stability

The asymptotic stability is an essential system property. It is defined as follows: A system is asymptotic stable, if the limit of the solution of the differential equations (see Eq. (3.54)) of the homogeneous system (see Eq. (3.53)) $\mathbf{x}(t \to \infty)$ is zero from any initial state $\mathbf{x}(0)$ [81].

From Eq. (3.53) it is obvious, that the homogeneous solution must contain self reproducing functions, e.g. exponential and sinusoidal functions. It can be shown that the homogeneous solution is described by linear combinations, where time exponential functions of the real parts of the eigenvalues of $\mathbf{A}$ represent factors [78]. The exponential functions of the imaginary parts of the eigenvalues of the system matrix $\mathbf{A}$ can be expressed as sinusoidal functions using Euler's formula [75].

Transferring these considerations to the definition of the asymptotic stability, the condition of the asymptotic stability reveals: A system is asymptotically stable, if all eigenvalues of the system matrix $\mathbf{A}$ have negative real parts, see Eq. (3.61) [79].

$$\text{Re}\{\lambda_i\} < 0 \tag{3.61}$$

If a system is asymptotically stable, the bounded-input bounded-output (BIBO) stability is also given [79]. The inversion of this statement is not true. That means that the asymptotic stability is a stronger criterion than the BIBO stability. In contrast to the BIBO stability, the asymptotic stability contains constraints on state variables.

A relation between the eigenvalues of continuous-time state-space models λ_i and discrete-time state-space models $\lambda_{d,i}$ can be derived based on the relation between $\mathbf{A}$ and $\mathbf{A}_d$ (see Eq. (3.58) and Eq. (3.55a)) according to Eq. (3.62) [79].

$$\lambda_{d,i} = e^{\lambda_i T_S} \tag{3.62}$$

Discrete state-space models are asymptotically stable, if the eigenvalues of $\mathbf{A}_d$ are inside the unit circle of the complex plane [79]. Merging this statement with Eq. (3.62) and the condition for asymptotic stability of continuous-time state-space models, it reveals, that asymptotic stability of continuous-time and discrete-time state-space models is equivalent.

3.3.3 Controllability

The complete state controllability is a further important system property. It is essential for the design of the controller of a VSC, see also Sect. 5.3.

A system is complete state controllable, if it can be moved from any initial state $\mathbf{x}(0)$ to any final state in a finite time interval T_C [79]. The final state may especially be zero.

To derive the condition of complete state controllability, the homogeneous system with the actuating variables $\mathbf{u}$ is regarded, see Eq. (3.63).

$$\frac{\mathrm{d}}{\mathrm{d}t}\mathbf{x}(t) = \mathbf{A}\mathbf{x}(t) + \mathbf{B}\mathbf{u}(t) \tag{3.63}$$

The solution of this system of differential equations is given in Eq. (3.64).

3.3 System Properties in the State-space

$$\mathbf{x}(t) = \mathbf{\Phi}(t)\mathbf{x}(0) + \int_0^t \mathbf{\Phi}(t-\tau)\mathbf{B}\mathbf{u}(\tau)d\tau \tag{3.64}$$

If the final state is zero, the initial state is calculated, see Eq. (3.65) [79].

$$\mathbf{x}(0) = -\int_0^{T_C} \mathbf{\Phi}(-\tau)\mathbf{B}\mathbf{u}(\tau)d\tau \tag{3.65a}$$

$$= -\int_0^{T_C} \sum_{k=0}^{\infty} \mathbf{A}^k \frac{(-\tau)^k}{k!} \mathbf{B}\mathbf{u}(\tau)d\tau \tag{3.65b}$$

Assuming, that the system is complete state controllable, the evaluation of the integral in Eq. (3.65b) leads to linear combinations with different sequences of the actuating variables. The factors of the linear combinations are $\mathbf{B}$, $\mathbf{AB}$, $\mathbf{A}^2\mathbf{B}$ and so on. [79]

Any initial state $\mathbf{x}(0)$ can be reached by this linear combinations, if the columns of the matrices $\mathbf{AB}$, $\mathbf{A}^2\mathbf{B}$... span a space with the dimension, which equals the order of the system n, see Sect. 3.2.1. According to the Cayley-Hamilton theorem, n powers of $\mathbf{A}$ have to be regarded, since higher powers of $\mathbf{A}$ can be represented by linear combinations of lower powers of $\mathbf{A}$. [79]

So, the controllability matrix $\mathbf{Q}_C$ is defined according to Eq. (3.66).

$$\mathbf{Q}_C = \begin{pmatrix} \mathbf{B} & \mathbf{AB} & \mathbf{A}^2\mathbf{B} & \cdots & \mathbf{A}^{n-1}\mathbf{B} \end{pmatrix} \tag{3.66}$$

A system is complete state controllable, if the controllability matrix $\mathbf{Q}_C$ has the rank of n [79].

$$\text{rank}\{\mathbf{Q}_C\} = n \tag{3.67}$$

If the sampling time T_S is suitably selected, a complete state controllable continuous-time system becomes a complete state controllable time-discrete system. Suitably selected means, that different eigenvalues of $\mathbf{A}$ lead to different eigenvalues of $\mathbf{A}_d$. [79]

The following examples demonstrate the complete state controllability. The continuous-time state-space model of a L-structure using $\alpha\beta$-components is investigated, see Sect. 3.2.2.1. The state-space model is described by the following systems of equations, see Eq. (3.68) and Eq. (3.69).

$$\underbrace{\frac{d}{dt}\begin{pmatrix}i_{c\alpha}\\i_{c\beta}\end{pmatrix}}_{\mathbf{x}} = \underbrace{\begin{pmatrix}-\frac{R_c}{L_c} & 0 \\ 0 & -\frac{R_c}{L_c}\end{pmatrix}}_{\mathbf{A}} \underbrace{\begin{pmatrix}i_{c\alpha}\\i_{c\beta}\end{pmatrix}}_{\mathbf{x}} + \underbrace{\begin{pmatrix}\frac{1}{L_c} & 0 \\ 0 & \frac{1}{L_c}\end{pmatrix}}_{\mathbf{B}} \underbrace{\begin{pmatrix}v_{c\alpha}\\v_{c\beta}\end{pmatrix}}_{\mathbf{u}} \quad (3.68)$$

$$+ \underbrace{\begin{pmatrix}-\frac{1}{L_c} & 0 \\ 0 & -\frac{1}{L_c}\end{pmatrix}}_{\mathbf{E}} \underbrace{\begin{pmatrix}v_{f\alpha}\\v_{f\beta}\end{pmatrix}}_{\mathbf{z}}$$

$$\underbrace{\begin{pmatrix}i_{c\alpha}\\i_{c\beta}\end{pmatrix}}_{\mathbf{y}} = \underbrace{\begin{pmatrix}1 & 0 \\ 0 & 1\end{pmatrix}}_{\mathbf{C}} \underbrace{\begin{pmatrix}i_{c\alpha}\\i_{c\beta}\end{pmatrix}}_{\mathbf{x}} \quad (3.69)$$

The state differential equations in Eq. (3.68) are illustrated via the two equivalent circuits in Fig. 3.8 and Fig. 3.9.

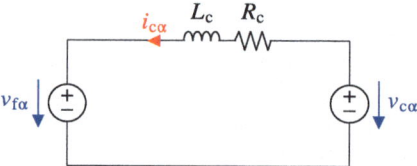

Fig. 3.8 Equivalent circuit of the α-differential equation of a L-structure

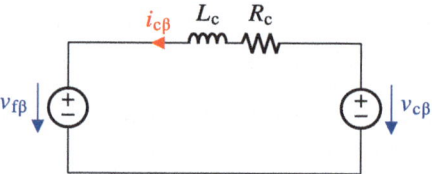

Fig. 3.9 Equivalent circuit of the β-differential equation of a L-structure

Both equivalent circuits are manipulated by the according actuating variable as a voltage source ($v_{c\alpha}$ and $v_{c\beta}$). This system is complete state controllable, since the condition for controllability in Eq. (3.67) is fulfilled.

3.3 System Properties in the State-space

Assuming, that the voltage source $v_{c\beta}$ does not exist, the following equations are valid, see Eq. (3.70).

$$\mathbf{Bu} = \begin{pmatrix} \frac{1}{L_c} \\ 0 \end{pmatrix} v_{c\alpha} \qquad (3.70)$$

The equivalent circuit of the β-components is changed from Fig. 3.9 to Fig. 3.10 in this case.

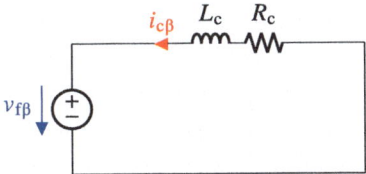

Fig. 3.10 Equivalent circuit of the β-differential equation of a L-structure without controllability

In this case, the system is not complete state controllable, since the rank of $\mathbf{Q_C}$ equals one and is smaller than the order of the system, which equals two. There is no actuating variable in the β-circuit (see Fig. 3.10) that can be manipulated.

3.3.4 Observability

In addition to the complete state controllability, the complete state observability is another important system property. It is essential for the developed model-based protection algorithm, see Chap. 8.

A system is complete state observable, if the initial state $\mathbf{x}(0)$ can be calculated from known trajectories of input variables and output variables over a finite time interval T_O [79].

To derive the condition of complete state observability, the homogeneous system and its output is investigated, see Eq. (3.53) and Eq. (3.54).

Using Eq. (3.55a) and Eq. (3.55b), the output equations in accordance with Eq. (3.71) reveal.

$$\mathbf{y}(t) = \mathbf{C}\mathbf{x}(t) \qquad (3.71a)$$
$$= \mathbf{C}\mathbf{\Phi}(t)\mathbf{x}(0) \qquad (3.71b)$$
$$= \mathbf{C}\sum_{k=0}^{\infty} \mathbf{A}^k \frac{t^k}{k!}\mathbf{x}(0) \qquad (3.71c)$$

However, since $\mathbf{y}(t)$ is known for the time interval T_O, this equation can be formulated for all points inside T_O. The questions to be examined are how many equations are necessary and whether the system of equations can be uniquely solved. [79]

The formulation of Eq. (3.71b) at several points in time leads to Eq. (3.72).

$$\begin{pmatrix} \mathbf{y}(T_1) \\ \mathbf{y}(T_2) \\ \vdots \\ \mathbf{y}(T_n) \end{pmatrix} = \begin{pmatrix} \mathbf{C}\mathbf{\Phi}(T_1) \\ \mathbf{C}\mathbf{\Phi}(T_2) \\ \vdots \\ \mathbf{C}\mathbf{\Phi}(T_n) \end{pmatrix} \mathbf{x}(0) \qquad (3.72)$$

The system of equations in Eq. (3.72) is solvable for $\mathbf{x}(0)$, if the matrix of the right side is invertible. Using Eq. (3.71c), each row of this matrix represents linear combinations with the factors $\mathbf{C}, \mathbf{CA}, \mathbf{CA}^2$ and so on. This consideration shows, that the initial state $\mathbf{x}(0)$ can be calculated from n measurement points of the output variables $\mathbf{y}(t)$, if the row vectors of Eq. (3.72) are linearly independent. According to the Cayley-Hamilton theorem, n powers of $\mathbf{A}$ have to be regarded. [79]

So, the observability matrix $\mathbf{Q_O}$ is defined according to Eq. (3.73).

$$\mathbf{Q_O} = \begin{pmatrix} \mathbf{C} \\ \mathbf{CA} \\ \mathbf{CA}^2 \\ \vdots \\ \mathbf{CA}^{n-1} \end{pmatrix} \qquad (3.73)$$

A system is complete state observable, if the observability matrix $\mathbf{Q_O}$ has the rank of n [79].

$$\text{rank}\{\mathbf{Q_O}\} = n \qquad (3.74)$$

3.3 System Properties in the State-space

If different eigenvalues of **A** lead to different eigenvalues of $\mathbf{A}_d$, a complete state observable continuous-time system becomes a complete state observable time-discrete system [79]. This condition is similar to the condition of complete state controllability, see Eq. (3.67).

The following example demonstrates the complete state observability. The continuous-time state-space model of a L-structure using $\alpha\beta$-components is regarded, see Sect. 3.2.2.1. The systems of equations are given in Eq. (3.68) and Eq. (3.69). The equivalent circuits are shown in the Fig. 3.8 and Fig. 3.9

In both equivalent circuits, the according output variable ($i_{c\alpha}$ and $i_{c\beta}$) is measured. This system is complete state observable, since the rank of the observability matrix $\mathbf{Q_O}$ equals the order of the system n. According to the condition for complete state observability in Eq. (3.74), this system is complete state observable.

Assuming, that the current $i_{c\beta}$ is not measured, Eq. (3.75) is valid.

$$\mathbf{Cx} = \begin{pmatrix} 1 & 0 \end{pmatrix} \begin{pmatrix} i_{c\alpha} \\ i_{c\beta} \end{pmatrix} \quad (3.75)$$

The equivalent circuit of the β-component is illustrated in Fig. 3.11.

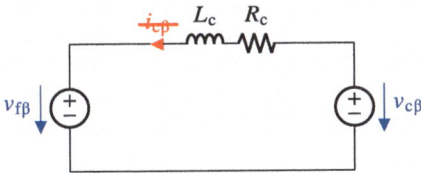

Fig. 3.11 Equivalent circuit of the β-differential equation of a L-structure without observability

In this case, the system is not complete state observable, since the rank of $\mathbf{Q_O}$ equals one and is smaller than the order of the system, which equals two. The initial state $\mathbf{x}(0)$ cannot be calculated.

3.3.5 Duality of Controllability and Observability

According to the investigations in Sect. 3.3.3 and Sect. 3.3.4, there are similarities between controllability and observability. These similarities are reflected in this section.

The state-space model according to Eq. (3.76a) and Eq. (3.76b) is considered.

$$\frac{d}{dt}\mathbf{x}(t) = \mathbf{A}\mathbf{x}(t) + \mathbf{B}\mathbf{u}(t) \tag{3.76a}$$

$$\mathbf{y}(t) = \mathbf{C}\mathbf{x}(t) \tag{3.76b}$$

Replacing the system matrix $\mathbf{A}$ with its transposed matrix $\mathbf{A}^T$, the input matrix $\mathbf{B}$ with $\mathbf{C}^T$ and the output matrix $\mathbf{C}$ with $\mathbf{B}^T$ using transformed variables (represented by the apostrophe), the state-space model according to Eq. (3.77a) and Eq. (3.77b) reveals.

$$\frac{d}{dt}\mathbf{x}'(t) = \mathbf{A}^T\mathbf{x}'(t) + \mathbf{C}^T\mathbf{u}'(t) \tag{3.77a}$$

$$\mathbf{y}'(t) = \mathbf{B}^T\mathbf{x}'(t) \tag{3.77b}$$

The systems of Eq. (3.76) and Eq. (3.77) are dual systems. The first system is called the primal system and the second system is called the dual system. [79]

Applying the conditions for controllability (see Eq. (3.67)) and observability (see Eq. (3.74)) it can be shown that the dual system is complete state controllable (observable), if the primal system is complete state observable (controllable) [79].

3.4 Summary

In this chapter, the fundamentals of the signal theory and the system theory used for the design of the control and protection algorithms in 100 % inverter-based power systems are presented.

Transient and steady states are described by space vectors. Space vectors are linked to phasors of symmetrical components in the steady state. Depending on the characteristics of a three-phase system, the space vector components in a synchronous reference frame (SRF) or a double synchronous reference frame (DSRF) reveal as time-constant quantities in the steady state. Models based on these components are used inside the developed control algorithms. The concept of SRF and DSRF is extended to polyfrequency three-phase four-wire systems, taking also the

3.4 Summary

zero sequence into account. Models based on space vector components in a stationary reference frame are used inside the developed protection algorithms. The representation of these models is simplified due to decoupled equations.

The relations between space vector components of different electrical quantities are described by differential equations. State-space models are one possibility to formulate these differential equations, considering also the algebraic output equations. State-space models of selected electrical circuit structures, which represent the fundament for the development of the control as well as the protection algorithms, are derived. Discrete-time state-space models based on the exact zero-order hold (ZOH) method or the approximate Forward-Euler method allow the realization of control and protection algorithms on digital computers.

Based on state-space models, the state controllability and the state observability represent essential system properties. They are explained using illustrative examples, which appear in filter applications of voltage source converters. Controllability is the basic prerequisite for the development of state control algorithms. Observability is the basic prerequisite for the novel model-based protection concept. Controllability and observability are dual properties from the system theory perspective. To follow the key aspects of the central theme of this thesis on a shortcut: The conclusion of the next chapter is provided in Sect. 4.6.

Open Access This chapter is licensed under the terms of the Creative Commons Attribution 4.0 International License (http://creativecommons.org/licenses/by/4.0/), which permits use, sharing, adaptation, distribution and reproduction in any medium or format, as long as you give appropriate credit to the original author(s) and the source, provide a link to the Creative Commons license and indicate if changes were made.

The images or other third party material in this chapter are included in the chapter's Creative Commons license, unless indicated otherwise in a credit line to the material. If material is not included in the chapter's Creative Commons license and your intended use is not permitted by statutory regulation or exceeds the permitted use, you will need to obtain permission directly from the copyright holder.

Synchronization of VSCs to the Grid 4

This chapter presents the synchronization of voltage source converters (VSCs) to the grid. In Sec. 4.1, the connection of a VSC to the grid based on the grid equivalent circuit and the energy transfer is explained. Sec. 4.2. gives an overview of different synchronization principles. Basic and advanced voltage-based synchronization concepts are shown and evaluated in Sec. 4.3. Power-based synchronization principles are the subject of Sec. 4.4. Beyond that, they are evaluated in the context of 100 % inverter-based power systems (IBPS). The adoption of physical properties of synchronous machines and the droop control are highlighted. In Sec. 4.5, the hybrid synchronization is presented and presented in a test scenario.

4.1 Grid Connection of a Voltage Source Converter

4.1.1 Grid Equivalent Circuit

4.1.1.1 Model of the Grid Equivalent Circuit

The grid connection of a voltage source converter (VSC) is represented by the equivalent circuit using space vectors (see Fig. 4.1) and the zero sequence component (see Fig. 4.2). The space vector equivalent circuit is valid for the α-component and the β-component.

The systems of equations are analogous to the ones in Sec. 3.2.2.1. The filter topologies are described in Sec. 3.2.2 and the fundamentals of VSCs are described in Sec. 2.1.2.

The voltage v_{gx} with $x \in \{1, 2, 3\}$ at the terminals of the grid-connected (GC) VSC represents the link between the grid and the VSC. The Tab. 4.1 shows the correspondences of physical quantities of the grid equivalent circuits and different filter structures.

© The Author(s) 2025
F. Mahr, *Control and Protection of 100% Inverter-based Power Systems*,
https://doi.org/10.1007/978-3-658-47217-7_4

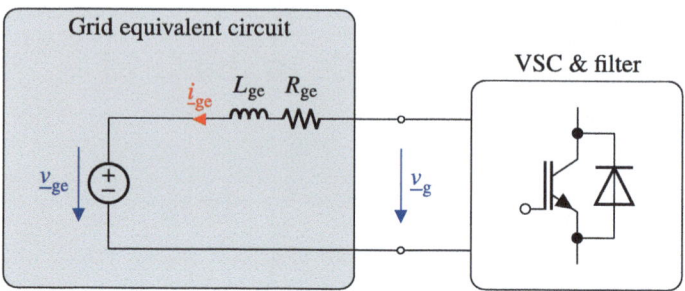

Fig. 4.1 Equivalent circuit of a grid-connected VSC using space vectors

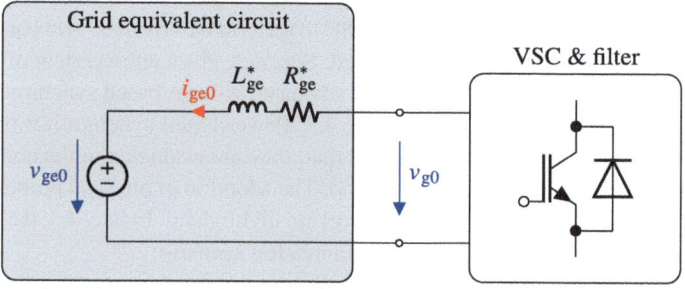

Fig. 4.2 Equivalent circuit of a grid-connected VSC using zero sequence components

Table 4.1 orrespondences of physical quantities of the grid equivalent circuits and the filter structures

	grid equivalent (Fig. 4.1 & Fig. 4.2)	
	i_{gex}	v_{gx}
L-filter (Fig. 3.4)	i_{cx}	v_{fx}
LC-filter (Fig. 3.5)	i_{gx}	v_{fx}
LCL-filter (Fig. 3.6)	i_{gx}	v_{gx}

The grid equivalent circuit reflects fundamental effects of the grid from the VSC perspective. It consists of a voltage source v_{gex} and an ohmic-inductive impedance. Capacitive effects are not included, since the effective capacitances are small compared to R_{ge} and L_{ge} and lead to big shunt impedances at the grid operating fre-

4.1 Grid Connection of a Voltage Source Converter

quency. In general, the voltage source and the impedance are not measurable directly since they do not appear physically.

Alternatively, the grid equivalent circuit can also represent a grid bus with the voltage v_{gex} and the line parameters R_{ge} and L_{ge}.

The monofrequency steady-state impedance $\underline{Z}_{ge}$ (see Eq. (4.1)) is valid for the positive sequence at the grid operating frequency.

$$\underline{Z}_{ge} = R_{ge} + j\omega L_{ge} \tag{4.1}$$

4.1.1.2 Grid Strength Assessment

The absolute value $|\underline{Z}_{ge}|$ of the monofrequency steady-state impedance (see Eq. (4.1)) is an essential, but not the only, parameter of the grid strength assessment.

$$|\underline{Z}_{ge}| = \sqrt{R_{ge}^2 + (\omega L_{ge})^2} \tag{4.2}$$

There are different definitions of the grid strength, and this term is in transition. Often, the ratio of the short circuit power at the infeed bus to the rated power (SCR) is used [82].

In 100 % inverter-based power systems (IBPS), the grid strength assessment via the SCR is not sufficient, since e.g. interactions between controllers are not considered [82]. Therefore, new parameters to characterize the grid strength of 100 % IBPS are defined. For example, the generalized short-circuit ratio (gSCR) interprets the grid strength as a sensitivity index of the terminal voltage of an inverter-based resource (IBR) to variations of its current injection [83].

Applying Kirchhoff's voltage law (KVL) to the equivalent circuits in Fig. 4.1 and Fig. 4.2 reveals, that $|\underline{Z}_{ge}|$ has significant influence on the terminal voltage v_{gx} during current injection. Consequently, $|\underline{Z}_{ge}|$ also influences the dimensioning of the DC-link voltage, see also Sec. 2.1.1.

Moreover, the synchronization of a VSC to the grid incorporates physical and controller laws, see Sec. 4.4. Since $|\underline{Z}_{ge}|$ is an important parameter of the physical equations, it also influences the synchronization of a VSC to the grid.

Regarding photovoltaic (PV) power plants and wind power plants (WPPs), the line impedance tends to be high, since they are often located in remote sites [50]. In these weak grids with high-impedance, problems of phase-locked loop (PLL)-based synchronization (see Sec. 4.3) are reported [84]. Power-based synchronization (see Sec. 4.4) is a promising solution for synchronizing VSCs to weak grids [50].

4.1.1.3 Grid Characteristics Assessment

The grid characteristics assessment is influenced by the angle of the monofrequency steady-state impedance (see Eq. (4.1)) according to Eq. (4.3).

$$\arg\{\underline{Z}_{ge}\} = \operatorname{atan2}(\omega L_{ge},\ R_{ge}) \tag{4.3}$$

Often, the term X/R-ratio is used for grid characteristics assessment. The X/R-ratio depends on the voltage level and the grid size. Especially low-voltage microgrids tend to a low X/R-ratio [46].

The X/R-ratio defines the ratio of the active and the reactive power as parts of the apparent power between two voltage nodes, see Eq. (4.18) and Eq. (4.19) in Sec. 4.1.2.2. For grids with a low X/R-ratio, the active and reactive power can be decoupled mathematically by introducing the modified active and reactive power, see Eq. (4.20) and [85]. Beyond that, the X/R-ratio can be adjusted from the perspective of the VSC by using the principle of the virtual impedance, see Sec. 5.2.4.2 and [46].

4.1.2 Energy Transfer between the Grid and the VSC

4.1.2.1 The p-q Theory

The p-q theory is a fundamental power theory, that is presented in [86]. It is used to calculate parts of the instantaneous power from voltages and currents in three-phase four-wire systems with a neutral conductor. In this section, the main aspects are summarized.

The p-q theory is based on the instantaneous complex power $\underline{s}$, see Eq. (4.4).

$$\underline{s} = \frac{3}{2}\underline{v}_{\alpha\beta}\underline{i}_{\alpha\beta}^* \tag{4.4a}$$

$$= p + jq \tag{4.4b}$$

It equals the product of the space vector of the phase-to-neutral voltages and the complex conjugate space vector of the phase currents using $\alpha\beta$-components. For consistency reasons, the definitions of the space vector and the zero sequence component from Eq. (3.13) and Eq. (3.14) are applied. Therefore, the factor 3/2 is used in Eq. (4.4a)—in contrast to [86]. This factor does not influence the p-q theory in general.

The real part of the instantaneous complex power $\underline{s}$ equals the instantaneous real power p, see Eq. (4.5).

4.1 Grid Connection of a Voltage Source Converter

$$p = \frac{3}{2}(v_\alpha i_\alpha + v_\beta i_\beta) \qquad (4.5)$$

The imaginary part of the instantaneous complex power $\underline{s}$ equals the instantaneous imaginary power q, see Eq. (4.6).

$$q = \frac{3}{2}(v_\beta i_\alpha - v_\alpha i_\beta) \qquad (4.6)$$

The instantaneous zero-sequence power p_0 equals the product of the zero sequence of the voltage and the current multiplied by 3, see Eq. (4.7).

$$p_0 = 3v_0 i_0 \qquad (4.7)$$

The three-phase instantaneous real power $p_{3\Phi}$ equals the sum of the instantaneous real power and the instantaneous zero-sequence, see Eq. (4.8).

$$p_{3\Phi} = p + p_0 \qquad (4.8)$$

The instantaneous real power p, the instantaneous imaginary power q and the instantaneous zero-sequence power p_0 contain average parts and oscillating parts. The mean values of p and q are given in Eq. (4.9) and Eq. (4.10) [86]. The notation from Eq. (3.26) is applied.

$$\overline{p} = \frac{3}{2}\sum_{n=1}^{\infty} \hat{V}_{(1)}^n \hat{I}_{(1)}^n \cos\left(\varphi_{v(1)}^n - \varphi_{i(1)}^n\right) + \frac{3}{2}\sum_{n=1}^{\infty} \hat{V}_{(2)}^n \hat{I}_{(2)}^n \cos\left(\varphi_{v(2)}^n - \varphi_{i(2)}^n\right) \qquad (4.9)$$

$$\overline{q} = \frac{3}{2}\sum_{n=1}^{\infty} \hat{V}_{(1)}^n \hat{I}_{(1)}^n \sin\left(\varphi_{v(1)}^n - \varphi_{i(1)}^n\right) - \frac{3}{2}\sum_{n=1}^{\infty} \hat{V}_{(2)}^n \hat{I}_{(2)}^n \sin\left(\varphi_{v(2)}^n - \varphi_{i(2)}^n\right) \qquad (4.10)$$

From that point of view, the well-known three-phase fundamental frequency active and reactive power of the positive sequence as the real part and the imaginary part of the complex power are special cases of Eq. (4.9) and Eq. (4.10), respectively. They are given in Eq. (4.11) and Eq. (4.12).

$$P = \frac{3}{2}\hat{V}_{(1)}\hat{I}_{(1)}\cos(\varphi_{v(1)} - \varphi_{i(1)}) \qquad (4.11)$$

$$Q = \frac{3}{2}\hat{V}_{(1)}\hat{I}_{(1)}\sin(\varphi_{v(1)} - \varphi_{i(1)}) \qquad (4.12)$$

The apparent power S equals the absolute value of the complex power, see Eq. (4.13).

$$S = \frac{3}{2}\hat{V}_{(1)}\hat{I}_{(1)} \qquad (4.13)$$

4.1.2.2 Energy Transfer between two Voltage Sources

The energy transfer between two voltage sources is essential for the normal operation of a GC VSC and its synchronization to the grid. The presented considerations are also applicable to describe the energy transfer between two grid nodes.

In general, the normal operation and the synchronization are related to the positive sequence. Based on Fig. 4.1, the model of the equivalent circuit of the positive sequence is used. This equivalent circuit is regarded as a monofrequency system in the steady state and described by the according root-mean-square (RMS) phasors, see also Eq. (3.15). For simplification and adaption to the literature, the names of the quantities are adjusted and the indices are reduced in Fig. 4.3.

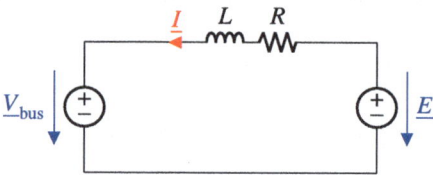

Fig. 4.3 General equivalent circuit of two voltage sources in the positive sequence connected by a RL-structure using phasors

Basically, the following considerations can also be adapted to the negative sequence and the zero sequence. In the zero sequence, the values of R and L have to be adjusted, see Sec. 3.2.2.1.

In Fig. 4.3, $\underline{E}$ represents the positive sequence RMS-phasor of the capacitor voltage v_{fx} of a LC-filter (see Fig. 3.5) or a LCL-filter (see Fig. 3.6). Due to the existence of the grid-side choke of the LCL-filter, R and L differ in both cases when assuming identical line parameters.

The phasor $\underline{V}_{bus}$ is the positive sequence RMS-phasor of the open circuit voltage v_{gex} of the grid equivalent or the bus voltage, see Fig. 4.1.

The circuit in Fig. 4.3 is described by a KVL according to Eq. (3.22).

4.1 Grid Connection of a Voltage Source Converter

$$\underline{E} = (R+j\omega L)\underline{I} + \underline{V}_{bus} \tag{4.14a}$$
$$= (R+jX)\underline{I} + \underline{V}_{bus} \tag{4.14b}$$

Applying Eq. (4.14) to a LCL-filter (see also Fig. 4.1 and Fig. 3.6) yields Eq. (4.15).

$$\underline{V}_{f(1)} = \left(R_g + R_{ge} + j\omega(L_g + L_{ge})\right)\underline{I}_{g(1)} + \underline{V}_{ge(1)} \tag{4.15}$$

According to Sec. 4.1.2.1, the complex power of the positive sequence is calculated in accordance with Eq. (4.16). Here, the definition of the symmetrical components according to Eq. (3.6) are applied.

$$\frac{1}{3}\underline{S} = \underline{V}_{bus}\underline{I}^* \tag{4.16a}$$
$$= -\frac{V_{bus}^2}{R-jX} + \frac{EV_{bus}}{R-jX}\left[\cos(\varphi_{V_{bus}} - \varphi_E) + j\sin(\varphi_{V_{bus}} - \varphi_E)\right] \tag{4.16b}$$

The angle δ equals voltage angle difference, see Eq. (4.17).

$$\delta = \varphi_E - \varphi_{V_{bus}} \tag{4.17}$$

The active and the reactive power according to Eq. (4.18) and Eq. (4.19) reveal. Additional calculation steps are shown in [143].

$$\frac{1}{3}P = -\frac{R}{R^2+X^2}V_{bus}^2 + \frac{R}{R^2+X^2}EV_{bus}\cos(\delta) \tag{4.18}$$
$$+ \frac{X}{R^2+X^2}EV_{bus}\sin(\delta)$$

$$\frac{1}{3}Q = -\frac{X}{R^2+X^2}V_{bus}^2 - \frac{R}{R^2+X^2}EV_{bus}\sin(\delta) \tag{4.19}$$
$$+ \frac{X}{R^2+X^2}EV_{bus}\cos(\delta)$$

4.1.2.3 P/Q-transformation

By introducing the modified active power P_t and the modified reactive power Q_t, the Eq. (4.18) and the Eq. (4.19) are decoupled and simplified. The following transformation according to Eq. (4.20) is applied [85]:

$$\begin{pmatrix} P_t \\ Q_t \end{pmatrix} = \begin{pmatrix} \frac{X}{Z} & -\frac{R}{Z} \\ \frac{R}{Z} & \frac{X}{Z} \end{pmatrix} \begin{pmatrix} P \\ Q \end{pmatrix} \tag{4.20}$$

The absolute value of the Impedance Z from Eq. (4.21) is used.

$$Z = \sqrt{R^2 + X^2} \tag{4.21}$$

Using the transformation according to Eq. (4.20) yields Eq. (4.22) and Eq. (4.23).

$$P_t = 3\frac{EV_{\text{bus}}}{Z}\sin(\delta) \tag{4.22}$$

$$Q_t = 3\frac{V_{\text{bus}}}{Z}\left[-V_{\text{bus}} + E\cos(\delta)\right] \tag{4.23}$$

Assuming dominant inductive grid characteristics ($R \ll X$), the Eq. (4.24) and Eq. (4.25) reveal.

$$P_t \approx 3\frac{EV_{\text{bus}}}{X}\sin(\delta) \tag{4.24}$$

$$Q_t \approx 3\frac{V_{\text{bus}}}{X}\left[-V_{\text{bus}} + E\cos(\delta)\right] \tag{4.25}$$

From Eq. (4.24) and the Eq. (4.25) is it seen, that the modified active power P_t is linked to the difference of the voltage angles. The modified reactive power Q_t is linked to the difference of the voltage RMS-values. Both powers are dependant from the frequency via X.

4.1.2.4 Linearising at an Operating Point

The considerations in this section are based on Fig. 4.3. The aim is to linearise the power equations in Eq. (4.24) and Eq. (4.25) at an operating point. By introducing time-variant quantities, linear differential equations appear. These physical equations are part of a state-space model, that can be investigated regarding small-signal power synchronization stability (see also Sec. 3.3.2).

In the following investigations, it is assumed for simplification, that $R \ll X$ is valid. Therefore, the modified powers of Sec. 4.1.2.3 equal the actual powers. The reactance X involves—at least approximately – all reactances between the voltage sources $\underline{E}$ and $\underline{V}_{\text{bus}}$. Consequently, especially the grid-side reactance of a LCL-filter and the line reactance has to be considered.

4.1 Grid Connection of a Voltage Source Converter

The operating point is defined by the RMS-voltage V_{base}, which is assumed to be equal for $\underline{E}$ and $\underline{V}_{\text{bus}}$ and the difference of the voltage angles δ_0. Deviations from the operating point are labelled by Δ. The principles of linearisation are given in [78].

Linearising Eq. (4.24) leads to Eq. (4.26). The difference of the voltage angles δ is given in Eq. (4.17). The synchronizing power coefficient K is used, see Eq. (4.27) [87].

$$\Delta P \approx K \Delta \delta \tag{4.26}$$

$$K = 3 \frac{V_{\text{base}}^2}{X} \cos(\delta_0) \tag{4.27}$$

The time-derivative of $\Delta\delta$ leads to the difference of the angular frequencies of the voltage sources $\underline{E}$ and $\underline{V}_{\text{bus}}$, see Eq. (4.28).

$$\frac{\mathrm{d}\Delta\delta}{\mathrm{d}t} = \frac{\mathrm{d}}{\mathrm{d}t}(\Delta\varphi_E - \Delta\varphi_{V_{\text{bus}}}) \tag{4.28a}$$

$$= \Delta\omega_{\text{m}} - \Delta\omega_{\text{bus}} \tag{4.28b}$$

From the Eq. (4.26), the derivative of ΔP after $\Delta\delta$ yields Eq. (4.29).

$$\frac{\mathrm{d}\Delta P}{\mathrm{d}\Delta\delta} = K \tag{4.29}$$

Combining Eq. (4.29) and Eq. (4.28b), a linear first-order differential equation for the active power according to Eq. (4.30) appears.

$$\frac{\mathrm{d}\Delta P}{\mathrm{d}t} = K(\Delta\omega_{\text{m}} - \Delta\omega_{\text{bus}}) \tag{4.30}$$

Eq. (4.30) is written for multiple VSCs (index i) on a common bus, see Eq. (4.31).

$$K_i \Delta\omega_{\text{bus}} = K_i \Delta\omega_{\text{m},i} - \frac{\mathrm{d}\Delta P_i}{\mathrm{d}t} \tag{4.31}$$

Multiple applications of Eq. (4.31) with identical $\Delta\omega_{\text{bus}}$ yield Eq. (4.32).

$$\Delta\omega_{\text{bus}} = \frac{1}{\sum_{i=1}^{n} K_i} \left(\sum_{i=1}^{n} K_i \Delta\omega_{\text{m},i} - \frac{\mathrm{d}}{\mathrm{d}t} \sum_{i=1}^{n} \Delta P_i \right) \tag{4.32}$$

Regarding the reactive power, the linearisation of Eq. (4.25) leads to Eq. (4.33).

$$\Delta Q \approx K_{QE} \Delta E + K_{QV} \Delta V_{bus} + K_{Q\delta} \Delta \delta \qquad (4.33)$$

The coefficients according to Eq. (4.34), Eq. (4.35) and Eq. (4.36) are used [88].

$$K_{QE} = 3 \frac{V_{base}}{X} \cos(\delta_0) \qquad (4.34)$$

$$K_{QV} = 3 \frac{V_{base}}{X} (\cos(\delta_0) - 2) \qquad (4.35)$$

$$K_{Q\delta} = -3 \frac{V_{base}^2}{X} \sin(\delta_0) \qquad (4.36)$$

Subsequently, the coefficient $K_{Q\delta}$ is set to zero, since the difference of the voltage angles δ_0 is nearly zero during most normal grid operations.

In contrast to the active power (see Eq. ((4.30))), the reactive power does not show an integral character (see Eq. (4.33)). Hence, the reactive power is usually filtered by a low-pass filter (LPF). The Eq. (4.37) shows the filtered reactive power Q_F by a first-order LPF with the time constant T_{FQ} in the frequency domain.

$$Q_F(s) = \frac{1}{1 + sT_{FQ}} Q(s) \qquad (4.37)$$

In the time domain, the Eq. (4.38) is valid.

$$\frac{dQ_F}{dt} = \frac{1}{T_{FQ}} (Q - Q_F) \qquad (4.38)$$

The linearisation of Eq. (4.38) yields Eq. (4.39).

$$\frac{d\Delta Q_F}{dt} = \frac{1}{T_{FQ}} (\Delta Q - \Delta Q_F) \qquad (4.39)$$

Substituting Eq. (4.33) into Eq. (4.39), a linear first-order differential equation for the reactive power appears, see Eq. (4.40).

$$\frac{d\Delta Q_F}{dt} = \frac{1}{T_{FQ}} K_{QE} \Delta E + \frac{1}{T_{FQ}} K_{QV} \Delta V_{bus} - \frac{1}{T_{FQ}} \Delta Q_F \qquad (4.40)$$

The Eq. (4.33) is written for multiple VSCs (index i) on a common bus, see Eq. (4.41).

$$K_{QV,i}\Delta V_{bus} = \Delta Q_i - K_{QE,i}\Delta E_i \qquad (4.41)$$

Multiple applications of Eq. (4.41) with an identical ΔV_{bus} leads to Eq. (4.42).

$$\Delta V_{bus} = \frac{1}{\sum_{i=1}^{n} K_{QV,i}} \left(\sum_{i=1}^{n} \Delta Q_i - \sum_{i=1}^{n} K_{QE,i}\Delta E_i \right) \qquad (4.42)$$

4.2 Synchronization Principles

Synchronization is the temporal adaption of waveforms of voltages and currents of the grid and the VSC. This adaption is linked to the fundamental frequency f of the positive sequence and the according angular frequency ω. The fundamental frequency f of the positive sequence reflects the time dependence of waveforms, see Eq. (3.16). Additionally, it defines the fundamental and harmonic frequencies of the positive, the negative and the zero sequence, see Eq. (3.26).

For the temporal adaption of voltages and currents in different sequences at different frequencies, all phase angles $\varphi_{(1)}^n$, $\varphi_{(2)}^n$ and $\varphi_{(0)}^n$ (see Eq. (3.26)) are defined in relation to a reference phase angle. Here, the reference phase angle is set to the phase angle $\varphi_{(1)}$ of the positive sequence at the fundamental frequency. The fundamental angular frequency ω and the phase angle $\varphi_{(1)}$ of the positive sequence lead to the time-dependent reference angle γ^+ in the steady state, see Eq. (4.43).

$$\gamma^+ = \omega t + \varphi_{(1)} \qquad (4.43)$$

The superscript $+$ is used, since γ^+ equals the time-dependent angle of the positive sequence space vector $\underline{v}^+$ at the fundamental frequency, see Eq. (3.17) and Eq. (3.15).

The objective of the synchronization is to define the reference angle γ^+, to which the VSC-controller is adapted. This reference angle and integer multiples of the reference angle are used as transformation angles γ_T, see Sec. 3.1.2.4, Sec. 3.1.2.5, Sec. 3.1.3 and Eq. (3.19). In this way, space vector components of different sequences at different frequencies are gained as DC quantities. Beyond that, this principle is necessary in order to decouple the positive and the negative sequence, see Sec. 4.3.3 and Sec. 4.3.4.1. In general, there are two different principles to define the reference angle γ^+.

The voltage-based principle is based on the temporal adaption to the voltages v_{gx} at the terminals of the VSC. These voltages represent the link between the grid and the VSC, see Fig. 4.1 and Fig. 4.2. Consequently, the angle of the positive sequence space vector $\underline{v}_g^+$ at the fundamental frequency is chosen as the reference angle γ^+. The angle of $\underline{v}_g^+$ is not measurable directly. Therefore, the voltages v_{gx} are measured and the angle of $\underline{v}_g^+$ is calculated by a phase-locked loop (PLL), see Sec. 4.3. This synchronization principle leads to the grid-following (GFL) mode, which is used for the current control of the VSC.

The power-based principle is based on the adaption to the value of the active power at the terminals of the VSC. The reference angle γ^+ is defined based on the active power, see Sec. 4.4. Here, the voltages v_{gx} and the currents i_{gex} (see Fig. 4.1 and Fig. 4.2) are measured and the active power is calculated according to Sec. 4.1.2.1. This synchronization principle leads to the grid-forming (GFM) mode, which is used for the voltage control by the VSC. Reactive power is used to adjust the reference value of the voltage amplitude. Sine the active and the reactive power are used as input variables to calculate the reference voltage angle and amplitude, the term "power-based principle" is used.

The voltage-based synchronization principle and the power-based synchronization principle can be combined and the hybrid synchronization principle reveals, see Sec. 4.5. Fig. 4.4 shows the synchronization principles of VSCs to the grid.

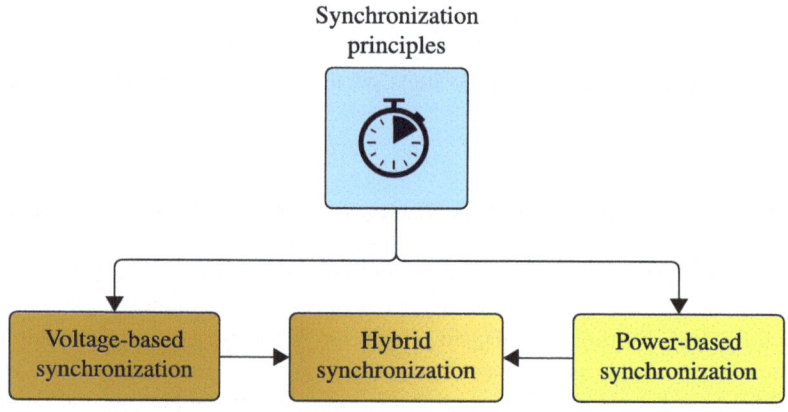

Fig. 4.4 Classification of synchronization principles of VSCs to the electrical power grid

The stability of the grid operation is influenced by the synchronization principle in combination with the grid strength, see Sec. 4.1.1.2. The voltage-based synchronization shows stability problems in weak grids [84, 89]. In contrast, the power-based synchronization principle is a promising solution for synchronizing VSCs to weak grids [50].

4.3 Voltage-based Synchronization

The voltage-based synchronization is state of the art in industrial applications and leads to the GFL-mode of the GC VSC. A PLL is applied to calculate the positive sequence space vector $\underline{v}_g^+$ at the fundamental frequency. The output angle of the PLL is denoted as γ_{PLL}^+ and is used as the reference angle γ^+. Integer multiples of γ_{PLL}^+ are used as transformation angles γ_T, see Eq. (3.19).

Using the transformations from Eq. (3.19), the positive and negative sequence space vector components of v_{gx} at the fundamental frequency are determined. Based on the according components (v_{gd}^+, v_{gq}^+, v_{gd}^- and v_{gq}^-), the positive and the negative sequence space vector components of i_{gx} are calculated as reference values of the current controller e.g. to inject the desired active and reactive power in the positive and the negative sequence, see Sec. 5.2.3.1. Consequently, the voltage-based synchronization is especially relevant for the GFL-mode, but it is also of great relevance for the steady state current limitation of GFM-VSCs, see Sec. 6.3.2.3.

Different PLL principles are presented. All of them include a PLL-controller (see Sec. 4.3.1). A decoupling cell is essential for the GFL-mode and the GFM-mode to decouple positive and negative sequence components—optionally at different frequencies. An overview of different PLL principles is given in [90].

4.3.1 PLL-controller

The objective of the PLL-controller is to calculate the time-dependent angle γ_{PLL}^+ and the according angular frequency ω_{PLL}. The PLL-controller is used in the presented PLL principles in the following sections, see Sec. 4.3.2, Sec. 4.3.3 and Sec. 4.3.4.

The PLL-controller controls the q-component v_q^+ of $\underline{v}_g^+$ to zero. The angle of $\underline{v}_g^+$ and the transformation angle γ_T^+ are equal, if the q-component v_q^+ is zero and $v_d^+ > 0$, see Sec. 3.1.2.4. The parametrization of the PLL-controller is usually done empirically. A guide for the parametrization is given in [90].

The general structure of the PLL-controller is shown in Fig. 4.5. The q-component v_q^+ is obtained from v_{gx}. A proportional-integral (PI) controller and a feedforward control of the nominal angular frequency ω_{nom} is applied. In Fig. 4.5, the entire system consisting of the VSC and the grid, including the feedback loop, is not shown to focus on the PLL-controller.

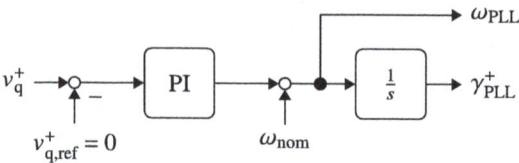

Fig. 4.5 General block diagram of a PLL-controller

4.3.2 SRF-PLL

The synchronous reference frame (SRF) PLL is a suitable solution for the voltage-based synchronization, if the voltages v_{gx} describe a monofrequency system in the positive sequence, see Sec. 3.1.2.4. It does not work properly, if either negative sequence components or harmonic frequency components appear in v_{gx}.

The principle of the SRF-PLL is to adjust the transformation angle γ_T^+ to the angle of the space vector $\underline{v}_g^+$ [90]. In the steady state, v_{gd}^+ is a constant value and v_{gq}^+ is zero. Therefore, a PLL-controller is used, see Sec. 4.3.1. The general structure of the SRF-PLL is shown in Fig. 4.6. It is especially valid using the voltages v_{gx} as input variables (index g). For clarity reasons, the index g is neglected in Fig. 4.6.

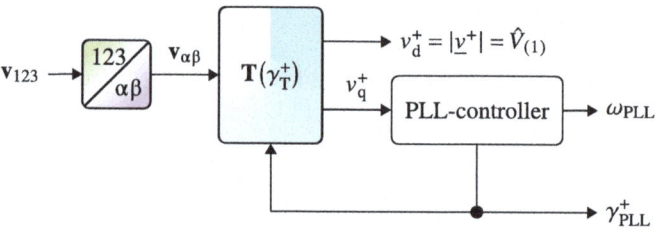

Fig. 4.6 General block diagram of a SRF-PLL

The PLL-controller (see Fig. 4.5) calculates ω_{PLL} and γ_{PLL}^+ from v_{gq}^+. The component v_{gq}^+ is calculated via a transformation block with γ_{PLL}^+ as the transformation angle γ_T^+. The component v_{gd}^+ equals the amplitude of the voltages v_{gx}. The input variables of the transformation block are calculated by the inverse transformation matrix in Eq. (3.12). A state-space model of the SRF-PLL for stability investigations is derived in [91].

4.3 Voltage-based Synchronization

4.3.3 DSRF-PLL

The double synchronous reference frame (DSRF) PLL is a suitable solution for the voltage-based synchronization, if the voltages v_{gx} describe a monofrequency system in the positive sequence or the negative sequence or both, see Sec. 3.1.2.5. It does not work properly, if harmonic frequency components appear in v_{gx}. The DSRF-PLL consists of a PLL-controller (see Sec. 4.3.1) and a decoupling cell [90].

The PLL-controller calculates ω_{PLL} and γ_{PLL}^+ from the mean value of the q-component $\overline{v}_{gq}^+$ of the positive sequence space vector $\underline{v}_g^+$. Furthermore, the amplitude of the positive sequence of the voltages v_{gx} is calculated.

The decoupling cell calculates the mean values of the positive and negative sequence space vector components $\overline{v}_{gd}^+, \overline{v}_{gq}^+, \overline{v}_{gd}^-, \overline{v}_{gq}^-$. Two SRFs are employed for the space vector $\underline{v}_g$, that is regarded as the sum of a positive sequence space vector $\underline{v}_g^+$ and a negative sequence space vector $\underline{v}_g^-$, see Eq. (3.25). The output angle of the PLL-controller γ_{PLL}^+ is used as the transformation angle γ_T^+.

These space vectors lead to DC-components in the according reference frame and to double frequency AC-components in the other reference frame. Consequently, low-pass filters (LPFs) are used for each reference frame in order to gain $\underline{v}_g^+$ and $\underline{v}_g^-$ exclusively in the according reference frame. Since the LPFs are not ideal, the subtraction of the space vectors in the other reference using cross-coupling feedback is necessary. The double transformation angle γ_T^+ with the according sign has to be applied.

The general structure of the decoupling cell of the DSRF-PLL is shown in Fig. 4.7. It is especially valid using the voltages v_{gx} as input variables (index g). For clarity reasons, the index g is neglected in Fig. 4.7.

This decoupling concept can also be described mathematically according to Eq. (4.44). The compact form is given in Eq. (4.45).

$$\begin{pmatrix} \mathbf{v}_{dq}^+ \\ \mathbf{v}_{dq}^- \end{pmatrix} = \begin{pmatrix} \mathbf{T}(\gamma_T^+) & 0 \\ 0 & \mathbf{T}(-\gamma_T^+) \end{pmatrix} \begin{pmatrix} \mathbf{v}_{\alpha\beta} \\ \mathbf{v}_{\alpha\beta} \end{pmatrix} - \begin{pmatrix} 0 & \mathbf{T}(2\gamma_T^+) \\ \mathbf{T}(-2\gamma_T^+) & 0 \end{pmatrix} \begin{pmatrix} \overline{\mathbf{v}}_{dq}^+ \\ \overline{\mathbf{v}}_{dq}^- \end{pmatrix}$$
(4.44)

$$\mathbf{v}_{dq}^\pm = \mathbf{T}(\pm \gamma_T^+)\mathbf{v}_{\alpha\beta} - \mathbf{T}(\pm 2\gamma_T^+)\overline{\mathbf{v}}_{dq}^\mp \quad (4.45)$$

The phase angle of the negative sequence is calculated applying the atan2-function to the pair $\overline{v}_{gd}^-, \overline{v}_{gq}^-$. The result of the negative sequence phase angle has to be regarded relatively to the positive sequence phase angle. The amplitude of the negative

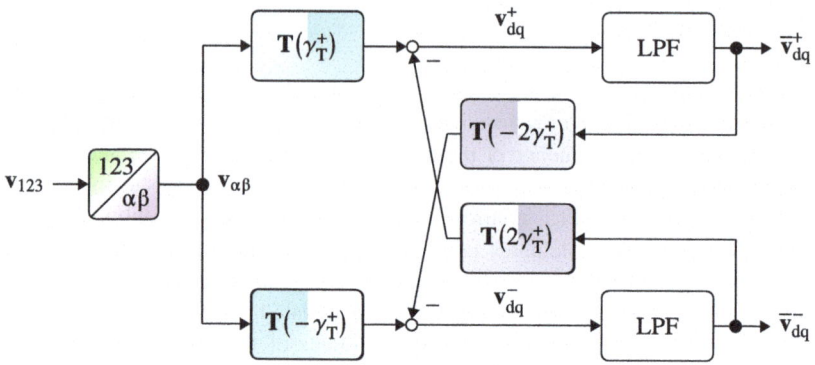

Fig. 4.7 General block diagram of the decoupling cell of a DSRF-PLL

sequence is determined by rotating $\underline{v}_g^-$ in the negative sequence reference frame by the negative sequence phase angle to the d^--axis.

This decoupling cell is an essential tool for the GFL-mode and for the GFM-mode to decouple positive and negative sequence components of measured values of voltages and currents. These components are used for the comparison of real and reference values or the feedforward control. In the GFM-mode, the transformation angle γ_T^+ is set to γ_m^+, which is not calculated by a PLL, but from the active power, see Sec. 5.2.2.1.

4.3.4 MSHDC-PLL

4.3.4.1 Principle of the MSHDC-PLL

The multi-sequence harmonic decoupling cell (MSHDC)-PLL [90] is a suitable solution for the voltage-based synchronization, if the voltages v_{gx} describe a polyfrequency system in the positive sequence or the negative sequence or both, see Sec. 3.1.3. Mean values are removed by using LPFs. The zero sequence component is removed by calculating it from Eq. (3.14) and subtracting it from v_{gx}.

In contrast to the DSRF-PLL (see Sec. 4.3.3), the MSHDC-PLL shows good results for waveforms including harmonic components. The consideration of harmonics is crucial for VSCs, since their operation principle evokes harmonics, see Sec. 2.1.1. Especially low-frequency harmonics have to be considered because it is difficult to attenuate their amplitude by passive filters [92]. The attention of harmonics is also important due to the transient current limiting, see Sec. 6.3.1.

4.3 Voltage-based Synchronization

The principle of the MSHDC-PLL is based on the PLL-controller (see Sec. 4.3.1) with $\overline{v}_{gq}^+$ of the fundamental frequency as input variable. The decoupling cell according to Fig. 4.7 is generalized by introducing an arbitrary number of reference frames. Here, LPFs also play an important role, since the principle of DC-quantities in the according reference frame is applied. The MSHDC is described by Eq. (4.46) [90]:

$$\mathbf{v}_{dq}^k = \mathbf{T}(k\gamma_T^+)\mathbf{v}_{\alpha\beta} - \sum_{k \neq m} \mathbf{T}((k-m)\gamma_T^+)\overline{\mathbf{v}}_{dq}^m \qquad (4.46)$$

In Eq. (4.46), the variable k equals the orders in the positive and the negative sequence. The variable m also equals the orders in the positive and the negative sequence, whereas $m = k$ is excluded. It is seen, that Eq. (4.45) is a special case of Eq. (4.46) with $k = 1$ and $k = -1$.

The phase angles of the positive and the negative sequence at all regarded frequencies (positive sequence fundamental frequency is excluded) are calculated using the atan2-function of the according dq-components. They have to be regarded relatively to the positive sequence phase angle at the fundamental frequency. The amplitudes of the individual systems are calculated by rotating the according space vectors in their reference frame by the phase angle to the according d-axis.

In addition to the voltages v_{gx}, the currents i_{gex} are analysed by an additional MSHDC. In this way, the active and the reactive power in the positive and the negative sequence at each frequency are calculated using Eq. (4.9) and Eq. (4.10).

4.3.4.2 Scenario: Distorted Voltage Conditions

In the following scenario, it is shown via simulations in MATLAB/Simulink, that voltage-based synchronization under distorted voltage conditions using the MSHDC-PLL (see Sec. 4.3.4.1) works properly. The 5th and 7th harmonic components of the voltages v_{gx} at the terminals of the VSC are considered. These frequency components often appear due to the modulation of VSCs [92].

In the investigated scenario, one VSC is synchronized to a 10 kV-grid by a MSHDC-PLL. Initially, the voltages v_{gx} contain a negative sequence at the fundamental frequency with 50 % of the amplitude of the positive sequence. At $t = 0.2\,\text{s}$, 5th and 7th harmonic components in the positive and the negative sequence with 20 % and 10 % of the amplitude of the positive sequence are added. The waveforms of v_{gx} are shown in Fig. 4.8.

The Fig. 4.9 shows the actual fundamental frequency of the positive sequence f and the calculated fundamental frequency of the positive sequence f_{PLL} of the MSHDC-PLL. It is demonstrated that the actual frequency is properly tracked in the interval $t < 0.2\,\text{s}$, despite the existence of the negative sequence at the fundamental

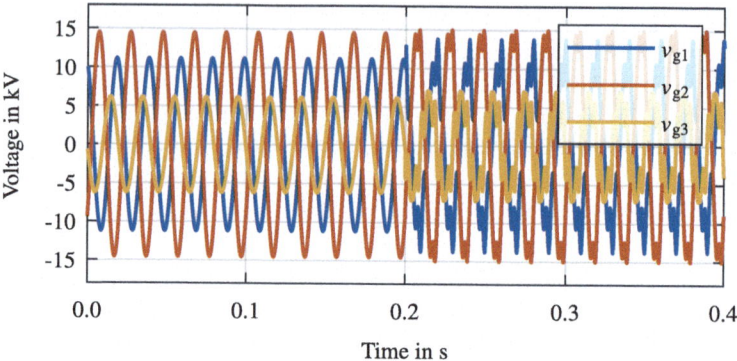

Fig. 4.8 Voltage signals to test the MSHDC-PLL with added harmonic signals at $t = 0.2\,\text{s}$

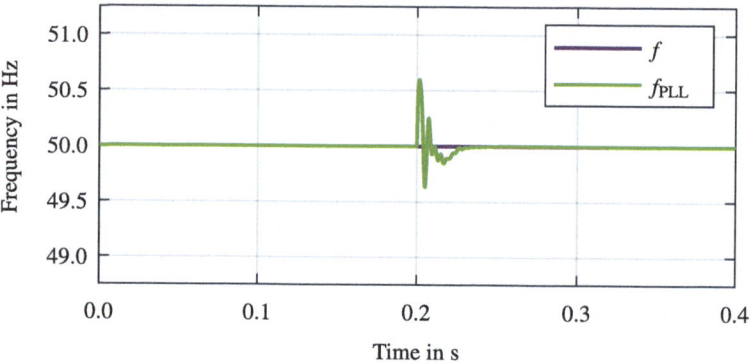

Fig. 4.9 Fundamental frequency of the positive sequence space vector f_{PLL} calculated by the MSHDC-PLL and its reference signal f tested with added harmonic signals at $t = 0.2\,\text{s}$

frequency. In the interval $t > 0.2\,\text{s}$ it is shown that the actual frequency is properly tracked—even under distorted voltage conditions—after a settling time of a few ms.

The Fig. 4.10 shows the actual fundamental frequency angle of the positive sequence γ^+ as the reference angle and the angle γ^+_{PLL} calculated by the MSHDC-PLL. The deviations between the reference angle and the calculated angle are negligible and the MSHDC-PLL allows the voltage-based synchronization under distorted voltage conditions.

4.3 Voltage-based Synchronization

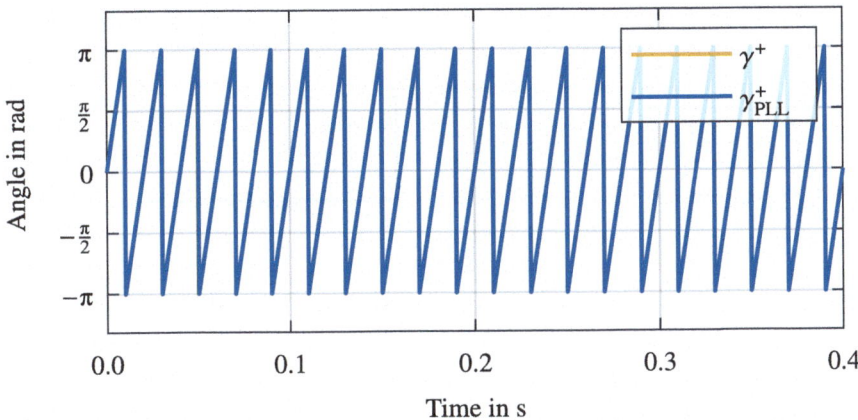

Fig. 4.10 Angle of the positive sequence space vector at the fundamental frequency γ^+_{PLL} calculated by the MSHDC-PLL and its reference signal γ^+ tested with added harmonic signals at $t = 0.2\,\text{s}$

4.3.4.3 Scenario: Frequency Dip

In the next scenario it is demonstrated via simulations in MATLAB/Simulink, that voltage-based synchronization using the MSHDC-PLL (see Sec. 4.3.4.1) works properly even a frequency dip occurs. The frequency dip can for example be caused by droop-controlled GFM-VSCs (see Sec. 4.4.1.2) due to changing load conditions or a grid fault.

The scenario equals the scenario of Sec. 4.3.4.2 in the interval $t < 0.2\,\text{s}$. At $t = 0.2\,\text{s}$, a frequency drop of 1 Hz in the fundamental frequency of the positive and the negative sequence is applied. The waveforms of v_{gx} correspond essentially to the waveforms in Fig. 4.8 in the interval $t < 0.2\,\text{s}$.

The Fig. 4.11 shows the actual fundamental frequency of the positive sequence f and the calculated fundamental frequency of the positive sequence f_{PLL} by the MSHDC-PLL. It is seen that the actual frequency is properly tracked after a settling of approx. 100 ms in the interval $t > 0.2\,\text{s}$ despite the frequency drop.

The Fig. 4.12 shows the actual fundamental frequency angle of the positive sequence γ^+ as the reference angle and the angle calculated by the MSHDC-PLL γ^+_{PLL}. The difference from Fig. 4.10 to Fig. 4.12 is induced by the frequency drop and seen by comparing γ^+_{PLL} at $t = 0.4\,\text{s}$. The deviations between the reference angle and the calculated angle is negligible and the MSHDC-PLL allows the voltage-based synchronization after a frequency dip.

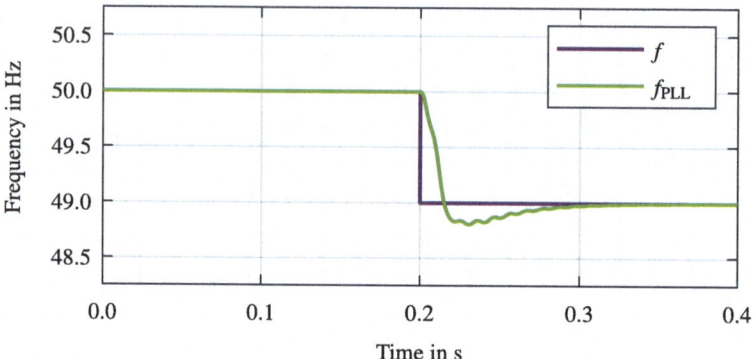

Fig. 4.11 Fundamental frequency of the positive sequence space vector f_{PLL} calculated by the MSHDC-PLL and its reference signal f tested with a frequency drop of 1 Hz at $t = 0.2\,\text{s}$

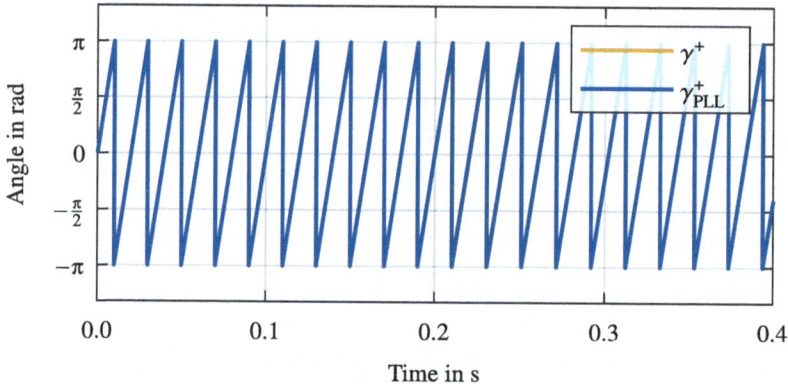

Fig. 4.12 Angle of the positive sequence space vector at the fundamental frequency γ^+_{PLL} calculated by the MSHDC-PLL and its reference signal γ^+ tested with a frequency drop of 1 Hz at $t = 0.2\,\text{s}$

4.4 Power-based Synchronization

The power-based synchronization is an alternative synchronization principle to the voltage-based synchronization (see Sec. 4.3) and leads to the GFM-mode of the GC-VSC. The GFM-mode is favourable over the GFL-mode in 100 % IBPS, due to the inherit properties (see Sec. 2.2.2), e.g. black-start capability and provision of

4.4 Power-based Synchronization

instantaneous power. Since power is composed by active and reactive power, two parameters for the synchronization appear.

There are numerous possibilities for active and reactive power synchronization [50, 51]. In this thesis, the configurable natural droop (CND) controller is applied for the active power synchronization (see Sec. 4.4.1.2) and a droop controller for the filtered reactive power (see Sec. 4.4.2) is used.

4.4.1 Active Power Synchronization

The active power synchronization as a part of the power synchronization is based on the calculation of the angle γ_m^+ from the active power P at the terminals of the VSC. The angle γ_m^+ is used as the reference angle γ^+ (see Eq. (4.43)) instead of γ_{PLL}^+ of the voltage-based synchronization, see Sec. 4.3.

The angle γ_m^+ is used as the reference angle of the capacitor voltages v_{fx} of the LC- and the LCL-filter, see Fig. 3.5 and Fig. 3.6. The capacitor voltages v_{fx} are controlled by the voltage controller, see Sec. 5.2.2.

According to Eq. (4.24), the active power depends on the voltage angle difference δ. The angle δ is adjusted using the integral of $\Delta\omega_m$ (see Eq. (4.28b)). The detailed correlation between γ_m^+ and $\Delta\omega_m$ in the transient state is given in Eq. (2.2c). There are several options to calculate ω_m from P. The transfer function can be selected freely and is called the active power synchronization controller. An overview of different transfer functions is given in [93]. In this thesis, the CND-controller is investigated due to its numerous beneficial design options and its adaption to the proposed grid control (see Sec. 2.2.3).

In contrast to the presented approach calculating ω_m from P, there are different approaches to directly manipulate δ by calculating $\varphi_{(1)}$ from P. A linear transfer function between $\varphi_{(1)}$ and P leads to the angle droop controller [94]. It is demonstrated in [95], that the angle droop controller equals the virtual impedance (see Sec. 5.2.4.2) and shows an analogy to the frequency droop with a high-pass filter (HPF). Using this concept exclusively, constant frequency regulation without communication is achieved [95]. Accurate power sharing cannot be guaranteed and Global Positioning System (GPS) signals to synchronize VSCs are needed [95]. The angle droop controller/virtual impedance can be added to the investigated CND-controller. The influence of the virtual impedance is discussed in Sec. 5.2.4.2.

4.4.1.1 Adoption of Physical Properties of Synchronous Machines to VSCs

One possibility to define the transfer function of the active power synchronization controller is to adopt Newton's second law from synchronous machines [42].

$$J\frac{\mathrm{d}}{\mathrm{d}t}\omega_m = M_m - M_e - M_d \quad (4.47)$$

Here, J is the total moment of inertia, ω_m is the rotor shaft velocity, M_m is the mechanical torque, M_e is the electromagnetic torque and M_d is the damping torque. The torques are approximated according to Eq. (4.48), Eq. (4.49) and Eq. (4.50) [42].

$$M_m \approx \frac{P_m}{\omega_{nom}} \quad (4.48)$$

$$M_e \approx \frac{P_e}{\omega_{nom}} \quad (4.49)$$

$$M_d \approx D\omega_m \quad (4.50)$$

In these equations, P_m represents the net shaft power, P_e represents the electrical air gap power, ω_{nom} is the nominal rotor shaft velocity and D is the damping coefficient. Inserting Eq. (4.48), Eq. (4.49) and Eq. (4.50) into Eq. (4.47) yields Eq. (4.51). The Eq. (4.51) contains the inertia J and the damping coefficient D as characteristic properties.

$$\frac{\mathrm{d}}{\mathrm{d}t}\omega_m = \frac{1}{J}\left(\frac{P_m - P_e}{\omega_{nom}} - D\omega_m\right) \quad (4.51)$$

4.4.1.2 Configurable Natural Droop Controller

The configurable natural droop (CND) controller is one possibility to define the transfer function of the active power synchronization controller. It is inspired by the properties of synchronous machines, see Sec. 4.4.1.1. The rotor shaft velocity ω_m and its nominal value ω_{nom} are now interpreted as the angular frequency and its nominal value of the voltages v_{fx}.

Assuming P_m as the reference active power P_0 and P_e as the actual active power P in Eq. (4.51) yields Eq. (4.52).

$$\frac{\mathrm{d}}{\mathrm{d}t}\omega_m = \frac{1}{J}\left(\frac{P_0 - P}{\omega_{nom}} - D\omega_m\right) \quad (4.52)$$

4.4 Power-based Synchronization

In the frequency domain the following equation using the angular frequency ω_0 at the nominal operating point (index: 0) is valid, see Eq. (4.53).

$$\omega_m(s) = \omega_0 + \frac{1}{\omega_{nom}(Js + D)}(P_0 - P(s)) \qquad (4.53)$$

This concept of active power synchronization controller is called synchronverter [51]. In contrast to a physical synchronous machine, the inertia J and the damping coefficient D are realized by adjusting the parameters of the synchronverter transfer function, instead of being predefined by the hardware design. The steady state characteristics in Eq. (4.53) are not adjustable independently of the damping coefficient.

The inertia, the damping coefficient and the droop constant define the frequency and active power characteristics after a grid fault or changing load conditions. To realize the desired inertia, damping coefficient and droop constant independently of each other, Eq. (4.53) is extended and leads to the transfer function of the CND-controller [93]. The variable $\Delta\omega_{m,PC}$ indicates the deviation of the angular frequency from the nominal angular frequency ω_0 in order to realize the primary grid control (index: PC), see Eq. (2.6). The secondary control and the tertiary control are neglected in this section. The angular frequency ω_m is calculated in accordance with Eq. (4.54).

$$\omega_m(s) = \omega_0 + \Delta\omega_{m,PC}(s) \qquad (4.54a)$$
$$= \omega_0 + G_{CND}(s)(P_0 - P(s)) \qquad (4.54b)$$

The transfer function of the CND-controller is given in Eq. (4.55) [93]. The adjustable parameters K_P, K_I and K_G are used.

$$G_{CND}(s) = \frac{sK_P + K_I}{s + K_G} \qquad (4.55)$$

In practical applications, the implementation of the transfer function in Eq. (4.55) on a digital computer can be realized using Tustin's method with the sampling time T_S [79, 96], see Eq. (4.56).

$$s \approx \frac{2}{T_S}\frac{z - 1}{z + 1} \qquad (4.56)$$

The Fig. 4.13 illustrates the calculation of the angular frequency ω_m as a block diagram. The structure is similar to the PLL-controller, see Fig. 4.5.

Analysing the steady state characteristics of Eq. (4.55) yields Eq. (4.57).

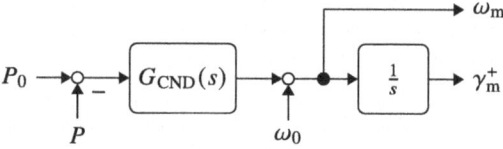

Fig. 4.13 Block diagram of the active power synchronization using a CND-controller

$$G_{\text{CND}}(s=0) = \frac{K_I}{K_G} \qquad (4.57)$$

The Eq. (4.57) shows that the CND-controller allows to adjust the droop constant independently of the inertia and the damping coefficient by an additional parameter—in contrast to the synchronverter transfer function, see Eq. (4.53) and also Eq. (2.3).

Regarding Eq. (4.54b) under steady state conditions (see Eq. (4.57)) shows the equivalence of the CND-controller to a droop-controller with the droop constant k_P, see Eq. (4.58).

$$\omega_m = \omega_0 + \frac{1}{k_P}(P_0 - P) \qquad (4.58)$$

The values P_0 and ω_0 define the nominal operating point given by the tertiary grid control, see Sec. 2.2.3. These values and the droop constant from Eq. (4.57) are used for the primary grid control, see Sec. 2.2.3 and Eq. (2.3).

The linearised state-space model of Eq. (4.54b) is derived using the controllable canonical form, see Eq. (4.59) and Eq. (4.60).

$$\frac{d}{dt}\Delta x_{\text{CND}} = A_{\text{CND}}\Delta x_{\text{CND}} + B_{\text{CND}}(\Delta P_0 - \Delta P) \qquad (4.59)$$

$$\Delta\omega_m = C_{\text{CND}}\Delta x_{\text{CND}} + D_{\text{CND}}(\Delta P_0 - \Delta P) \qquad (4.60)$$

The artificial state variable x_{CND} is introduced. The following state-space parameters are obtained from the CND-parameters, see Eq. (4.61), Eq. (4.62), Eq. (4.63) and Eq. (4.64).

$$A_{\text{CND}} = -K_G \qquad (4.61)$$

$$B_{\text{CND}} = 1 \qquad (4.62)$$

$$C_{\text{CND}} = K_I - K_G K_P \qquad (4.63)$$

$$D_{\text{CND}} = K_P \qquad (4.64)$$

4.4 Power-based Synchronization

Substituting the output equation (see Eq. (4.60)) into the linear first-order differential equation for the active power (see Eq. ((4.30))) yields Eq. (4.65).

$$\frac{d}{dt}\Delta P = KC_{CND}\Delta x_{CND} + KD_{CND}\Delta P_0 - KD_{CND}\Delta P - K\Delta\omega_{bus} \quad (4.65)$$

Combining the physical differential equation from Eq. (4.65) and the controller differential equation from Eq. (4.59), the state-space model of the active power synchronization using a CND-controller is gained, see Eq. (4.66) and Eq. (4.67).

$$\frac{d}{dt}\underbrace{\begin{pmatrix}\Delta x \\ \Delta P\end{pmatrix}}_{\mathbf{x}_{SP}} = \underbrace{\begin{pmatrix}A_{CND} & -B_{CND} \\ KC_{CND} & -KD_{CND}\end{pmatrix}}_{\mathbf{A}_{SP}}\underbrace{\begin{pmatrix}\Delta x \\ \Delta P\end{pmatrix}}_{\mathbf{x}_{SP}} + \underbrace{\begin{pmatrix}0 & B_{CND} \\ -K & KD_{CND}\end{pmatrix}}_{\mathbf{B}_{SP}}\underbrace{\begin{pmatrix}\Delta\omega_{bus} \\ \Delta P_0\end{pmatrix}}_{\mathbf{u}_{SP}} \quad (4.66)$$

$$\underbrace{\begin{pmatrix}\Delta\omega_m \\ \Delta P\end{pmatrix}}_{\mathbf{y}_{SP}} = \underbrace{\begin{pmatrix}C_{CND} & -D_{CND} \\ 0 & 1\end{pmatrix}}_{\mathbf{C}_{SP}}\underbrace{\begin{pmatrix}\Delta x \\ \Delta P\end{pmatrix}}_{\mathbf{x}_{SP}} + \underbrace{\begin{pmatrix}0 & D_{CND} \\ 0 & 0\end{pmatrix}}_{\mathbf{D}_{SP}}\underbrace{\begin{pmatrix}\Delta\omega_{bus} \\ \Delta P_0\end{pmatrix}}_{\mathbf{u}_{SP}} \quad (4.67)$$

Analogue to Eq. (3.34b), the matrix of the transfer functions for all input variables in $\mathbf{u}_{SP}$ to all output variables in $\mathbf{y}_{SP}$ is calculated, see Eq. (4.68).

$$\mathbf{G}_{SP,u}(s) = \mathbf{C}_{SP}(s\mathbf{I} - \mathbf{A}_{SP})^{-1}\mathbf{B}_{SP} + \mathbf{D}_{SP} \quad (4.68)$$

The denominator is the same for all transfer functions, see Eq. (4.69).

$$\text{denom}_{SP} = \frac{1}{KK_I}s^2 + \left(\frac{K_P}{K_I} + \frac{K_G}{KK_I}\right)s + 1 \quad (4.69)$$

The denominator of a second order (SO) lag element is given in Eq. (4.70).

$$\text{denom}_{SO} = T_{SO}^2 s^2 + 2D_{SO}T_{SO}s + 1 \quad (4.70)$$

Comparing the denominator of the transfer functions (see Eq. (4.69)) with the denominator of a SO lag element (see Eq. (4.70)), the characteristics of the SO lag element are adopted by the transfer functions using the CND-controller by calculating the parameters K_P, K_I and K_G, see Eq. (4.55). The synchronizing power coefficient K (see Eq. (4.27)) has to be considered.

In the first step, the time constant of the transfer functions using the CND-controller (see Eq. (4.69)) is realized by tuning K_I. In the second step, the droop constant (see Eq. (4.57)) is realized by tuning K_G. In the third step, the damping constant of the transfer functions using the CND-controller (see Eq. (4.69)) is realized by adjusting K_P.

4.4.2 Reactive Power Synchronization

The reactive power synchronization as a part of the power synchronization is based on the calculation of the RMS-value (or the amplitude respectively) of the voltages v_{fx} from the active power Q at the terminals of the VSC. According to Eq. (4.25), the reactive power depends on the RMS-value E.

There are several options to calculate E from Q. The transfer function can be selected freely and is called the reactive power synchronization controller. An overview of different transfer functions is given in [51]. In this thesis, the droop-controller using the filtered reactive power (see also Sec. 4.1.2.4) is investigated due to its simplicity. Alternatively, E can also be set independently of Q to reach filter capacitor voltages showing a constant RMS-value, see Sec. 5.2.2.2.

The droop-controller using the actual reactive power Q and the filtered reactive power Q_F is described in accordance with Eq. (4.71). The transfer function of the droop-controller is given in Eq. (4.72).

$$E(s) = E_0 + G_D(s)(Q_0 - Q_F(s)) \tag{4.71}$$

$$G_D(s) = \frac{1}{k_Q} \tag{4.72}$$

The structure of the reactive power droop-controller using the filtered reactive power is shown as a block diagram in Fig. 4.14.

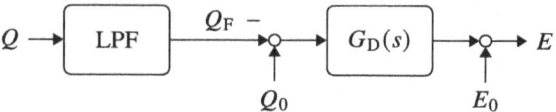

Fig. 4.14 Block diagram of the reactive power synchronization using a droop-controller

4.4 Power-based Synchronization

The values Q_0 and E_0 define the nominal operating point given by the tertiary grid control (see Sec. 2.2.3). These values and the droop constant from Eq. (4.72) are used for the primary grid control. In this way, reactive power-dependent voltage amplitude is realized.

Linearising Eq. (4.71) yields Eq. (4.73).

$$\Delta E = \frac{1}{k_Q}(\Delta Q_0 - \Delta Q_F) \tag{4.73}$$

Substituting the transfer function of the controller (see Eq. (4.73)) into the linear differential equation of the reactive power (see Eq. (4.40)) reveals Eq. (4.74). This equation is the state-space differential equation of the reactive power synchronization using a first-order LPF. The variable T_{FQ} equals the time constant of the LPF.

$$\frac{d}{dt} \underbrace{\Delta Q_F}_{x_{SQ}} = \underbrace{\left(-\frac{K_{QE}}{T_{FQ}k_Q} - \frac{1}{T_{FQ}}\right)}_{A_{SQ}} \underbrace{\Delta Q_F}_{x_{SQ}} + \underbrace{\left(\frac{K_{QV}}{T_{FQ}} \quad \frac{K_{QE}}{T_{FQ}k_Q}\right)}_{B_{SQ}} \underbrace{\begin{pmatrix} \Delta V_{bus} \\ \Delta Q_0 \end{pmatrix}}_{u_{SQ}} \tag{4.74}$$

4.4.3 Small-signal Power Synchronization Stability

The description of the power synchronization is based on physical and controller equations, that are combined into state differential equations, see Eq. (4.66) and Eq. (4.74). The subordinate V/I-controller (see Sec. 5.3) is neglected due to its fast dynamics. The power synchronization is linked to the grid operation principle by the droop-characteristics in the steady-state.

To investigate the small-signal power synchronization stability, the active and reactive power synchronization models from Eq. (4.66) and Eq. (4.74) are merged, see Eq. (4.75).

$$\frac{d}{dt}\begin{pmatrix} x_{SP} \\ x_{SQ} \end{pmatrix} = \begin{pmatrix} A_{SP} & 0 \\ 0 & A_{SQ} \end{pmatrix} \begin{pmatrix} x_{SP} \\ x_{SQ} \end{pmatrix} + \begin{pmatrix} B_{SP} & 0 \\ 0 & B_{SQ} \end{pmatrix} \begin{pmatrix} u_{SP} \\ u_{SQ} \end{pmatrix} \tag{4.75}$$

These holistic state differential equations are investigated concerning asymptotic stability according to Sec. 3.3.2. Therefore, the eigenvalues of the system matrix in Eq. (4.75) are calculated. In general, these eigenvalues depend on the synchronizing active and reactive power coefficients, the CND-controller parameters, the reactive power LPF time constant and the reactive power droop constant.

The small-signal power synchronization stability is one aspect of the converter-driven stability as a superordinate stability category [89]. The model from Eq. (4.75) can in general be extended for a system of multiple VSCs by using Eq. (4.32) and Eq. (4.42).

Small-signal power synchronization stability analyses are essential for the operation of a 100 % IBPS. They also help to define expedient controller parameters in Eq. (4.55) and Eq. (4.72).

4.4.4 Scenario: Different CND-control Parameters

In the following scenario, it is demonstrated via simulations in MATLAB/Simulink, that active power-based synchronization using a CND-controller (see Sec. 4.4.1.2) works properly and the desired transient and the steady state characteristics of ω_m are realized. The time constant works analogue to the inertia of synchronous machines (SMs) and is essential for the operation of 100 % IBPS to avoid sharp frequency steps caused by load changes or grid faults. The droop constant is fundamental for accurate active power sharing according to the grid operation principle after load changes or grid faults, see Sec. 2.2.2.

In the investigated scenario, a VSC forms a 10 kV-grid at the fundamental frequency of 50 Hz via a CND-controller and a droop-controller using the filtered reactive power (see Sec. 4.4.2). The symmetrical load L at the busbar is changed from 10 MW to 12 MW at $t = 1.0$ s via the switch S. The scenario is shown in Fig. 4.15.

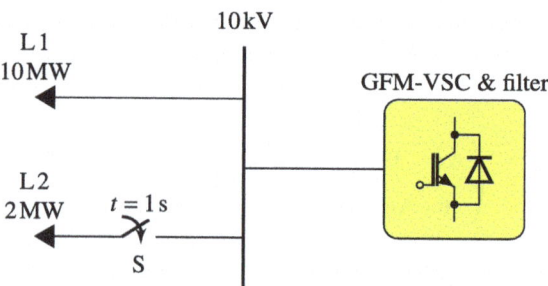

Fig. 4.15 Scenario to investigate the frequency characteristics of a single GFM-VSC under changing load conditions

4.4 Power-based Synchronization

The inverse droop constant of 0.1 Hz/2 MW is chosen and realized by adjusting the CND-controller parameters. The product of the time constant and the damping constant according to a SO lag element (see Sec. 4.4.1.2) is kept constant by adjusting the CND-controller parameters, while the time constant is varied. So three different dynamic characteristics are realized, which are defined as slow, medium and fast.

The Fig. 4.16 shows the actual fundamental frequency of the positive sequence space vector f_m of the filter voltages v_{fx} of the VSC (see Fig. 3.6).

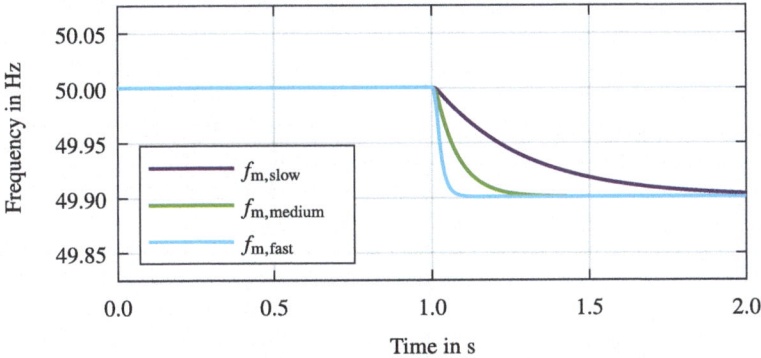

Fig. 4.16 Fundamental frequencies f_m of the positive sequence space vector $\underline{v}_f$ at different parameter configurations as a response to a load step of 2 MW at $t = 1.0\,\text{s}$

It is demonstrated that the VSC works as the only grid-forming element in the grid and no further SM is necessary. Beyond that, the desired transient state characteristics of the VSC are realized by the time constant. The desired steady state characteristics are realized by the droop constant.

4.4.5 Scenario: Two GFM-VSCs and their Interactions

This scenario demonstrates, via simulations in MATLAB/Simulink, that the active power-based synchronization using a CND-controller (see Sec. 4.4.1.2) of two VSCs works properly. The transient and the steady state characteristics of ω_m and P are investigated. The damping constants are of great importance for the parallel operation of multiple VSCs inside a 100 % IBPS. Power swings of high amplitude and unstable grid operation caused by load changes or grids faults are avoided.

The droop constants are essential for accurate active power sharing respecting the nominal power of the individual VSCs.

In the investigated scenario, two VSCs form a 10 kV-grid at the fundamental frequency of 50 Hz via individual CND-controllers and individual droop-controllers using the filtered reactive power (see Sec. 4.4.2). The symmetrical load L with a power of 10 MW is equally supplied by each VSC. At $t = 1.0$ s, the load is changed from 10 MW to 12 MW via the switch S. The additional load shall be shared unequally on the two VSCs by the ratio 2:1. The scenario is shown in Fig. 4.17.

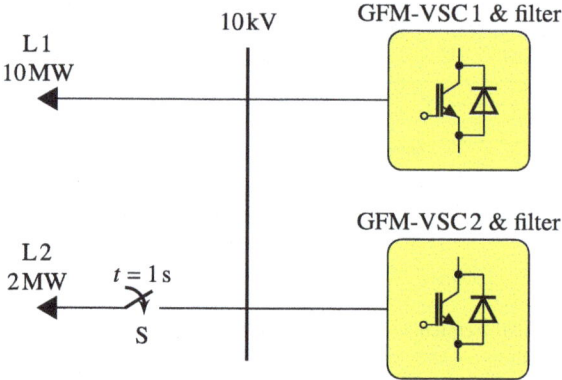

Fig. 4.17 Scenario to investigate the frequency characteristics of two GFM-VSCs under changing load conditions

The inverse droop constant of VSC 1 is chosen to 0.1 Hz/2 MW. The inverse droop constant of VSC 2 is chosen to 0.05 Hz/2 MW. The time constants are equal for each VSC. The setting for slow characteristics of Sec. 4.4.4 is chosen. Moreover, the damping constants are equal for both VSCs.

In the first test case, poor damping of both VSCs is realized. The Fig. 4.18 shows the actual fundamental frequencies of the positive sequence space vectors of the filter voltages v_{fx} of both VSCs. In the steady state, the frequencies of both VSCs are equal and differ from the steady state frequency in Fig. 4.16 of Sec. 4.4.4, since the added load of 2 MW is shared upon two VSCs.

The Fig. 4.19 shows the active powers of both VSCs. It is seen, that power swings with high amplitude appear in the transient state. They are caused by poor damping

4.4 Power-based Synchronization

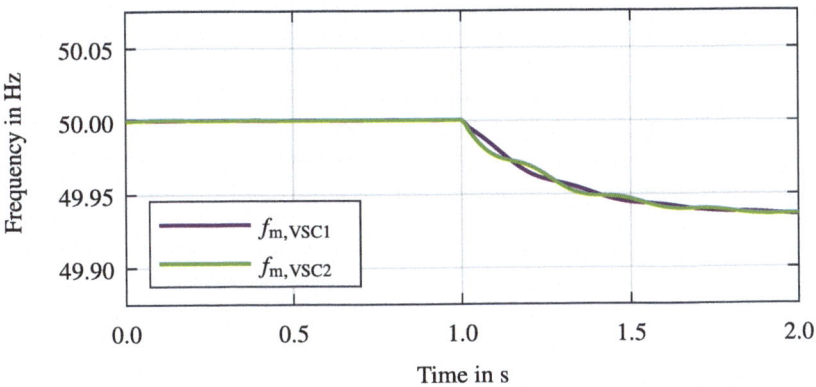

Fig. 4.18 Fundamental frequencies $f_{m,VSCi}$ of the positive sequence space vectors $\underline{v}_{f,VSCi}$ calculated by the CND-controllers as a response to a load step of 2 MW at $t = 1.0$ s with poor damping

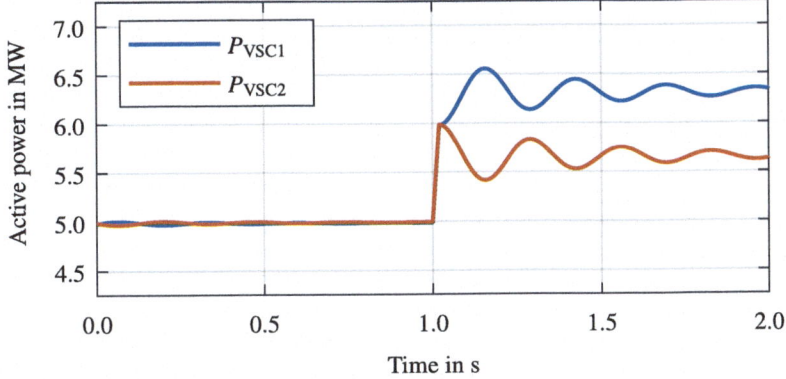

Fig. 4.19 Active power P_{VSCi} of the positive sequence at fundamental frequency as a response to a load step of 2 MW at $t = 1.0$ s with poor damping

of both VSCs. In the steady state, the added load of 2 MW is shared correctly between the two VSCs according to their droop constants by the ratio of 2:1.

In the second test case, good damping of both VSCs is realized. The actual fundamental frequencies of the positive sequence space vectors of the filter voltages v_{fx} of both VSCs is shown in The Fig. 4.20.

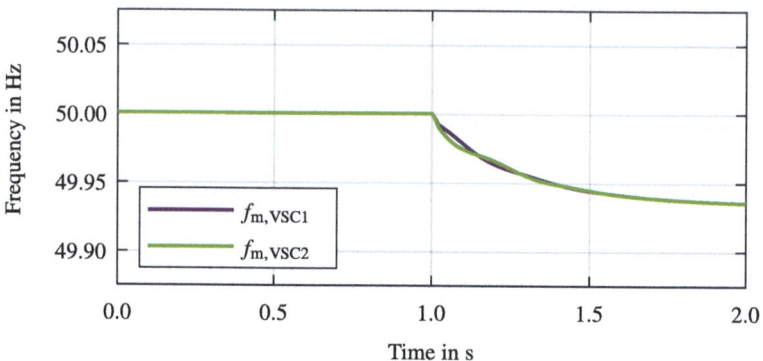

Fig. 4.20 Fundamental frequencies $f_{m,VSCi}$ of the positive sequence space vectors $\underline{v}_{f,VSCi}$ calculated by the CND-controllers as a response to a load step of 2 MW at $t = 1.0$ s with good damping

The active powers of both VSCs are shown in Fig. 4.21. Fig. 4.21 demonstrates good damping in the transient state—in contrast to the first test case realizing poor damping, see Fig. 4.19. In the steady state, the added load of 2 MW is also shared correctly between the two VSCs.

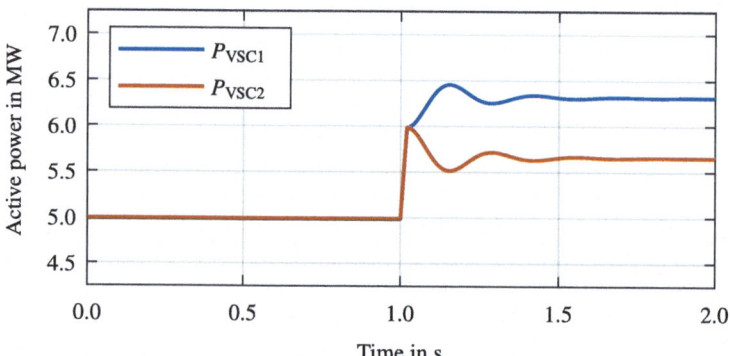

Fig. 4.21 Active power P_{VSCi} of the positive sequence at fundamental frequency as a response to a load step of 2 MW at $t = 1.0$ s with good damping

It is demonstrated, that GFM-VSCs work properly in parallel inside a 100 % IBPS. Good damping characteristics are realized independently of the time constants and the droop constants. Beyond that, the desired steady state characteristics are realized for accurate active power sharing.

4.5 Hybrid Synchronization

4.5.1 Principle of the Hybrid Synchronization

The hybrid synchronization unites the voltage-based synchronization (see Sec. 4.3) and the active power synchronization (see Sec. 4.4.1). The reactive power synchronization (see Sec. 4.4.2) is not affected. In [97] a similar concept is proposed. In contrast to [97], in this thesis a CND-controller and a MSHDC-PLL are combined. Consequently, the presented concept profits from the advantages of the CND-controller (see Sec. 4.4.1.2) and the MSHDC-PLL (see Sec. 4.3.4). In [98, 99] another similar concept is presented. In contrast to these publications, the angle of the PLL is separated from the angle of the active power control in this thesis.

The hybrid synchronization is based on the calculation of the angle γ_h^+ from the active power P at the terminals of the VSC and the calculation of the positive sequence space vector $\underline{v}_g^+$ at the fundamental frequency via a PLL. The angle γ_h^+ is used as the reference angle γ^+ instead of γ_m^+ or γ_{PLL}^+. It is also used as the reference angle of the capacitor voltages v_{fx} of the LC- or the LCL-filter, see Fig. 3.5 and Fig. 3.6.

The principle idea of the hybrid synchronization is to replace the angular frequency ω_0 defining a nominal operating point by ω_{PLL}. The Eq. (4.76) shows the calculation of ω_h based on ω_{PLL} (see Fig. 4.5) and ω_m (see e.g. Eq. (4.54b)). The calculation of γ_h^+ is shown in the block diagram in Fig. 4.22.

$$\omega_h = \omega_{PLL} + \omega_m - \omega_0 \qquad (4.76)$$

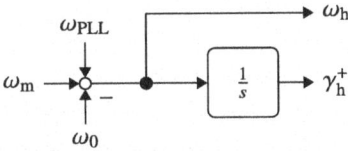

Fig. 4.22 Block diagram of the hybrid active power synchronization

In this way, the droop-characteristics disappear in favour of an accurate control of P_0 as P_{ref} using the hybrid synchronization. In this way, frequency-independent active power injection is realized. The steady state characteristics of the CND-controller are interpreted as a proportional controller. The transient characteristics of the CND-controller are the same as in Sec. 4.4.1.2.

Droop-characteristics are an essential grid-supporting feature concerning active power sharing of energy storage (ES) based renewable energy sources (RES). In contrast, the control of P_{ref} is crucial for the control of the DC-link voltage of wind power plants (WPP) and photovoltaic (PV) power plants. Unlike the GFL-mode, the hybrid synchronization includes fundamental GFM-properties, such as the capability to form an island grid and the provision of instantaneous power, that are favourable for the operation of 100 % IBPS.

Different PLL-principles and transfer functions of the active power control can be used in general. In this thesis, the MSHDC-PLL and the CND-controller are used.

4.5.2 Scenario: Grid Behaviour of a GFM-VSC and a H-VSC

In this scenario, the interaction of active power-based synchronization (see Sec. 4.4.1.2) and the hybrid synchronization (see Sec. 4.5.1) is investigated via simulations in MATLAB/Simulink. A 100 % IBPS with two differently synchronized VSCs is investigated. The transient and the steady state characteristics of ω_m, ω_h and P are presented. The hybrid synchronization allows the frequency-independent active power injection, which is favourable for VSCs that are not storage-based to ensure a constant DC-link voltage (see also Sec. 2.1.1). The active power-based synchronization provides active power sharing by active power-dependent frequency, see Sec. 4.4.5. In this way, a 100 % IBPS using ES-based VSCs and VSCs without an energy storage system (ESS) is realized.

In the investigated scenario, two VSCs are synchronized to a 10 kV-grid at the fundamental frequency of 50 Hz. The symmetrical load of 10 MW is equally supplied by each VSC. At $t = 1.0$ s, the load is changed from 10 MW to 12 MW. The additional active power shall be provided by the ES-based, power-synchronized VSC 1 using a CND-controller. VSC 2 is not equipped with an ESS and hybrid synchronization using as MSHDC-PLL and a CND-controller is applied. From the hardware perspective, this scenario is equal to Fig. 4.17.

The time constants are equal for each VSC. The setting for slow characteristics of Sec. 4.4.4 is chosen. Besides, the good damping characteristics of Sec. 4.4.5 are

4.5 Hybrid Synchronization

applied for both VSCs. The inverse droop constant of 0.1 Hz/2 MW is chosen for VSC 1.

The Fig. 4.23 presents the actual fundamental frequencies of the positive sequence space vectors $\underline{v}_{fx}$ of both VSCs. In the steady state, the frequencies of both VSCs are equal and correspond to the steady state frequency in Fig. 4.16, since the added load power of 2 MW is solely provided by VSC 1. Thus, its frequency is reduced corresponding to the droop-constant.

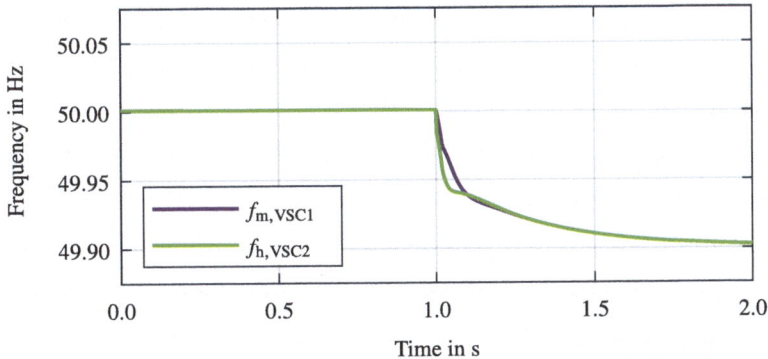

Fig. 4.23 Fundamental frequencies of the positive sequence space vectors $\underline{v}_{f,VSCi}$ calculated as a response to a load step of 2 MW at $t = 1.0$ s using power-based synchronization for VSC 1 and hybrid synchronization for VSC 2

The Fig. 4.24 shows the active powers of both VSCs. It is seen that the added load power is solely provided by VSC 1 in the steady state. The active power of VSC 2 is same at $t = 0.0$ s and $t = 2.0$ s as a feature of GFL-VSCs. In the transient state at $t = 1.0$ s, VSC 2 rises its active power and provides instantaneous active power injection as a feature of GFM-VSCs. This instantaneous active power injection decreases the active power overshoot and avoids the disconnection of VSC 1 due to overcurrent limitation. Consequently, VSC 2 combines favourable GFL- and GFM-features.

It is demonstrated, that an active power synchronized-VSC and a hybrid synchronized VSC work properly in parallel inside a 100 % IBPS. While VSC 1 provides variable frequency grid-operation for active power sharing, VSC 2 ensures frequency-independent active power injection. Moreover, VSC 2 provides instantaneous active power injection and stabilizes the grid operation by instantaneous

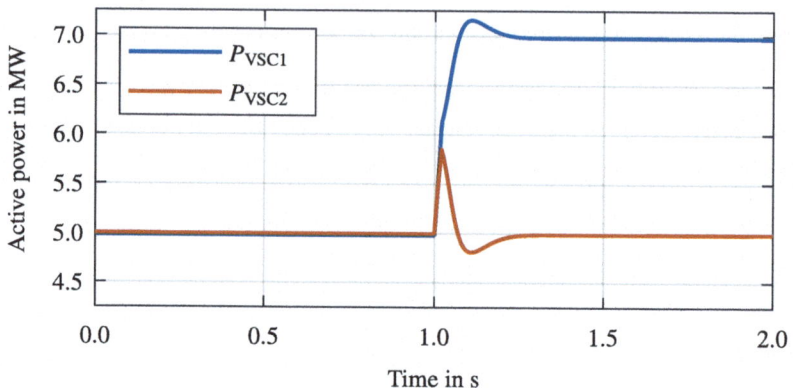

Fig. 4.24 Active power P_{VSCi} of the positive sequence at fundamental frequency as a response to a load step of 2 MW at $t = 1.0$ s using power-based synchronization for VSC 1 and hybrid synchronization for VSC 2

power injection. In this way, ES-based VSCs and VSCs without ES are combined inside a 100 % IBPS.

4.6 Summary

In this chapter, the synchronization of grid-connected voltage source converters (VSCs) and the electrical power grid is presented. This interdependency and related effects are the basis for the operation of a 100 % inverter-based power system (IBPS). This chapter is closely linked to the grid dynamics and the grid stability.

The properties of the grid equivalent circuit define the energy transfer between the VSC and the grid essentially. The p-q-theory represents a universal power theory and allows the description of power in all grid states. The P/Q-transformation and the linearisation at an operating point allow a simplified mathematical description of the physical effects.

Basically, there are two different principles for the synchronization of a VSC to the grid. The voltage-based synchronization uses a phase-locked loop (PLL) and leads to the grid-following (GFL) mode of the VSC. Current source characteristics are achieved. The multi-sequence harmonic decoupling cell (MSHDC) PLL allows the voltage-based synchronization in multi-sequence and distorted systems. The included decoupling method of space vector components is an important tool for the GFL-mode as well as for the grid-forming (GFM) mode.

4.6 Summary

The power-based synchronization leads to the GFM-mode of the VSC. Voltage source characteristics are achieved. The configurable natural droop (CND) controller enables the realization of the desired frequency inertia, the power damping coefficient and the frequency droop constant independently of each other. In contrast to a synchronous machine, these parameters are independent of the hardware design of the VSC and are realized via software parameters by the VSC controller.

The hybrid synchronization combines the voltage-based and the power-based synchronization. Especially, power-synchronized and hybrid-synchronized VSCs are expedient to operate 100 % IBPS. Several simulation scenarios are investigated in this context.

Derived state-space models allow the analysis of the small-signal power synchronization stability of VSCs. Synchronization stability also depends on the design of the VSC controller and its parameter selection.

The synchronization of a VSC is closely connected to the grid control, representing one layer in a hierarchical control structure. This control layer calculates the reference variables of the voltage/current control layer. This hierarchical structure is the starting point of Chap. 5. To follow the key aspects of the central theme of this thesis on a shortcut: The conclusion of the next chapter is provided in Sec. 5.4.

Open Access This chapter is licensed under the terms of the Creative Commons Attribution 4.0 International License (http://creativecommons.org/licenses/by/4.0/), which permits use, sharing, adaptation, distribution and reproduction in any medium or format, as long as you give appropriate credit to the original author(s) and the source, provide a link to the Creative Commons license and indicate if changes were made.

The images or other third party material in this chapter are included in the chapter's Creative Commons license, unless indicated otherwise in a credit line to the material. If material is not included in the chapter's Creative Commons license and your intended use is not permitted by statutory regulation or exceeds the permitted use, you will need to obtain permission directly from the copyright holder.

Grid-side Characteristics and V/I-Control of VSCs

This chapter presents the grid-side characteristics and the voltage/current (V/I) control of voltage source converters (VSCs). In Sec. 5.1, the hierarchical control structure of VSCs and the functional chain to realize desired grid-side characteristics are explained. After discussing possible options of grid-side characteristics in a generalized symmetrical components equivalent circuit, the calculation of reference variables is presented in Sec. 5.2. The design of voltage sources and currents sources as well as the realization of V/I-characteristics is shown. The injection of reactive current in the negative sequence is evaluated. Sec. 5.3 describes the design of the state-feedback V/I-controller. The control law is derived and aspects for the analysis of the closed control loop are given. Moreover, the feedforward control of reference and disturbance variables is presented. The principle of the state-feedback controller based on the linear–quadratic regulator (LQR) algorithm is shown. Furthermore, the implementation of integral behaviour is described.

5.1 Hierarchical Control Structure and Functional Chain

There are different requirements for grid-connected (GC) voltage source converters (VSCs) from the grid perspective during normal operation conditions and during grid faults. To fulfil these requirements, VSCs must realize different grid-side characteristics in the positive, in the negative and optionally in the zero sequence. This chapter presents how to realize required grid-side characteristics in the positive and the negative sequence, as well as the voltage/current (V/I) control of VSCs. The realization of grid-side characteristics in the zero sequence is shown in Sec. 6.2.2.

A hierarchical control structure is used to reflect the causal order from grid control (see Sec. 2.2.3) to the modulator. Regarding grid-forming (GFM) VSCs, the reference operating point (calculated from the nominal operating point and the droop characteristics) in the steady state and the transient behaviour (e.g. inertia and damping) are defined by the active and reactive power synchronization (see Sec. 4.4.1 and Sec. 4.4.2). In this thesis, a configurable natural droop (CND) controller (see Sec. 4.4.1.2) and a filtered reactive power droop controller (see Sec. 4.4.2) are used. The V/I-controller calculates actuating variables **u** from reference variables **w**, see Fig. 5.1. The modulator activates suitable switching states of the VSC to synthesize the actuating variables, see Sec. 2.1.1.

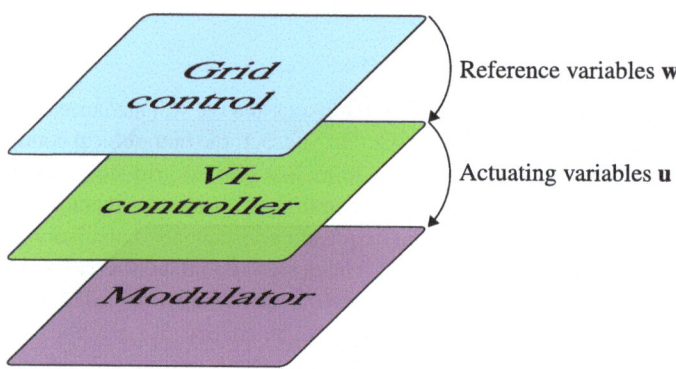

Fig. 5.1 Hierarchical control structure of a grid-connected VSC

The Fig. 5.2 shows the functional chain to realize the desired grid-side characteristics using a state-feedback V/I-controller.

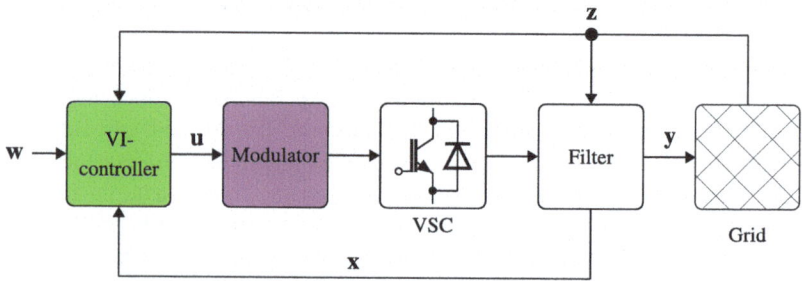

Fig. 5.2 Functional chain of a grid-connected VSC using a state-feedback V/I-controller and a modulator visualized as a block diagram

5.1 Hierarchical Control Structure and Functional Chain

The filter (see Sec. 3.2.2) represents the control plant. In the following investigations, LCL-filters (see Sec. 3.2.2.3) are applied due to the superior filter performance compared to L- and LC-filters. Further benefits of LCL-filters over L- and LC-filters are discussed in [143]. The combination of the VSC and the modulator represents the actuator. From the control design perspective, the actuator is assumed to be ideal. That means that the transfer function of the actuator equals 1. The interaction between the controlled variables **y** and the disturbance variables **z** is also affected by the grid strength, see Sec. 4.1.1.2.

There are different possibilities for the design of the V/I-controller. Each filter type (especially the LCL-filter) leads to a multiple-input multiple output (MIMO) (regarding **y**, **z** and the actuating variables **u**) model of higher order (regarding the state variables **x**). Using space vector components in rotating reference frames, the d- and q- state differential equations are coupled, see Sec. 3.2.2. These models even grow, if negative and zero sequence components are included or multiple frequencies are regarded.

To handle these complex models by an appropriate V/I-controller, state-feedback controllers are used in this thesis. The characteristic feature of state-feedback controllers is the feedback of state variables instead of controlled variables, e.g. when using proportional-integral (PI) controllers. The applied state-feedback controllers use space vector components in rotating reference frames to gain DC-quantities in the steady state. State-feedback controllers are particularly suitable for GFM-VSCs, since no cascaded control loops with slow outer loops are necessary—in contrast to PI-controllers. Beyond that, the decoupling of d- and q- control loops is not needed. Consequently, state-feedback V/I-controller will be the backbone of 100% inverter-based power system (IBPS).

The actuating variables **u** are adjustable input variables and are calculated by the V/I-controller. For each filter type, the pair v_{cd}^+/v_{cq}^+ equals the actuating variables of the positive sequence and the pair v_{cd}^-/v_{cq}^- equals the actuating variables of the negative sequence. After the calculation, each pair is transformed into the stationary reference frame (see Sec. 3.1.2.4 and Sec. 3.1.2.5) and then into the natural reference frame using Eq. (3.12). Afterwards, the positive and negative sequence components are added and sent to the modulator, which generates gate signals for the power devices.

Disturbance variables **z** are non-adjustable input variables. The pair of disturbance variables differs for each filter type, see Sec. 3.2.2. Using a LCL-filter the pair v_{gd}^+/v_{gq}^+ equals the disturbance variables of the positive sequence and the pair v_{gd}^-/v_{gq}^- equals the disturbance variables of the negative sequence. The components are gained by measuring the variables in the natural reference frame and transforming them by a decoupling cell, see Sec. 4.3.3 and Sec. 4.3.4.1. It is seen, that decoupling cells are necessary for grid-following (GFL) and GFM VSCs. The treatment of disturbance variables inside the V/I-controller is described in Sec. 5.3.3.

State variables **x** are inner system variables, see Sec. 3.2.1. Different filter models lead to different state variables, see Sec. 3.2.2. As disturbance variables, state variables are measured in the natural reference frame and transformed by a decoupling cell.

Controlled variables **y** are variables, that should equal certain values and therefore represent output variables of the control plant. In this thesis, the controlled variables are a subset of the state variables. They are chosen by the definition of the matrix **C**, see Eq. (3.30). The number of selected controlled variables must not exceed the number of actuating variables of the positive and the negative sequence due to the controllability of the system, see Sec. 3.3.3. For example, one pair out of i_{gd}^+/i_{gq}^+, v_{fd}^+/v_{fq}^+ and i_{cd}^+/i_{cq}^+ is chosen as the controlled variables in the positive sequence using a LCL-Filter. At the same time, one pair out of i_{gd}^-/i_{gq}^-, v_{fd}^-/v_{fq}^- and i_{cd}^-/i_{cq}^- is chosen as the controlled variables in the negative sequence. The choice of currents components in the positive sequence leads to the GFL-mode, see Sec. 4.2. In contrast, the choice of voltage components in the positive sequence leads to the GFM-mode. The chosen pair can be different in the positive and the negative sequence.

Reference variables are labelled by **w**. Assuming that **y** = **w** is ensured by the V/I-controller, the reference variables define the steady state grid-side characteristics of the VSC. They are calculated in the grid control layer, which is connected to the synchronization principle.

5.2 Calculation of Reference Variables

5.2.1 Symmetrical Components Equivalent Circuit and Overview of Grid-Side Characteristics

The steady state grid-side characteristics are described by phasors. It is favourable to transform these phasors into symmetrical components to evaluate the positive, the negative and optionally the zero sequence individually, see Sec. 3.1.2.2.

Conventional VSCs represent three-phase three-wire systems. Due to Kirchhoff's current law (KCL), zero sequence currents are not possible and the VSC represents an open circuit in the zero sequence. From the fundamentals of VSCs connecting different DC-link potentials to the AC-side (see Sec. 2.1), a voltage source in the positive sequence and a voltage source in the negative sequence appear in general. The monofrequency symmetrical components equivalent circuit (see Fig. 5.3) of a three-phase three-wire VSC is derived from Fig. 2.4.

5.2 Calculation of Reference Variables

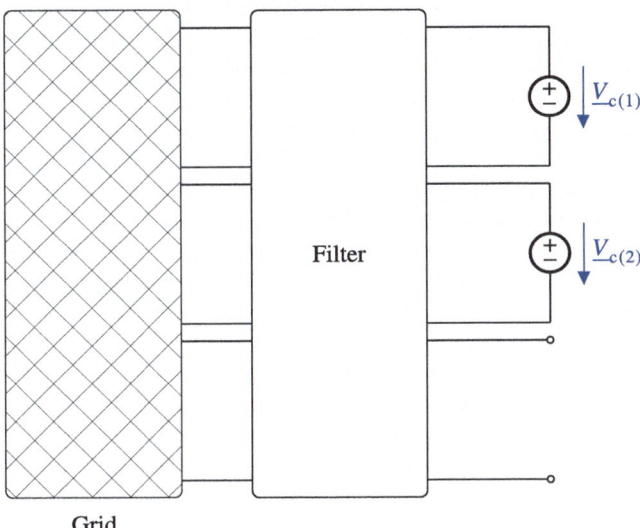

Fig. 5.3 Steady state monofrequency symmetrical components equivalent circuit of a three-phase three-wire grid-connected (GC) VSC representing a voltage source in the positive sequence (1), a voltage source in the negative sequence (2) and an open circuit in the zero sequence (0), see also Fig. 2.4

If the negative sequence of the VSC voltages v_{cx} is set to zero, the negative sequence voltage phasor (see Eq. (3.22)) is zero and the VSC represents a short circuit in the negative sequence, see Fig. 5.4.

Connecting a VSC to the grid via a filter, filter voltages or currents are controlled. From the grid perspective, the VSC and the filter merge. The generalized steady state symmetrical components equivalent circuit of the GC VSC is shown in Fig. 5.5.

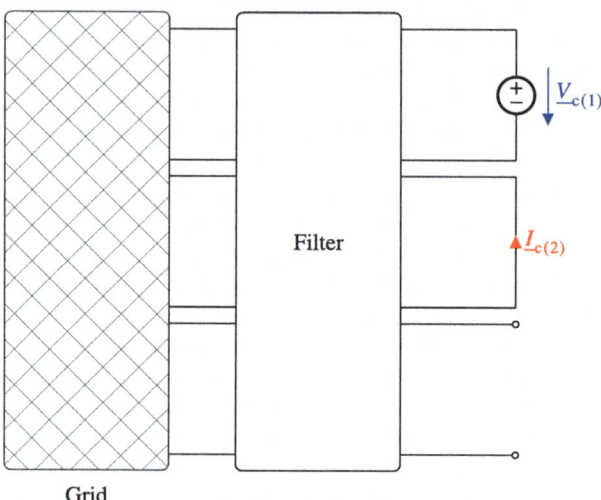

Fig. 5.4 Steady state monofrequency symmetrical components equivalent circuit of the three-phase three-wire GC VSC, if a positive sequence (1) voltage phasor is realized and the negative sequence (2) voltage phasor is set to zero

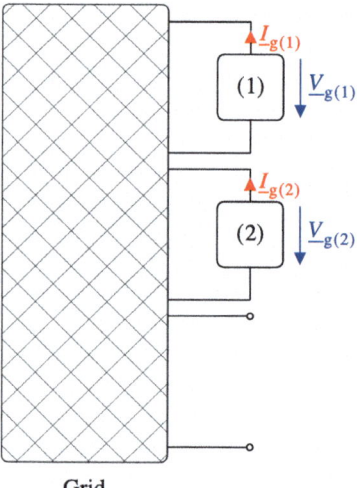

Fig. 5.5 Generalized steady state symmetrical components equivalent circuit of the GC VSC regarding the positive sequence (1), the negative sequence (2) and an open circuit in the zero sequence from the grid perspective

5.2 Calculation of Reference Variables

Various options for the grid characteristics reveal in the positive and the negative sequence, that are realized by proper calculation of the according reference variables. The Fig. 5.6 and the Fig. 5.7 give an overview of different possibilities of grid-side characteristics in the positive and the negative sequence and the according controlled variables.

In general, the characteristics of the positive and the negative sequence are independent of each other, since the according variables are decoupled from each other. In both sequences, the characteristics of a voltage source, a current source and V/I-characteristics are possible. Characteristics of sources and V/I-characteristics can be combined.

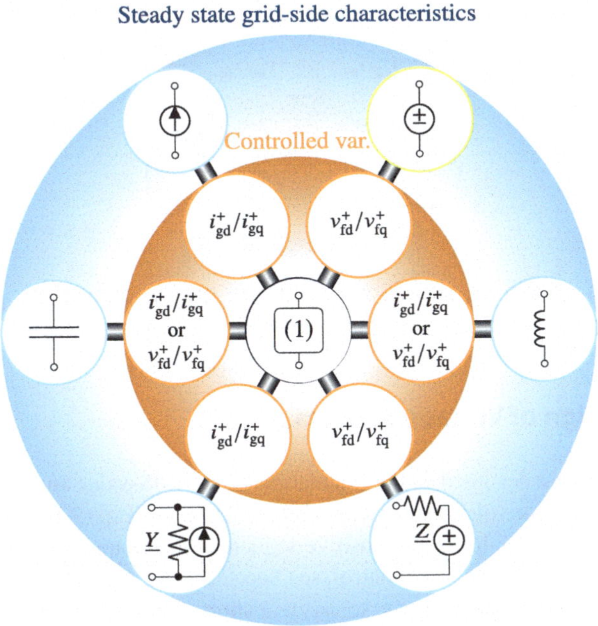

Steady state grid-side characteristics

Fig. 5.6 Options of grid-side characteristics of the GC VSC in the positive sequence (1) in the steady state and the according controlled variables

Steady state grid-side characteristics

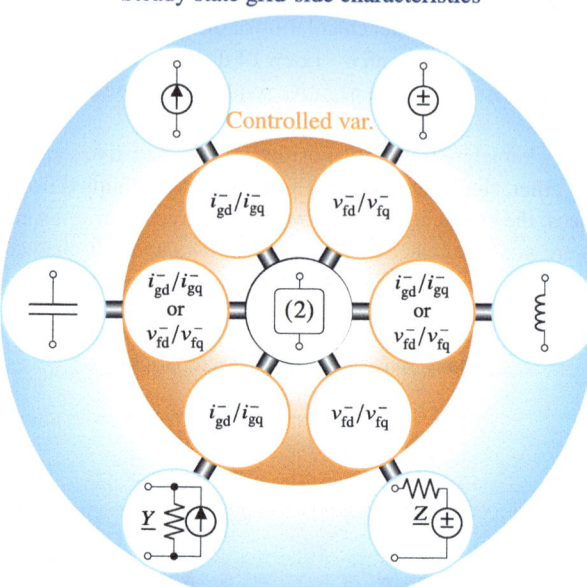

Fig. 5.7 Options of grid-side characteristics of the GC VSC in the negative sequence (2) in the steady state and the according controlled variables

5.2.2 Design of Voltage Sources

Voltage sources are designed by the calculation of the RMS-value and the frequency of the filter capacitor voltages v_{fx} with $x \in \{1, 2, 3\}$, see Eq. (3.16) or Eq. (3.23) and Sec. 3.2.2. Especially in the positive sequence, the characteristics of voltage sources are relevant for the grid control, see Sec. 2.2.3. Voltage sources can also be used in the negative sequence, e.g. to create a short circuit to realize symmetrical filter capacitor voltages.

For the design of a voltage source, the pairs v_{fd}^+/v_{fq}^+ or v_{fd}^-/v_{fq}^- have to be chosen as controlled variables. The inverse transformation from Eq. (3.12) has to be applied and the principles of rotation reference frames, see Sec. 3.1.2.4 or Sec. 3.1.2.5, have to be used.

Subsequently, the variables ω_m, E, P and Q are related to the positive or the negative sequence. For better readability, the indices (1) or (2) are neglected. The active and the reactive power in the positive or the negative sequence are calculated from decoupled space vector components, see Sec. 5.2.3.1.

5.2.2.1 f/P-Characteristics

The angular frequency ω_m of the filter capacitor voltages v_{fx} is linked to the active power P, see Eq. (4.22), Eq. (4.17) and Eq. (4.28b). The time-independent part of the angle of v_{fx} (see Eq. (3.16) or Eq. (3.23)) adjusts according to the operating point.

Three different mathematical relations between ω_m and P are discussed. They are visualized in Fig. 5.8.

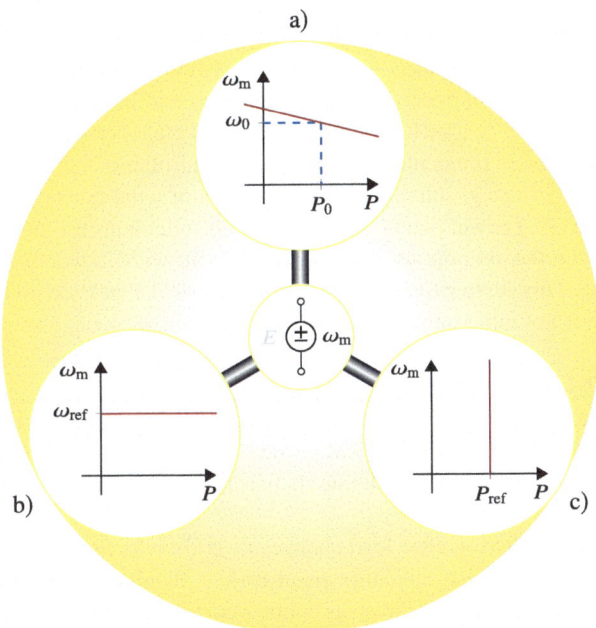

Fig. 5.8 Design options of the angular frequency ω_m of the capacitor voltage v_{fx} of a GFM-VSC depending on the active power P

A droop control (see Eq. (2.3)) allows the active power sharing after load changes or grid faults (option a) in Fig. 5.8. The idea is to change the voltage angle difference δ (see Eq. (4.17)) to adjust the active power (see Eq. (4.22)) by realizing $\omega_m \neq \omega_{vg}$ (see also Eq. (4.28b)). In the steady state, ω_m and ω_{vg} are equal but can deviate from the nominal operating frequency. The nominal operating point (index 0) is given by the tertiary grid control (see Sec. 2.2.3). The droop control in the steady state is realized by the CND-controller, see Eq. (4.54b) and Eq. (4.55). The droop control configuration is suitable for GC VSCs, preferably equipped with a storage device, or an active power source at the DC-link.

The angular frequency ω_m can be set independently of the active power P to ω_{ref} (option b) in Fig. 5.8). The time-independent part of the angle of v_{fx} must also be defined. This configuration realizes slack characteristics concerning the frequency and the angle and is e.g. suitable for load-side VSCs—in contrast to GC VSCs.

The active power P can be set independently of the angular frequency ω_m to P_{ref} (option c) in Fig. 5.8). These characteristics are realized by a PI-controller (see [93]) or the hybrid synchronization, see Sec. 4.5.1. Using a PI-controller, the inertia and the damping coefficient are fixed by the PI-controller parameters. In contrast—using the hybrid synchronization—the droop constant, which can be regarded as a proportional controller, is independent of the inertia and the damping coefficient. This configuration can be used for VSCs in the parallel grid operation. Optionally, the reference active power is calculated by a $P(v_{DC})$-controller or a $P(f_{vg})$-controller. The v_{DC}-controller is e.g. suitable for the operation of wind power plants (WPPs) or photovoltaic (PV) power plants. The $P(f_{vg})$-controller aims for the active power sharing based on the assumption, that f_{vg} is a system-wide indicator of the active power balance. This perspective results from the physical characteristics of rotating machines and is not inherently valid for 100% IBPS.

5.2.2.2 E/Q-Characteristics

The RMS-value E of the filter capacitor voltages v_{fx} is linked to the reactive power Q, see Eq. (4.23). By calculating the RMS-value, the amplitude of v_{fx} is also defined.

Three different mathematical relations between E and Q are discussed. They are visualized in Fig. 5.9.

Reactive power sharing after load changes or grid faults is realized by a droop control using the reference operating point (index: 0) given by the tertiary grid control (see Sec. 2.2.3). This is shown as option a) in Fig. 5.9. The Eq. (4.71) and Eq. (4.72) show the droop equation using the filtered reactive power. This configuration can be used for GC VSCs, preferably equipped with an energy storage device.

The RMS-value E can be set independently of the reactive power Q to E_{ref} (option b) in Fig. 5.9). Slack characteristics concerning the RMS-value of the voltages v_{fx} are gained. Consequently, this configuration is suitable for the load-side VSCs.

The reactive power Q can be set independently of the RMS-value E to Q_{ref} (option c) in Fig. 5.9). These characteristics are realized by a PI-controller [51]. This configuration can be used for VSCs in the parallel grid operation.

5.2 Calculation of Reference Variables

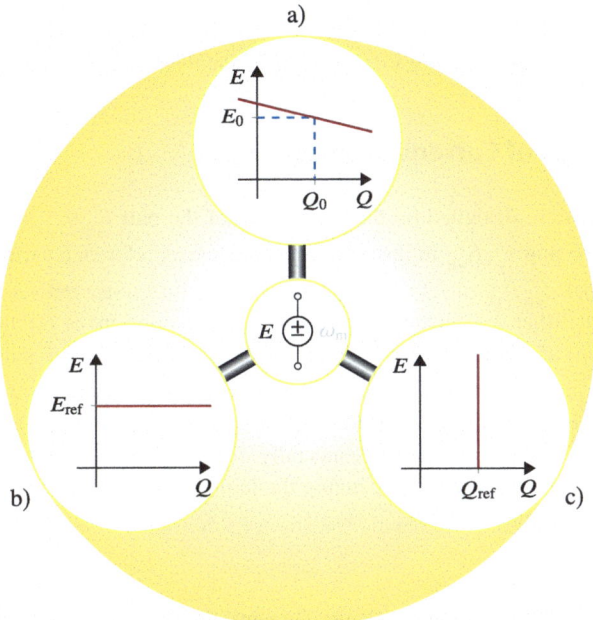

Fig. 5.9 Design options of the RMS-value E of the capacitor voltage v_{fx} of a GFM-VSC depending on the reactive power Q

5.2.2.3 Selected Combinations for the Positive Sequence

Based on Sec. 5.2.2.1 and Sec. 5.2.2.2, three different combinations to design voltage sources based on the calculation of the angular frequency ω_m and the RMS-value E are presented.

Combining the angular frequency ω_m droop control (Fig. 5.8 a)) and the RMS-value E droop control (Fig. 5.9 a)) leads to grid-supporting characteristics. Both the active and the reactive power are shared after load changes or grid faults.

Alternatively, slack characteristics can be realized. The angular frequency ω_m and the RMS-value E are set independently of the active power P and the reactive power Q (Fig. 5.8 b) and Fig. 5.9 b)). In this configuration, the slack-VSC compensates all active and reactive power imbalances. Only one slack-VSC can be used inside the grid.

Beyond that, generator characteristics (see also [71]) can be realized. In this case, the angular frequency ω_m droop control (Fig. 5.8 a)) and a fixed RMS-value E

(Fig. 5.9 b)) are combined. This configuration shows slack characteristics regarding reactive power and can be used for VSCs in the parallel grid operation.

5.2.3 Design of Current Sources

Current sources are designed by the calculation of the pair i_{gd}^+/i_{gq}^+ in the positive sequence or the pair i_{gd}^-/i_{gq}^- in the negative sequence as reference variables. Alternatively, the pair i_{cd}^+/i_{cq}^+ or i_{cd}^-/i_{cq}^- is calculated. The difference between these two options is, that either the currents on the grid side (first option) or on the VSC side (second option) are controlled, see also Fig. 3.6. Subsequently, the pair i_{gd}^+/i_{gq}^+ or the pair i_{gd}^-/i_{gq}^- is controlled, since these pairs represent the currents i_{gx}, that directly affect the grid and contain less harmonic distortions.

In the following investigations, especially current sources in the negative sequence play an important role influencing the fault characteristics. In this case, the pair i_{gd}^-/i_{gq}^- is controlled.

5.2.3.1 P/Q-Control

Subsequently, the voltages v_{gx} and the currents i_{gx} are regarded to characterize the point of interconnection of the VSC to the grid, see Sec. 4.1.1.1. For better readability, the index g is neglected.

One possibility is to define the pair i_d^+/i_q^+ or the pair i_d^-/i_q^- independently of other variables. From the grid perspective, especially the active power P and the reactive power Q are relevant. Consequently, the relations between the space vector components and the active and the reactive power are derived.

The starting point is Eq. (4.5) describing the instantaneous real power p and Eq. (4.6) describing the instantaneous imaginary power q. Using the vectorial notation instead of complex numbers, these equations are written as dot products, see Eq. (5.1) and Eq. (5.2).

$$p = \frac{3}{2}(\mathbf{v}_{\alpha\beta} \cdot \mathbf{i}_{\alpha\beta}) \tag{5.1}$$

$$q = \frac{3}{2}(\mathbf{v}_{\perp\alpha\beta} \cdot \mathbf{i}_{\alpha\beta}) \tag{5.2}$$

In Eq. (5.2), the rotated orthogonal voltage space vector $\mathbf{v}_{\perp\alpha\beta}$ (see Eq. (5.3)) is used [77]. This space vector equals the original voltage space vector $\mathbf{v}_{\alpha\beta}$ rotated by 90 degrees in the mathematical negative direction. The original and the orthogonal voltage space vectors have the same length.

5.2 Calculation of Reference Variables

$$\mathbf{v}_{\perp\alpha\beta} = \begin{pmatrix} 0 & 1 \\ -1 & 0 \end{pmatrix} \mathbf{v}_{\alpha\beta} \tag{5.3}$$

This principle is also valid for the orthogonal positive sequence voltage space vector $\mathbf{v}_{\perp\alpha\beta}^+$ and the orthogonal negative sequence voltage space vector $\mathbf{v}_{\perp\alpha\beta}^-$. The following space vectors according to Eq. (5.4), Eq. (5.5) and Eq. (5.6) are obtained.

$$\mathbf{v}_{\alpha\beta} = \mathbf{v}_{\alpha\beta}^+ + \mathbf{v}_{\alpha\beta}^- \tag{5.4}$$

$$\mathbf{i}_{\alpha\beta} = \mathbf{i}_{\alpha\beta}^+ + \mathbf{i}_{\alpha\beta}^- \tag{5.5}$$

$$\mathbf{v}_{\perp\alpha\beta} = \mathbf{v}_{\perp\alpha\beta}^+ + \mathbf{v}_{\perp\alpha\beta}^- \tag{5.6}$$

The rotation of the orthogonal voltage space vector is also valid for space vectors in rotating reference frames.

The active power P and the reactive power Q equal the mean values of the instantaneous real power p and the instantaneous imaginary power q, see Sec. 4.1.2.1. They are calculated according to Eq. (5.7) and Eq. (5.8) [77, 86].

$$P = \frac{3}{2}\left(\mathbf{v}_{\alpha\beta}^+ \mathbf{i}_{\alpha\beta}^+ + \mathbf{v}_{\alpha\beta}^- \mathbf{i}_{\alpha\beta}^-\right) \tag{5.7}$$

$$Q = \frac{3}{2}\left(\mathbf{v}_{\perp\alpha\beta}^+ \mathbf{i}_{\alpha\beta}^+ + \mathbf{v}_{\perp\alpha\beta}^- \mathbf{i}_{\alpha\beta}^-\right) \tag{5.8}$$

The current space vector is separated into the active current space vector $\mathbf{i}_{Pdq}$ and reactive current space vector $\mathbf{i}_{Qdq}$ [77], see Eq. (5.9). The parts are calculated in accordance with Eq. (5.9) and Eq. (5.10) [77].

$$\mathbf{i}_{dq} = \mathbf{i}_{Pdq} + \mathbf{i}_{Qdq} \tag{5.9}$$

$$\mathbf{i}_{Pdq} = \frac{2}{3}\left(\frac{P_{(1)}}{|\mathbf{v}_{dq}^+|^2}\mathbf{v}_{dq}^+ + \frac{P_{(2)}}{|\mathbf{v}_{dq}^-|^2}\mathbf{v}_{dq}^-\right) \tag{5.10}$$

$$\mathbf{i}_{Qdq} = \frac{2}{3}\left(\frac{Q_{(1)}}{|\mathbf{v}_{\perp dq}^+|^2}\mathbf{v}_{\perp dq}^+ + \frac{Q_{(2)}}{|\mathbf{v}_{\perp dq}^-|^2}\mathbf{v}_{\perp dq}^-\right) \tag{5.11}$$

The active current space vector $\mathbf{i}_{Pdq}$ consists of projections on the original positive sequence voltage space vector $\mathbf{v}_{dq}^{+}$ and the original negative sequence voltage space vector $\mathbf{v}_{dq}^{-}$. The reactive current space vector $\mathbf{i}_{Qdq}$ consists of projections on the orthogonal positive sequence voltage space vector $\mathbf{v}_{\perp dq}^{+}$ and the orthogonal negative sequence voltage space vector $\mathbf{v}_{\perp dq}^{-}$.

A positive value of $P_{(1)}$ ($P_{(2)}$) leads to active power injection from the VSC into the grid in the positive (negative) sequence based on the active sign convention. A positive value of $Q_{(1)}$ leads to reactive power injection from the VSC into the grid in the positive sequence based on the active sign convention. Assuming a dominant inductive grid (see also Sec. 4.1.1.3), the voltage amplitude in the positive sequence is raised. In contrast, a positive value of $Q_{(2)}$ leads to reactive power injection from the VSC into the grid in the negative sequence based on the active sign convention. Assuming a dominant inductive grid, the voltage amplitude in the negative sequence is decreased. In this way, symmetrical grid voltages are realized.

Active currents ($I_{P(1)}$ and $I_{P(2)}$) and reactive currents ($I_{Q(1)}$ and $I_{Q(2)}$) are introduced as projections of the active and reactive current space vector on the original and the orthogonal voltage space vectors based on Eq. (5.10) and Eq. (5.11). They are given in Eq. (5.12), Eq. (5.13), Eq. (5.14) and Eq. (5.15).

$$I_{P(1)} = \frac{2}{3} \frac{P_{(1)}}{|\mathbf{v}_{dq}^{+}|} \tag{5.12}$$

$$I_{P(2)} = \frac{2}{3} \frac{P_{(2)}}{|\mathbf{v}_{dq}^{-}|} \tag{5.13}$$

$$I_{Q(1)} = -\frac{2}{3} \frac{Q_{(1)}}{|\mathbf{v}_{dq}^{+}|} \tag{5.14}$$

$$I_{Q(2)} = -\frac{2}{3} \frac{Q_{(2)}}{|\mathbf{v}_{dq}^{-}|} \tag{5.15}$$

If the d^{+}-axis is aligned to $\mathbf{v}_{dq}^{+}$ and the d^{-}-axis is aligned to $\mathbf{v}_{dq}^{-}$ of rotating reference frames (see Fig. 3.2), the active (reactive) currents are equal to the d- (q-) components of the current space vectors. The following equations reveal, see Eq. (5.16), Eq. (5.17), Eq. (5.18)) and Eq. (5.19).

$$i_{d}^{+} = \frac{2}{3} \frac{P_{(1)}}{v_{d}^{+}} \tag{5.16}$$

5.2 Calculation of Reference Variables

$$i_d^- = \frac{2}{3}\frac{P_{(2)}}{v_d^-} \tag{5.17}$$

$$i_q^+ = -\frac{2}{3}\frac{Q_{(1)}}{v_d^+} \tag{5.18}$$

$$i_q^- = -\frac{2}{3}\frac{Q_{(2)}}{v_d^-} \tag{5.19}$$

These equations demonstrate, that the active power and the reactive power in the positive and the negative sequence can be adjusted, if the components of the voltage space vectors of the positive and the negative sequence (v_d^+, v_q^+, v_d^- and v_q^-) are known. To gain these components, a decoupling cell (see Sec. 4.3.3 and Sec. 4.3.4.1) are used. The components of the current space vectors of the positive and the negative sequence (i_d^+, i_q^+, i_d^- and i_q^-) represent the controlled variables. They are calculated from Eq. (5.16), Eq. (5.17), Eq. (5.18) and Eq. (5.19). The P/Q-transformation (see Sec. 4.1.2.3) is used to adapt to the grid characteristics.

The reference values of the active and the reactive power can optionally be calculated from a $P(f_{vg})$- or $Q(V_g)$-droop controller. Alternatively, a $P(v_{DC})$-controller for the operation of wind turbines or photovoltaic systems can be used. Moreover, a $P(df_{vg}/dt)$-controller can be used exclusively or additionally.

The use of the f_{vg}-based active power controllers is based on the assumption, that f_{vg} is a system-wide indicator of the active power balance. This perspective results from the physical characteristics of rotating machines and is not inherently valid for 100% IBPS.

In contrast to $P(f_{vg})$-controllers, $f_m(P)$-controllers (see Sec. 5.2.2.1) aim for active power sharing based on the locally indicated active power and not on a system-wide indicator. They are based on the frequencies of different voltages.

5.2.3.2 Reactive Current Injection in the Negative Sequence

The injection of reactive current in the negative sequence reduces the voltage amplitude in the negative sequence of dominant inductive grids. In [47], the injection of reactive current in the negative sequence of 50 Hz medium voltage grids is standardized. The scope of application of this norm is further described in [47]. This grid connection guideline is state-of-the-art.

The demand on reactive current of an inverter-based resource (IBR) is calculated in accordance with Eq. (5.20).

$$I_{Q(2)} = -k_{(2)}\frac{V_{(2)}}{V_{\text{ref}}}I_{\text{nom}} \tag{5.20}$$

The variable $k_{(2)}$ is often chosen between 2 and 6. The reference voltage V_{ref} depends on the grid voltage. The nominal current I_{nom} refers to the nominal apparent power of the IBR. In this way, IBRs with higher nominal apparent power inject more reactive current in the negative sequence than IBRs with lower nominal apparent power.

The Eq. (5.20) represents a special case of the P/Q-control in the negative sequence (see Sec. 5.2.3.1). Reactive current $I_{Q(2)}$ is drawn from the grid, if $V_{(2)}$ (concerning v_{gx}) is greater than zero and additional constraints are fulfilled, such as the positive sequence voltage falls below a certain value [47]. In this case, the injection of reactive power $Q_{(2)}$ into the grid leads to the reduction of $V_{(2)}$ in order to balance the grid voltage, see also Sec. 4.1.2.2.

Using Eq. (5.20), the reference value of $I_{Q(2)}$ is calculated. Afterwards, the reference value of the reactive power in the negative sequence $Q_{(2)}$ is calculated from Eq. (5.15). If the d^--axis is aligned to $\mathbf{v}_{dq}^-$ (see Fig. 3.2), the reference value of i_q^- is known from Eq. (5.19). Since the active power in the negative sequence is set to zero in most cases, the reference value of i_d^- is zero, see Eq. (5.17). The reference values of the current space vector components in the negative sequence are known, and the according current source in the negative sequence is realized independently of the characteristics of the VSC in the positive sequence. The P/Q-transformation (see Sec. 4.1.2.3) can optionally be used to adapt to the grid characteristics. Decoupling cells (see Sec. 4.3.3 and Sec. 4.3.4.1) are essential to gain the components of the negative sequence voltage space vector.

It is shown in Sec. 5.2.4.1, that this type of current source is equivalent to the characteristics of a virtual capacitance. The virtual capacitance appears in series to the line inductance and therefore compensates the effect of the line inductance. The reduction of the imaginary part of the impedance inside the negative sequence explains the reduction of $V_{(2)}$ in an alternative way.

The injection of reactive current in the negative sequence is important to manage asymmetrical faults in the steady state. A grid-forming (GFM) VSC is assumed, that is characterized by a voltage source in the positive sequence taking the grid-side filter impedance $\underline{Z}_{g(1)}$ into account and a current source in the negative sequence, see Fig. 5.10 that is based on Fig. 5.5. After a phase-to-phase fault occurs, the voltage source in the positive sequence of the VSC causes the fault current. The fault current leads the negative sequence voltage $V_{(2)}$. This voltage drop is calculated by a voltage divider and leads to asymmetrical voltages v_{gx}. The interconnection of the positive and the negative sequence circuit after the phase-to-phase fault at the fault location is shown in Fig. 5.12 and [71].

To balance the voltages v_{gx}, the VSC must reduce the voltage $V_{(2)}$ by reactive current injection. Since the imaginary part of the impedance inside the negative

5.2 Calculation of Reference Variables

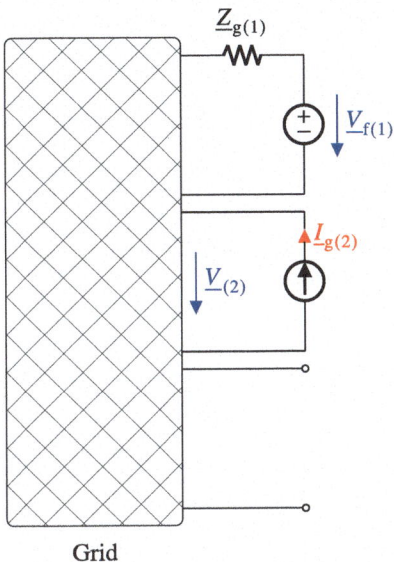

Fig. 5.10 Steady state symmetrical components equivalent circuit of the GC VSC applying a voltage source in the positive sequence (1), a current source in the negative sequence (2) and an open circuit in the zero sequence

sequence is reduced, the voltage $V_{(2)}$ is reduced. In contrast to a virtual short circuit of the VSC in the negative sequence using the characteristics of a voltage source, the reactive current injection leads to a non-zero impedance in the negative sequence.

5.2.3.3 Scenario: Voltage Balancing After a Phase-To-Phase Fault

In the following scenario, it is demonstrated via simulations in MATLAB/Simulink, that reactive current injection in the negative sequence (see Sec. 5.2.3.2) leads to symmetrical voltages v_{gx} at the point of interconnection after a phase-to-phase fault. In this way, the GC VSC adds resilience to the grid and the normal grid operation under symmetrical voltages is held up.

In the investigated scenario, a VSC is connected via an overhead line between the busbars A and B to a 10 kV-grid. The overhead line has a length of 10 km. The relays of this grid section are named R 1 and R 2. A phase-to-phase fault between phase 2 and phase 3 is applied at $t = 0.1$ s. The scenario is shown in Fig. 5.11.

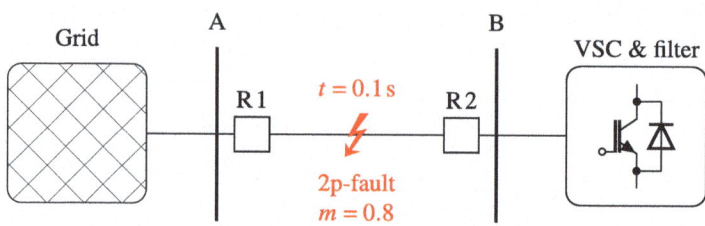

Fig. 5.11 Topology for evaluating the fault characteristics of a GC VSC during a phase-to-phase fault between phase 2 and phase 3

The electrical distance between the VSC and the fault equals 2 km. Therefore, the relative fault distance m equals 0.8 seen from busbar A. The fault resistance R_F equals 2 Ω.

The VSC is controlled to realize slack characteristics (see Fig. 5.8 b) and Fig. 5.9 b)) to investigate the negative sequence characteristics exclusively. Reactive current injection in the negative sequence is realized according to Sec. 5.2.3.2. The voltage source in the positive sequence and the current source in the negative sequence are realized via a linear–quadratic (LQ) controller according to Sec. 5.3.4. The reactive current injection in the negative sequence is activated at $t = 0.3$ s. The time interval between the fault occurrence and the activation of the reactive current injection of 200 ms is chosen to clearly demonstrate the effect of the reactive current injection. Consequently, the differences between the activated injection and no injection characteristics are revealed. In practical applications, this time delay can be reduced. The VSC is connected as a three-phase three-wire system, see Fig. 2.3. Consequently, it represents an open circuit in the zero sequence. The steady state symmetrical components equivalent circuit is shown in Fig. 5.12, which is based on Fig. 5.10.

In general, the influence of the capacitors in Fig. 5.12 after the fault occurs is negligible since the characteristics of the overhead line are predominantly inductive in this case, see [100, 143]

The Fig. 5.13 shows the negative sequence voltages $v_{gx(2)}$ at the point of interconnection at busbar B. The amplitude of the negative sequence voltages $v_{gx(2)}$ rises to 1 kV after the fault occurs due to the coupling of the positive and negative sequence. This circuit equals a voltage divider circuit, with the positive sequence voltage source representing the circuit source. After the reactive current injection is activated at $t = 0.3$ s, the amplitude of $v_{gx(2)}$ is significantly reduced and reaches a value of 0.4 kV in the steady state.

The voltages v_{gx} at the point of interconnection are shown in Fig. 5.14. It is seen, that the voltages v_{gx} become asymmetrical after the fault at $t = 0.1$ s. Furthermore,

5.2 Calculation of Reference Variables

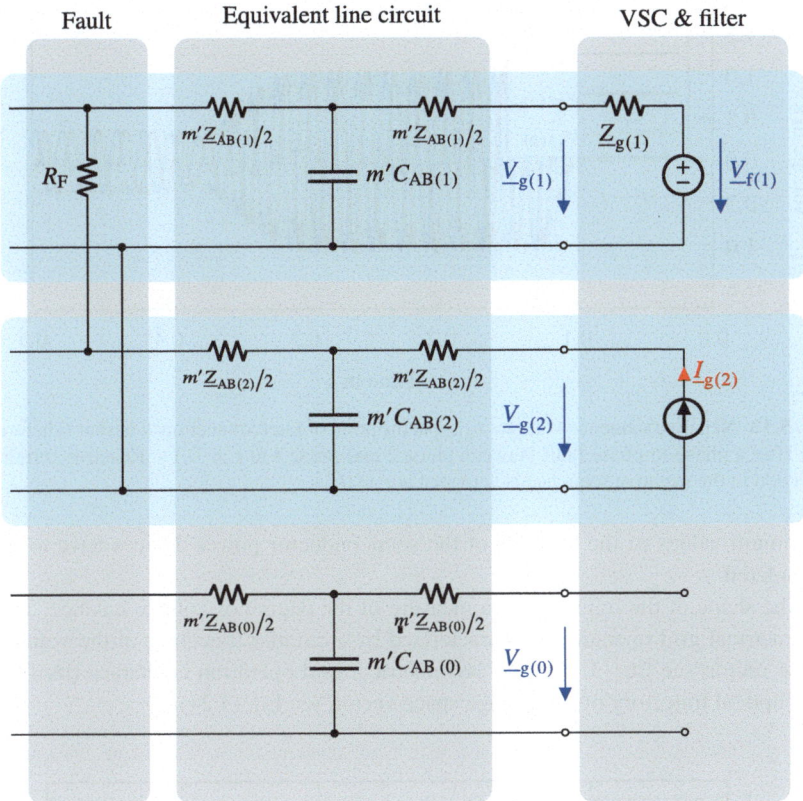

Fig. 5.12 Steady state symmetrical components equivalent circuit of the three-phase three-wire VSC applying a voltage source in the positive sequence (1) and a current source in the negative sequence (2) during a phase-to-phase fault between phase 2 and phase 3. The relative fault distance $m' = (1 - m)$ seen from the point of interconnection at busbar B is used

the balancing by the reactive current injection at $t = 0.3\,\text{s}$ is obvious. In general, the amplitude of the positive sequence voltage source has to be raised additionally after the fault. This effect is not realized in this scenario for demonstration issues of the negative sequence characteristics.

The Fig. 5.15 shows the phase currents i_{gx} at the point of interconnection at busbar B. The amplitude of i_{g2} rises to 2 kA after the fault. The injection of reactive current in the negative sequence slightly increases this value. For practical applications, the

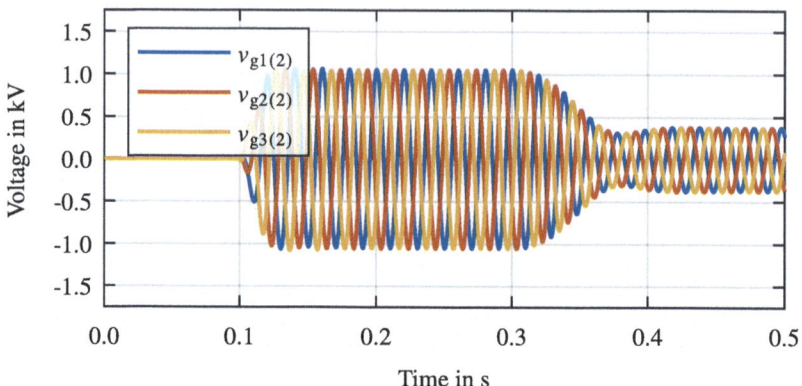

Fig. 5.13 Negative sequence voltages $v_{gx(2)}$ at the point of interconnection at busbar B before and after a phase-to-phase fault between phase 2 and phase 3 at $t = 0.1$ s. Reactive current injection in the negative sequence is activated at $t = 0.3$ s

maximum values of the currents of the semiconductor power devices have to be considered.

The shape of the trajectory is a measure of the degree of voltage balance. The symmetrical grid operation is characterized by a circular trajectory of the voltage space vector, see Eq. (3.17). The asymmetrical grid operation is characterized by an elliptical trajectory of the voltage space vector, see Eq. (3.24).

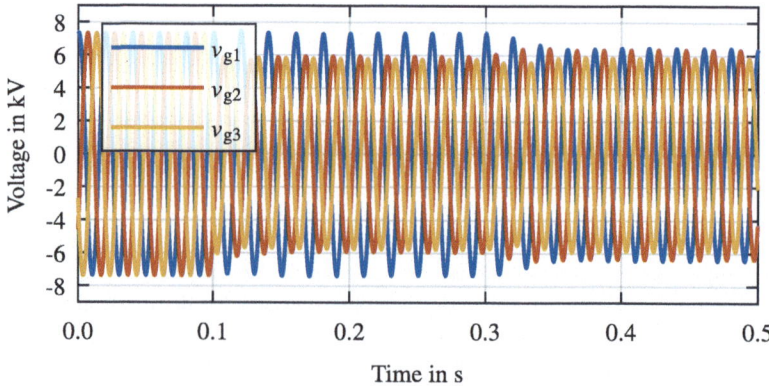

Fig. 5.14 Voltages v_{gx} at the point of interconnection at busbar B before and after a phase-to-phase fault between phase 2 and phase 3 at $t = 0.1$ s. Reactive current injection in the negative sequence is activated at $t = 0.3$ s

5.2 Calculation of Reference Variables

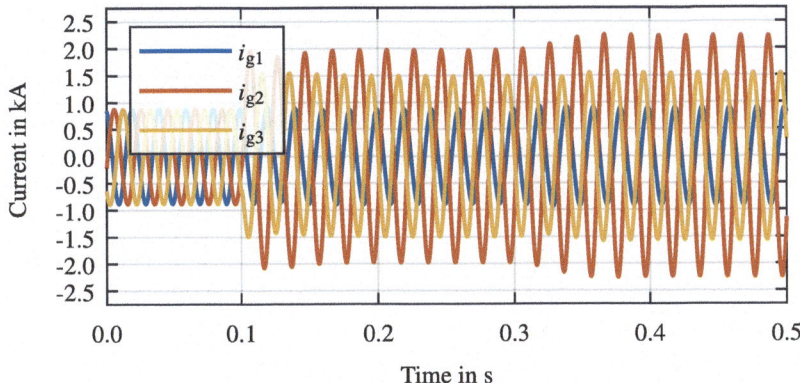

Fig. 5.15 Phase currents i_{gx} at the point of interconnection before and after a phase-to-phase fault between phase 2 and phase 3 at $t = 0.1\,\text{s}$. Reactive current injection in the negative sequence is activated at $t = 0.3\,\text{s}$

The trajectory in the complex plane of the space vector $\underline{v}_g$ at the point of interconnection at busbar B is shown in Fig. 5.16. The time is visualized by the colour progression from dark blue at $t = 0\,\text{s}$ to dark red at $t = 0.5\,\text{s}$. By the reactive current injection in the negative sequence, the elliptical shape is converged to a circular shape. The radius of the trajectory after the fault can be increased by raising the amplitude of the positive sequence voltage source. In this case, the maximum values of the phase currents have to be considered to not exceed their maximum ratings.

In this scenario, it is demonstrated, that the reactive current injection in the negative sequence leads to symmetrical voltages v_{gx} at the point of interconnection after a phase-to-phase fault. The grid operation under symmetrical voltages can be continued and the resiliency against grid faults is increased.

5.2.4 Realization of V/I-Characteristics

Voltage/current-characteristics (V/I-characteristics) are realized exclusively or in addition to the source characteristics in the positive sequence (see Fig. 5.6) and the negative sequence (see Fig. 5.7). The V/I-characteristics are realized in the steady state. As special cases, the characteristics of a resistor, an inductor, a capacitor or combinations can be realized.

The selection of measured variables and controlled variables leads to admittance or impedance characteristics. If the voltages v_{gx} are measured and the currents i_{gx}

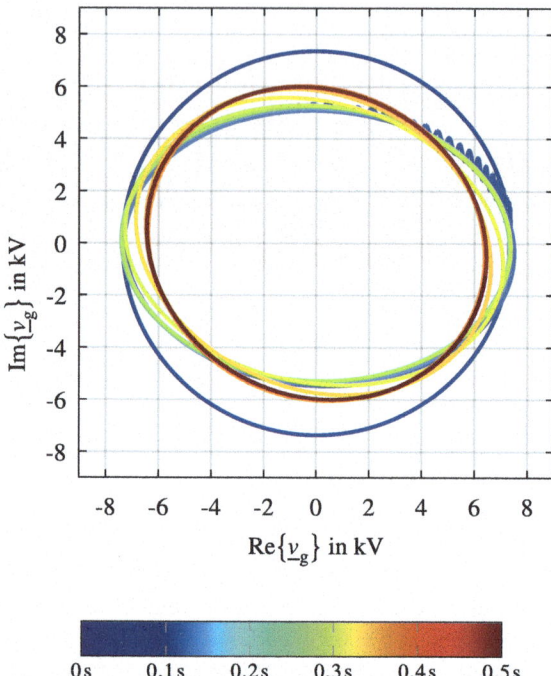

Fig. 5.16 Trajectory in the complex plane of the space vector $\underline{v}_g$ at the point of interconnection at busbar B before and after a phase-to-phase fault between phase 2 and phase 3 at $t = 0.1$ s. Reactive current injection in the negative sequence is activated at $t = 0.3$ s

(see Fig. 3.6) are controlled, admittance characteristics are realized at the point of interconnection via the current controller. This principle is called virtual admittance. If the currents i_{gx} are measured and the voltages v_{fx} are controlled, impedance characteristics are realized via the voltage controller right of the filter inductance L_g, see Fig. 3.6. This principle is called virtual impedance.

5.2.4.1 Virtual Admittance

The principle of the virtual admittance (VA) is derived using the active sign convention. The VA is used exclusively or in parallel to a current source, see Sec. 5.2.3. The Fig. 3.6 and the 5th and 6th row of Eq. (3.42) are the starting point of the following derivation. Additionally, a resistor R_f in parallel to the capacitor C_f is assumed, see Eq. (5.21).

5.2 Calculation of Reference Variables

$$\frac{d}{dt}\begin{pmatrix} v_{fd}^{\pm} \\ v_{fq}^{\pm} \end{pmatrix} = \frac{1}{C_f}\left[\begin{pmatrix} i_{cd}^{\pm} \\ i_{cq}^{\pm} \end{pmatrix} - \begin{pmatrix} i_{gd}^{\pm} \\ i_{gq}^{\pm} \end{pmatrix} + \frac{1}{R_f}\begin{pmatrix} v_{fd}^{\pm} \\ v_{fq}^{\pm} \end{pmatrix}\right] + \begin{pmatrix} 0 & \pm\omega \\ \mp\omega & 0 \end{pmatrix}\begin{pmatrix} v_{fd}^{\pm} \\ v_{fq}^{\pm} \end{pmatrix} \tag{5.21}$$

Rearranging these equations yields Eq. (5.22).

$$\begin{pmatrix} i_{gd}^{\pm} \\ i_{gq}^{\pm} \end{pmatrix} = \frac{1}{R_f}\begin{pmatrix} v_{fd}^{\pm} \\ v_{fq}^{\pm} \end{pmatrix} \pm \omega C_f \begin{pmatrix} v_{fq}^{\pm} \\ -v_{fd}^{\pm} \end{pmatrix} - C_f \frac{d}{dt}\begin{pmatrix} v_{fd}^{\pm} \\ v_{fq}^{\pm} \end{pmatrix} + \begin{pmatrix} i_{cd}^{\pm} \\ i_{cq}^{\pm} \end{pmatrix} \tag{5.22}$$

In the steady state, the time derivations are zero. Furthermore, the space vector components $i_{cd}^{\pm}$ and $i_{cq}^{\pm}$ are set to zero. The Eq. (5.23) is obtained from Eq. (5.21).

$$\begin{pmatrix} i_{gd}^{\pm} \\ i_{gq}^{\pm} \end{pmatrix} = \begin{pmatrix} \frac{1}{R_f} & \pm\omega C_f \\ \mp\omega C_f & \frac{1}{R_f} \end{pmatrix}\begin{pmatrix} v_{fd}^{\pm} \\ v_{fq}^{\pm} \end{pmatrix} \tag{5.23}$$

In the next step, the space vector components of v_{fx} are formally substituted by the space vector components of v_{gx} with negative sign to realize the active sign convention regarding the relevant variables. Moreover, R_f and C_f are substituted by the variables of the VA, see Eq. (5.24).

$$\begin{pmatrix} i_{gd}^{\pm} \\ i_{gq}^{\pm} \end{pmatrix} = \begin{pmatrix} -\frac{1}{R_{VA}} & \mp\omega C_{VA} \\ \pm\omega C_{VA} & -\frac{1}{R_{VA}} \end{pmatrix}\begin{pmatrix} v_{gd}^{\pm} \\ v_{gq}^{\pm} \end{pmatrix} \tag{5.24}$$

The admittance matrix $\mathbf{Y}_{VA}$ is introduced to generalize these equations, see Eq. (5.25).

$$\begin{pmatrix} i_{gd}^{\pm} \\ i_{gq}^{\pm} \end{pmatrix} = \underbrace{\begin{pmatrix} -G_{VA} & \mp B_{VA} \\ \pm B_{VA} & -G_{VA} \end{pmatrix}}_{\mathbf{Y}_{VA}}\begin{pmatrix} v_{gd}^{\pm} \\ v_{gq}^{\pm} \end{pmatrix} \tag{5.25}$$

In this way, the admittance $\underline{Y}_{VA}$ (see Eq. (5.26)) is realized in the positive or the negative sequence in the steady state by controlling the pair i_{gd}^{+}/i_{gq}^{+} or the pair i_{gd}^{-}/i_{gq}^{-}.

$$\underline{Y}_{VA} = G_{VA} + jB_{VA} \tag{5.26}$$

The Fig. 5.17 illustrates the realization of the virtual admittance in the negative sequence as an example based on Fig. 5.5. The parallel connection of the virtual resistance R_{VA} and the virtual capacitance C_{VA} appears at the point of interconnection.

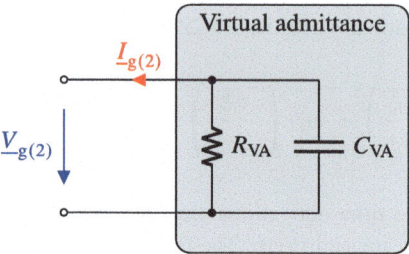

Fig. 5.17 Equivalent circuit of the negative sequence (2) of a GC VSC in the steady state applying the principle of the virtual admittance (VA)

Subsequently, the negative sequence is regarded under the assumption, that v_{gq}^- is set to zero by aligning the d^--axis to $\mathbf{v}_{gdq}^-$. The conductance G_{VA} is set to zero.

$$G_{VA} \stackrel{!}{=} 0 \quad (5.27)$$

Applying these assumptions, the formal equivalence of the 2th row of Eq. (5.25) regarding the negative sequence components and Eq. (5.20) is seen. In this way, the variable $k_{(2)}$ (see Sec. 5.2.3.2) can be interpreted as the value of a virtual capacitance C_{VA}. A low imaginary part of the impedance in the negative sequence based on a series connection of the virtual capacitance and the line inductance is realized. During an asymmetrical fault and a voltage source in the positive sequence, the voltage amplitude in the negative sequence is low based on the principle of a voltage divider.

Alternatively, if the virtual conductance G_{VA} is set to zero, a virtual inductance L_{VA} is realized by B_{VA} at the point of interconnection, see Eq. (5.28).

$$B_{VA} \stackrel{!}{=} \frac{1}{\omega L_{VA}} \quad (5.28)$$

The Eq. (5.25) reveals a further interpretation. The active power in the positive or negative sequence (see Eq. (5.16) and Eq. (5.17)) can be regarded as a virtual resistor R_{VA} using the active sign convention in the according symmetrical component circuit. Therefore, B_{VA} and $v_{gq}^{\pm}$ are set to zero.

5.3 Design of the State-Feedback V/I-Controller

5.2.4.2 Virtual Impedance

The principle of the virtual impedance (VI) is analogue to the principle of the virtual admittance. The VI is used exclusively or in series to a voltage source, see Sec. 5.2.3. The Fig. 3.6 and the 3th and 4th row of Eq. (3.42) are the starting point of the derivation, which is shortened here. The time derivations and the space vector components $v_{\text{gdq}}^{\pm}$ are set to zero. The impedance matrix $\mathbf{Z}_{\text{VI}}$ reveals in Eq. (5.29). The series connection of the virtual resistance R_{VI} and the virtual inductance L_{VI} appears right of the filter inductance L_{g}, see Fig. 3.6.

$$\begin{pmatrix} v_{\text{fd}}^{\pm} \\ v_{\text{fq}}^{\pm} \end{pmatrix} = \underbrace{\begin{pmatrix} R_{\text{VI}} & \mp \omega L_{\text{VI}} \\ \pm \omega L_{\text{VI}} & R_{\text{VI}} \end{pmatrix}}_{\mathbf{Z}_{\text{VI}}} \begin{pmatrix} i_{\text{gd}}^{\pm} \\ i_{\text{gq}}^{\pm} \end{pmatrix} \tag{5.29}$$

There are four main applications of the virtual impedance, which is often used in the positive sequence. The grid-side filter inductance L_{g} (see Fig. 3.6) can be compensated to zero or the electrical distance can be adjusted for reactive power sharing, see Sec. 4.1.2.2 and [101]. Furthermore, the virtual impedance can influence the small signal synchronization stability, see Eq. (4.27) and Sec. 4.4.3. Moreover, the X/R-ratio can be influenced, see Sec. 4.1.1.3. Alternatively, the P/Q-transformation from [85] and Sec. 4.1.2.3 can be applied. Beyond that, a purely resistive virtual impedance can be used to limit the converter currents during faults in the steady state [102]. The problem is, that the fault location and the grid impedance have to known previously to properly calculate the value of R_{VI}.

5.3 Design of the State-Feedback V/I-Controller

In this section, the design of the state-feedback voltage/current (V/I) controller is presented. The objective of this controller is to calculate the actuating variables in a way, that the controlled variables adjust to the reference variables, see Fig. 5.1.

The filter of the VSC represents the control plant, see Fig. 5.2. The design of the state-feedback V/I-controller is based on a state-space model of the control plant. It is assumed that the state-space model meets the characteristics of the control plant. In this way, the behaviour of the controlled model is mirrored to the control plant.

In the following investigations, the state-space models from Sec. 3.2.2 and especially the state-space model of the LCL-filter in Sec. 3.2.2.3 are used. The space vector components in appropriate rotating reference frames represent DC-quantities.

For the practical implementation of the state-feedback V/I-controller on a digital computer, the discretization from Sec. 3.2.3 is applied. The control law of the state-feedback V/I-controller is designed in the time domain. In contrast, proportional-integral (PI) controllers are designed in the frequency domain.

Furthermore, the following assumptions are made: The transfer function of the actuator consisting of the modulator and the VSC equals 1, see also Fig. 5.2. A state observer is not necessary, since all state variables of the filter structures are measured. Furthermore, the disturbance variables are measured.

5.3.1 Derivation of the Control Law and Analysis of the Closed Control Loop

The philosophy of the design of the control law to calculate the actuating variables is divided into three sections. The desired influence of the reference variables and the disturbance variables on the controlled variables is realized by two individual feedforward controllers. At the same time, the reference state is shifted out of the origin. The state-feedback controller adjusts the initial state $x(0)$ to the reference state. If the reference state is reached, the controlled variables equal the reference variables. Using this philosophy, the feedforward controllers and the feedback controller are designed independently of each other. In this way, a control law with multiple degrees of freedom is realized.

The Fig. 5.18 illustrates the realization of this control design, see also [81]. A continuous-time state-space model of the control plant is shown in Fig. 3.3. A discrete-time state-space model of the control plant is shown in Fig. 3.7.

The reference variables w evoke the part of the actuating variables u_w using the feedforward controller M_u. The disturbance variables z evoke the part of the actuating variables u_z using the feedforward controller N_u. The part of the actuating variables u_R is calculated by the state-feedback controller R from the difference of the reference state x_s and the actual state x. The reference state is calculated from the matrices M_x and N_x regarding w and z.

According to Fig. 5.18, the actuating variables u are calculated in accordance with Eq. (5.30).

$$\begin{align} \mathbf{u}(k) &= \mathbf{u}_w + \mathbf{u}_z + \mathbf{u}_R \tag{5.30a} \\ &= \mathbf{M}_u \mathbf{w}(k) + \mathbf{N}_u \mathbf{z}(k) + \mathbf{R}\left[\mathbf{M}_x \mathbf{w}(k) + \mathbf{N}_x \mathbf{z}(k) - \mathbf{x}(k)\right] \tag{5.30b} \\ &= \left[\mathbf{M}_u + \mathbf{R}\mathbf{M}_x\right]\mathbf{w}(k) + \left[\mathbf{N}_u + \mathbf{R}\mathbf{N}_x\right]\mathbf{z}(k) - \mathbf{R}\mathbf{x}(k) \tag{5.30c} \end{align}$$

5.3 Design of the State-Feedback V/I-Controller

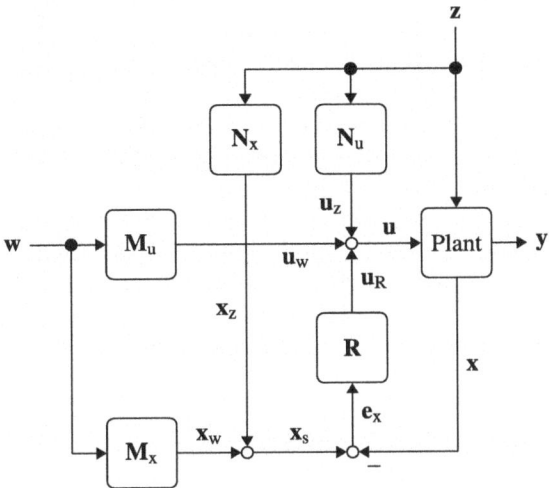

Fig. 5.18 Block diagram of the state-feedback V/I-controller using the feedback controller **R** of the difference of the reference state $\mathbf{x}_s$ and the actual state $\mathbf{x}$ and feedforward controllers ($\mathbf{M}_u$ and $\mathbf{N}_u$) of the reference variables **w** and the disturbance variables **z** calculating the actuating variables **u**

The state difference equations of the closed control loop yield Eq. (5.31). The system matrix $(\mathbf{A}_d - \mathbf{B}_d \mathbf{R})$ of the closed control loop reveals.

$$\mathbf{x}(k+1) = \mathbf{A}_d \mathbf{x}(k) + \mathbf{B}_d \mathbf{u}(k) + \mathbf{E}_d \mathbf{z}(k) \tag{5.31a}$$
$$= [\mathbf{A}_d - \mathbf{B}_d \mathbf{R}]\mathbf{x}(k) + \mathbf{B}_d[\mathbf{M}_u + \mathbf{R}\mathbf{M}_x]\mathbf{w}(k) + \{\mathbf{B}_d[\mathbf{N}_u + \mathbf{R}\mathbf{N}_x] + \mathbf{E}_d\}\mathbf{z}(k) \tag{5.31b}$$

The matrix **C** is designed according to Sec. 5.2 to select the desired control variables. The output equations are given in Eq. (5.32).

$$\mathbf{y}(k) = \mathbf{C}\mathbf{x}(k) \tag{5.32}$$

The asymptotic stability (see Sec. 3.3.2) and the dynamics of the controlled system are determined by the system matrix $(\mathbf{A}_d - \mathbf{B}_d \mathbf{R})$. If the state-feedback controller **R** is designed based on the linear–quadratic regulator (LQR) algorithm (see Sec. 5.3.4), the stability and the dynamics are examined after the calculation of **R**. Instead—using the pole placement concept to calculate **R**—the matrix is designed to ensure the stability and the desired dynamics inherently [81]. Asymptotic stability (see Sec. 3.3.2) is gained, if all eigenvalues of the system matrix $(\mathbf{A}_d - \mathbf{B}_d\mathbf{R})$ are inside the unit circle of the complex plane [78].

The continuous-time transfer functions from the reference variables **w** and the disturbance variables **z** to the state variables **x** are obtained according to Sec. 3.2.1 by adapting Eq. (3.32) to Eq. (5.31b). Based on the poles of the transfer functions, the time constants and therefore the dynamics of the closed control loop are calculated.

In general, the poles represent a subset of the eigenvalues of the system matrix [81]. If the system is complete state controllable and observable, the set of poles equals the set of eigenvalues [78]. The eigenvalues of continuous-time state-space models and discrete-time state-space models are converted using Eq. (3.62).

5.3.2 Feedforward Control of Reference Variables

The design of the feedforward control of reference variables is based on [81]. Subsequently, these principles are adapted to the discrete-time model in Eq. (5.31a). The objective is the calculation of the feedforward matrices $\mathbf{M}_u$ and $\mathbf{M}_x$, see Fig. 5.18.

The starting point is the model of the undisturbed subsystem without disturbance-feedforward controller and state-feedback-controller in Eq. (5.33). Consequently, the influence of the reference variables (index: w) on the system, induced by their feedforward controllers, is investigated.

$$\mathbf{x}_w(k+1) = \mathbf{A}_d \mathbf{x}_w(k) + \mathbf{B}_d \mathbf{u}_w(k) \tag{5.33}$$

The controlled variables must equal the reference variables, see Eq. (5.34b). It is assumed that the reference variables **w** are DC-quantities.

$$\mathbf{y}_w(k) = \mathbf{C}\mathbf{x}_w(k) \tag{5.34a}$$

$$\stackrel{!}{=} \mathbf{w}(k) \tag{5.34b}$$

In the steady state, the state variables do not vary. The following state-space model according to Eq. (5.35) and Eq. (5.36) reveals.

$$\mathbf{x}_w(k) = \mathbf{A}_d \mathbf{x}_w(k) + \mathbf{B}_d \mathbf{u}_w(k) \tag{5.35}$$

$$\mathbf{w}(k) = \mathbf{C}\mathbf{x}_w(k) \tag{5.36}$$

5.3 Design of the State-Feedback V/I-Controller

Using the matrix notation of Eq. (5.35) and Eq. (5.36) yields Eq. (5.37).

$$\begin{pmatrix} 0 \\ \mathbf{w}(k) \end{pmatrix} = \begin{pmatrix} \mathbf{A}_d - \mathbf{I} & \mathbf{B}_d \\ \mathbf{C} & 0 \end{pmatrix} \begin{pmatrix} \mathbf{x}_w(k) \\ \mathbf{u}_w(k) \end{pmatrix} \qquad (5.37)$$

The matrices $\mathbf{M}_u$ and $\mathbf{M}_x$ are introduced to characterize the relations between $\mathbf{w}$ and $\mathbf{x}_w$ and $\mathbf{u}_w$, respectively. The Eq. (5.38) reveals.

$$\begin{pmatrix} 0 \\ \mathbf{I} \end{pmatrix} \mathbf{w}(k) = \begin{pmatrix} \mathbf{A}_d - \mathbf{I} & \mathbf{B}_d \\ \mathbf{C} & 0 \end{pmatrix} \begin{pmatrix} \mathbf{M}_x \\ \mathbf{M}_u \end{pmatrix} \mathbf{w}(k) \qquad (5.38)$$

Solving these equations yields Eq. (5.39).

$$\begin{pmatrix} \mathbf{M}_x \\ \mathbf{M}_u \end{pmatrix} = \begin{pmatrix} \mathbf{A}_d - \mathbf{I} & \mathbf{B}_d \\ \mathbf{C} & 0 \end{pmatrix}^{-1} \begin{pmatrix} 0 \\ \mathbf{I} \end{pmatrix} \qquad (5.39)$$

5.3.3 Feedforward Control of Disturbance Variables

The design of the feedforward control of disturbance variables is similar to the design of the feedforward control of reference variables (see Sec. 5.3.2) and is also based on [81]. The principles are adapted to the discrete-time model in Eq. (5.31a). The objective is the calculation of the feedforward matrices $\mathbf{N}_u$ and $\mathbf{N}_x$, see Fig. 5.18.

The disturbed subsystem without reference-feedforward controller and state-feedback-controller in Eq. (5.40) is regarded. Consequently, the influence of the disturbance variables and their feedforward controller on the system (index: z) is investigated, see Eq. (5.40).

$$\mathbf{x}_z(k+1) = \mathbf{A}_d \mathbf{x}_z(k) + \mathbf{B}_d \mathbf{u}_z(k) + \mathbf{E}_d \mathbf{z}(k) \qquad (5.40)$$

The controlled variables are set to zero, since the disturbance variable must not influence the controlled variables, see Eq. (5.41b). It is assumed that the disturbance variables $\mathbf{z}$ are DC-quantities.

$$y_z(k) = Cx_z(k) \tag{5.41a}$$

$$\stackrel{!}{=} 0 \tag{5.41b}$$

In the steady state, the state variables do not vary. The state-space model according to Eq. (5.42) and Eq. (5.43) appears.

$$x_z(k) = A_d x_z(k) + B_d u_z(k) + E_d z(k) \tag{5.42}$$

$$0 = Cx_z(k) \tag{5.43}$$

The matrix representation of Eq. (5.42) and Eq. (5.43) is given in Eq. (5.44).

$$\begin{pmatrix} 0 \\ 0 \end{pmatrix} = \begin{pmatrix} A_d - I & B_d \\ C & 0 \end{pmatrix} \begin{pmatrix} x_z(k) \\ u_z(k) \end{pmatrix} + \begin{pmatrix} E_d \\ 0 \end{pmatrix} z(k) \tag{5.44}$$

The matrices N_u and N_x are introduced in Eq. (5.45) to characterize the relations between z and x_z and u_z, respectively.

$$\begin{pmatrix} A_d - I & B_d \\ C & 0 \end{pmatrix} \begin{pmatrix} N_x \\ N_u \end{pmatrix} z(k) = - \begin{pmatrix} E_d \\ 0 \end{pmatrix} z(k) \tag{5.45}$$

Solving these equations reveals the matrices N_u and N_x, see Eq. (5.46).

$$\begin{pmatrix} N_x \\ N_u \end{pmatrix} = - \begin{pmatrix} A_d - I & B_d \\ C & 0 \end{pmatrix}^{-1} \begin{pmatrix} E_d \\ 0 \end{pmatrix} \tag{5.46}$$

5.3.4 State-Feedback Controller Based on the LQR-Algorithm

5.3.4.1 Feedback of State Variables

According to the philosophy of the design of the control law (see [81]), the feedforward controllers of the reference variables (see Sec. 5.3.2) and the disturbance variables (see Sec. 5.3.3) calculate the reference state x_s, see Fig. 5.18. The objective of the state-feedback controller R is to adjust the initial state $x(0)$ to the reference state x_s by the appropriate feedback of the difference of the reference state and the

5.3 Design of the State-Feedback V/I-Controller

actual state $\mathbf{x}$. In this way, the controlled variables automatically equal the reference variables. The state controllability (see Sec. 3.3.3) of the control plant is a prerequisite.

First, the state error $\mathbf{e}_x$ is calculated, see Eq. (5.47).

$$\mathbf{e}_x(k) = \mathbf{x}_s(k) - \mathbf{x}(k) \tag{5.47a}$$
$$= \mathbf{x}_w(k) + \mathbf{x}_z(k) - \mathbf{x}(k) \tag{5.47b}$$
$$= \mathbf{M}_x \mathbf{w}(k) + \mathbf{N}_x \mathbf{z}(k) - \mathbf{x}(k) \tag{5.47c}$$

The feedforward controllers of the reference variables and the disturbance variables are neglected to solely investigate the influence of the state-feedback controller (index: R). The state-space model in Eq. (5.48) and Eq. (5.49) is derived from Eq. (5.31a).

$$\mathbf{e}_x(k+1) = \mathbf{A}_d \mathbf{e}_x(k) + \mathbf{B}_d \mathbf{u}_R(k) \tag{5.48}$$

$$\mathbf{y}_R(k) = \mathbf{C} \mathbf{e}_x(k) \tag{5.49}$$

The feedback law is chosen in a way, that the part of the actuating variables $\mathbf{u}_R$ is calculated from $\mathbf{e}_x$ (see Eq. (5.47a)). The state-feedback matrix $\mathbf{R}(k)$ is time-dependent in general.

$$\mathbf{u}_R(k) = \mathbf{R}(k) \mathbf{e}_x(k) \tag{5.50}$$

One possibility to design $\mathbf{R}(k)$ is to formulate a cost function and to solve the resulting optimization problem. The solution of the optimization problem (optimal trajectories of the actuating variables) is linked to $\mathbf{R}(k)$ by the choice of the feedback law in Eq. (5.50). If a quadratic cost function (see Eq. (5.51) in Sec. 5.3.4.2) is applied for linear state difference equations according to Eq. (5.48), the term linear-quadratic regulator (LQR) algorithm is used [103].

5.3.4.2 Formulation of the Optimization Problem

There are multiple options to define the cost function J_{LQR}. Subsequently, the quadratic cost function according to Eq. (5.51) is used [103]. This concept using weighting matrices allows an intuitive way of designing the controller according to the desired properties.

$$J_{LQR} = \frac{1}{2} \sum_{k=0}^{K-1} \left[\mathbf{e}_x^T(k) \mathbf{Q}_e(k) \mathbf{e}_x(k) + \mathbf{u}_R^T(k) \mathbf{Q}_u(k) \mathbf{u}_R(k) \right] + \frac{1}{2} \mathbf{e}_x^T(K) \mathbf{Q}_{eK} \mathbf{e}_x(K) \tag{5.51}$$

The factors 0.5 do not have any influence on the minimization of the cost function J_{LQR}. They improve the representation of the gradient.

In the first addend in Eq. (5.51), the square values of the elements of the state error $\mathbf{e}_x(k)$ are penalized by the elements of the weighting matrix $\mathbf{Q}_e(k)$ to track the reference state $\mathbf{x}_s$. Furthermore, the square values of the elements of the parts of the actuating variables $\mathbf{u}_R(k)$ are penalized by the elements of the weighting matrix $\mathbf{Q}_u(k)$ to reduce the control energy.

In the second addend in Eq. (5.51), the optimization horizon is described by the variable K. The square values of the elements of the final state error $\mathbf{e}_x(K)$ are penalized by the elements of the weighting matrix $\mathbf{Q}_{eK}$. Consequently, the reference state $\mathbf{x}_s$ is reached as good as possible at the end of the optimization horizon.

The weighting matrices $\mathbf{Q}_e(k)$ and $\mathbf{Q}_{eK}$ are assumed to be positive-semidefinite, whereas the matrix $\mathbf{Q}_u(k)$ is assumed to be positive-definite [103].

For the clear weighting of each element of $\mathbf{e}_x(k)$, $\mathbf{u}_R(k)$ and $\mathbf{e}_x(K)$, the weighting matrices $\mathbf{Q}_e(k)$, $\mathbf{Q}_u(k)$ and $\mathbf{Q}_{eK}$ are designed as diagonal matrices. Since the eigenvalues of diagonal matrices are located on the main diagonal, the requirements on the quadratic form are ensured by selecting positive values [75]. The elements of the weighting matrices $\mathbf{Q}_e(k)$, $\mathbf{Q}_u(k)$ and $\mathbf{Q}_{eK}$ are design parameters of the LQR-algorithm.

5.3.4.3 Solution of the Optimization Problem

The minimization of the cost function J_{LQR} from Eq. (5.51) leads to a dynamic optimization problem. The objective is to calculate the trajectories of the actuating variables $\mathbf{u}_R(k)$, which minimize J_{LQR}. The state difference equations according to Eq. (5.48) have to be considered.

The solution of the dynamic optimization problem is presented in [103]. The time-dependent state-feedback matrix $\mathbf{R}(k)$ is linked to the optimal trajectories of the actuating variables by the feedback law in Eq. (5.50). $\mathbf{R}(k)$ is calculated in accordance with Eq. (5.52).

$$\mathbf{R}(k) = \left[\mathbf{B}_d^T(k)\mathbf{P}(k+1)\mathbf{B}_d(k) + \mathbf{Q}_u(k)\right]^{-1} \mathbf{B}_d^T(k)\mathbf{P}(k+1)\mathbf{A}_d(k) \qquad (5.52)$$

The Riccati-Matrix $\mathbf{P}(k)$ appears. It characterizes the relation between $\mathbf{e}_x(k)$ and the Lagrange multipliers, which are introduced to solve the optimization problem [103]. $\mathbf{P}(k)$ is described by the discrete-time Riccati equation as a special case of matrix difference equations, see Eq. (5.53) [103].

$$\mathbf{P}(k) = \mathbf{A}_d^T(k)\mathbf{P}(k+1)\mathbf{A}_d(k) + \mathbf{Q}_e(k) - \mathbf{R}^T(k)\mathbf{B}_d^T(k)\mathbf{P}(k+1)\mathbf{A}_d(k) \qquad (5.53)$$

5.3 Design of the State-Feedback V/I-Controller

The Riccati-matrix at the end of the optimization horizon $\mathbf{P}(K)$ equals $\mathbf{Q}_{eK}$ as a constraint of the solution of the dynamic optimization problem, see Eq. (5.54) [103].

$$\mathbf{P}(K) = \mathbf{Q}_{eK} \tag{5.54}$$

After calculating $\mathbf{P}(K)$ from Eq. (5.54), the state-feedback matrix $\mathbf{R}(K-1)$ is calculated from Eq. (5.52). The Riccati-Matrix $\mathbf{P}(K-1)$ is calculated using $\mathbf{P}(K)$ and $\mathbf{R}(K-1)$ from Eq. (5.53) and so on. This iterative algorithm equals the integration backwards in time of Eq. (5.53) starting from $\mathbf{P}(K)$ and yields $\mathbf{P}(k)$ and $\mathbf{R}(k)$ inside the optimization horizon K at each time step.

Subsequently, several constraints are introduced [103]. The matrices $\mathbf{A}_d(k)$, $\mathbf{B}_d(k)$, $\mathbf{Q}_e(k)$ and $\mathbf{Q}_u(k)$ are assumed to be time-invariant. Furthermore, the system described by $\mathbf{A}_d$ and $\mathbf{B}_d$ is assumed to be complete state controllable and $\mathbf{Q}_e = \mathbf{C}^T\mathbf{C}$. Moreover, the optimization horizon K is set to infinity, and therefore the second addend in Eq. (5.51) describing the final state, is obsolete.

Regarding these constraints, the solution of the optimization problem leads to the discrete-time algebraic Riccati equation with the positive-definite time-independent Riccati-matrix $\overline{\mathbf{P}}$. The time-independent state-feedback matrix $\mathbf{R}$ is calculated according to Eq. (5.55).

$$\mathbf{R} = \left[\mathbf{B}_d^T\overline{\mathbf{P}}\mathbf{B}_d + \mathbf{Q}_u\right]^{-1}\mathbf{B}_d^T\overline{\mathbf{P}}\mathbf{A}_d \tag{5.55}$$

The state-feedback yields in this special case Eq. (5.56).

$$\mathbf{u}_R(k) = \mathbf{R}\mathbf{e}_x(k) \tag{5.56}$$

For practical applications, Eq. (5.55) might be useful. Although, regarding time-dependent grid frequency and therefore a time-dependent system matrix $\mathbf{A}_d(k)$ (see also Eq. (3.42)), the solution from Eq. (5.52) for $\mathbf{R}(k)$ provides better characteristics.

5.3.5 Implementation of Integral Characteristics

The implementation of integral characteristics into the control philosophy according to Fig. 5.18 is necessary, if the model parameters are not known accurate enough and stationary accuracy is crucial.

The principle to implement integral characteristics is to extend the state-feedback according to Eq. (5.50) by the integration of the difference of the reference variables and the controlled variables, see Eq. (5.57) [104].

$$\mathbf{u}_R(k) = \mathbf{R}_x[\mathbf{x}_s(k) - \mathbf{x}(k)] - \frac{1}{1-z^{-1}}\mathbf{R}_y[\mathbf{w}(k) - \mathbf{y}(k)] \tag{5.57}$$

The transfer function of the minuend in Eq. (5.57) equals the discrete-time integration of the difference of the reference variables and the controlled variables according to the Backward-Euler method [80].

The introduction of Δ-variables using the shift-operator z^{-1} yields Eq. (5.58), Eq. (5.59) and Eq. (5.60).

$$\Delta\mathbf{x}(k) = \mathbf{x}(k) - \mathbf{x}(k-1) \tag{5.58a}$$
$$= (1 - z^{-1})\mathbf{x}(k) \tag{5.58b}$$

$$\Delta\mathbf{u}_R(k) = \mathbf{u}_R(k) - \mathbf{u}_R(k-1) \tag{5.59a}$$
$$= (1 - z^{-1})\mathbf{u}_R(k) \tag{5.59b}$$

$$\Delta\mathbf{z}(k) = \mathbf{z}(k) - \mathbf{z}(k-1) \tag{5.60a}$$
$$= (1 - z^{-1})\mathbf{z}(k) \tag{5.60b}$$

Reformulating Eq. (5.57) by using the Δ-variables yields Eq. (5.61). The Eq. (5.61) is simplified in Eq. (5.62). In Eq. (5.62b) it is assumed, that the reference states $\mathbf{x}_s(k-1)$ and $\mathbf{x}_s(k)$ are equal.

$$(1 - z^{-1})\mathbf{u}_R(k) = (1 - z^{-1})\mathbf{R}_x[\mathbf{x}_s(k) - \mathbf{x}(k)] - \mathbf{R}_y[\mathbf{w}(k) - \mathbf{y}(k)] \tag{5.61}$$

$$\Delta\mathbf{u}_R(k) = \mathbf{R}_x\left[(1 - z^{-1})\mathbf{x}_s(k) - (1 - z^{-1})\mathbf{x}(k)\right] - \mathbf{R}_y[\mathbf{w}(k) - \mathbf{y}(k)] \tag{5.62a}$$
$$= -\mathbf{R}_x\Delta\mathbf{x}(k) - \mathbf{R}_y[\mathbf{w}(k) - \mathbf{y}(k)] \tag{5.62b}$$

The part of the actuating variables $\mathbf{u}_R(k)$ are calculated from Eq. (5.59a), see Eq. (5.63).

$$\mathbf{u}_R(k) = \mathbf{u}_R(k-1) + \Delta\mathbf{u}_R(k) \tag{5.63}$$

Subsequently, a model for the calculation of $\mathbf{R}_x$ and $\mathbf{R}_y$ to realize the feedback of $\mathbf{x}(k)$ and $(\mathbf{w}(k) - \mathbf{y}(k))$ is derived. The state difference equations are evaluated at k and $(k+1)$, see Eq. (5.64) and Eq. (5.65).

$$\mathbf{x}(k+1) = \mathbf{A}_d\mathbf{x}(k) + \mathbf{B}_d\mathbf{u}(k) + \mathbf{E}_d\mathbf{z}(k) \tag{5.64}$$

5.3 Design of the State-Feedback V/I-Controller

$$\mathbf{x}(k) = \mathbf{A}_d\mathbf{x}(k-1) + \mathbf{B}_d\mathbf{u}(k-1) + \mathbf{E}_d\mathbf{z}(k-1) \tag{5.65}$$

Subtracting Eq. (5.65) from Eq. (5.64) yields Eq. (5.66).

$$\Delta\mathbf{x}(k+1) = \mathbf{x}(k+1) - \mathbf{x}(k) \tag{5.66a}$$
$$= \mathbf{A}_d\Delta\mathbf{x}(k) + \mathbf{B}_d\Delta\mathbf{u}(k) + \mathbf{E}_d\Delta\mathbf{z}(k) \tag{5.66b}$$

The output equations according to Eq. (5.67) are valid.

$$\mathbf{y}(k+1) - \mathbf{y}(k) = \mathbf{C}\Delta\mathbf{x}(k+1) \tag{5.67a}$$
$$= \mathbf{C}\big[\mathbf{A}_d\Delta\mathbf{x}(k) + \mathbf{B}_d\Delta\mathbf{u}(k) + \mathbf{E}_d\Delta\mathbf{z}(k)\big] \tag{5.67b}$$

Reformulating Eq. (5.67b) yields Eq. (5.68).

$$\mathbf{y}(k+1) = \mathbf{C}\big[\mathbf{A}_d\Delta\mathbf{x}(k) + \mathbf{B}_d\Delta\mathbf{u}(k) + \mathbf{E}_d\Delta\mathbf{z}(k)\big] + \mathbf{y}(k) \tag{5.68}$$

Subtracting Eq. (5.68) from the time-independent reference variables $\mathbf{w}$ yields Eq. (5.69).

$$\mathbf{w} - \mathbf{y}(k+1) = -\mathbf{C}\big[\mathbf{A}_d\Delta\mathbf{x}(k) + \mathbf{B}_d\Delta\mathbf{u}(k) + \mathbf{E}_d\Delta\mathbf{z}(k)\big] + \big[\mathbf{w} - \mathbf{y}(k)\big] \tag{5.69}$$

By combining Eq. (5.66b) and Eq. (5.69), the extended state-space model in accordance with Eq. (5.70) and Eq. (5.71) appears.

$$\begin{pmatrix}\Delta\mathbf{x}(k+1)\\ \mathbf{w}-\mathbf{y}(k+1)\end{pmatrix} = \begin{pmatrix}\mathbf{A}_d & \mathbf{0}\\ -\mathbf{C}\mathbf{A}_d & \mathbf{I}\end{pmatrix}\begin{pmatrix}\Delta\mathbf{x}(k)\\ \mathbf{w}-\mathbf{y}(k)\end{pmatrix} + \begin{pmatrix}\mathbf{B}_d\\ -\mathbf{C}\mathbf{B}_d\end{pmatrix}\Delta\mathbf{u}(k) + \begin{pmatrix}\mathbf{E}_d\\ -\mathbf{C}\mathbf{E}_d\end{pmatrix}\Delta\mathbf{z}(k) \tag{5.70}$$

$$\mathbf{w} - \mathbf{y}(k) = \begin{pmatrix}\mathbf{0} & \mathbf{I}\end{pmatrix}\begin{pmatrix}\Delta\mathbf{x}(k)\\ \mathbf{w}-\mathbf{y}(k)\end{pmatrix} \tag{5.71}$$

The controller matrices $\mathbf{R}_x$ and $\mathbf{R}_y$ are calculated from the system matrix and the input matrix in Eq. (5.70). The following matrices appear according to the solution of the discrete-time algebraic Riccati equation (see also Sec. 5.3.4.3), see Eq. (5.72).

$$\mathbf{R} = \begin{pmatrix}\mathbf{R}_x & \mathbf{R}_y\end{pmatrix} \tag{5.72}$$

The weighting of the proportional state-feedback and the integration of $(\mathbf{w}-\mathbf{y}(k))$ is realized by the choice of the proportion of the according elements of the $\mathbf{Q}_e$-matrix of the extended model from Eq. (5.70). The higher $\Delta \mathbf{x}(k)$ is weighted compared to $(\mathbf{w}-\mathbf{y}(k))$, the slower is the operation of the state-feedback controller with integral characteristics. Via $\mathbf{Q}_u$, the elements of $\Delta \mathbf{u}_R(k)$ are weighted.

In this thesis—in contrast to [104]—the feedforward controllers of the reference variables $\mathbf{M}_u$ (see Sec. 5.3.2) and the disturbance variables $\mathbf{N}_u$ (see Sec. 5.3.3) are added to the principle of the state-feedback controller with integral characteristics. The control structure according to Fig. 5.19 is used.

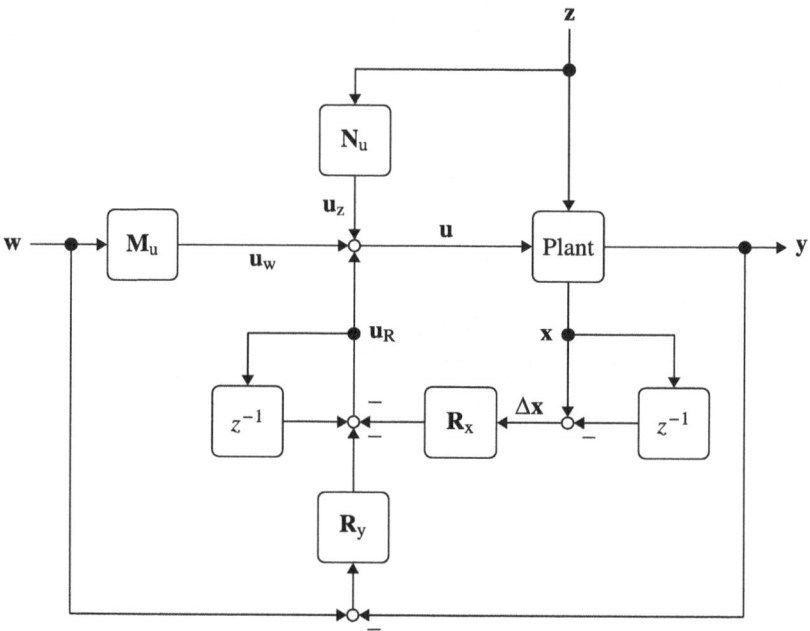

Fig. 5.19 Block diagram of the state-feedback V/I-controller with integral characteristics using the feedback controllers $\mathbf{R}_x$ and $\mathbf{R}_y$ and unit delay blocks z^{-1}

5.4 Summary

In this chapter, the voltage/current (V/I) control of voltage source converters (VSCs) and their grid-side characteristics are presented. The grid-side characteristics influence the interaction between the VSC and the electrical power grid, especially during the steady state of a grid fault.

The calculation of the reference variables of the V/I-control layer is linked to the synchronization of VSCs to the grid (see Chap. 4) in a hierarchical control structure. The VSC is identified as the actuator inside a functional chain to control and dispense active and reactive power.

Assuming that the V/I-controller works correctly, the reference variables define the steady state grid-side characteristics. Various options for the grid-side characteristics in the positive and negative sequence reveal. The characteristics of voltage sources and current sources as well as defined V/I-characteristics can be realized. The injection of reactive current in the negative sequence to achieve symmetrical voltages after a phase-to-phase fault is investigated via simulations.

The V/I-controller is the basic prerequisite to realize reference values of voltage or currents. It reacts to changing grid conditions or transitions of the operating point. The stability and the dynamics of the V/I-controller are dependant on the type of V/I-controller. In this thesis, a state-feedback controller based on the linear-quadratic regulator (LQR) algorithm added by two feedforward controllers is applied. This type of controller shows significant advantages over conventional proportional-integral (PI) controllers. High dynamic performance in the grid-forming (GFM) mode is reached, inner system properties reveal and control parameters are tuned intuitively using weighting matrices. The state-space models based on space vector components and the state controllability and asymptotic stability from Chap. 3 are used.

The investigation and optimization of selected grid-side characteristics of the VSC in the positive, the negative and the zero sequence, especially during grid faults, is presented in Chap. 6. To follow the key aspects of the central theme of this thesis on a shortcut: The conclusion of the next chapter is provided in Sec. 6.4.

Open Access This chapter is licensed under the terms of the Creative Commons Attribution 4.0 International License (http://creativecommons.org/licenses/by/4.0/), which permits use, sharing, adaptation, distribution and reproduction in any medium or format, as long as you give appropriate credit to the original author(s) and the source, provide a link to the Creative Commons license and indicate if changes were made.

The images or other third party material in this chapter are included in the chapter's Creative Commons license, unless indicated otherwise in a credit line to the material. If material is not included in the chapter's Creative Commons license and your intended use is not permitted by statutory regulation or exceeds the permitted use, you will need to obtain permission directly from the copyright holder.

Enhanced Grid Fault Characteristics of VSCs

This chapter presents the enhanced grid fault characteristics of voltage source converters (VSCs). In Sect. 6.1, the resilience of 100% inverter-based power systems is discussed. The resilience is subdivided into the resilience of the grid operation and the resilience of the VSC. The resilience of the grid operation is discussed in Sect. 6.2. It contains the presentation of the necessary hardware and software configuration of direct-connected VSCs. Different control modes of these VSCs are discussed. The resonant grounding system (RGS) mode of VSCs based on a virtual inductance realized in the zero sequence is introduced. The RGS-mode is analytically developed and evaluated in simulations. Sect. 6.3 describes the resilience of the VSC. It consists of the transient state current limiting and the steady state current limiting of grid-forming VSCs. Both current limiting methods are analytically developed. The transient current limiting represents an additional feature of the RGS-mode and is only activated if overcurrents are predicted. Furthermore, the realization of a constant switching frequency is shown. This concept is also evaluated in simulations.

6.1 Resilience of 100% IBPS Against Grid Faults

Resilience is the ability of a system to return to the initial state after a disturbing event without crucial damage in general. Resilience of the electrical power system includes the ability to restore the grid operation after a grid fault and to avoid destruction of equipment and components. Consequently, resilience is necessary for the reliable energy supply.

The resilience of a 100% inverter-based power system (IBPS) is separated into the resilience of the grid operation and the resilience of the voltage source converter (VSC). Both forms of resilience are linked to each other and are realized by the software-based protection induced by the VSC. The Fig. 6.1 shows this correlation and the different aspects.

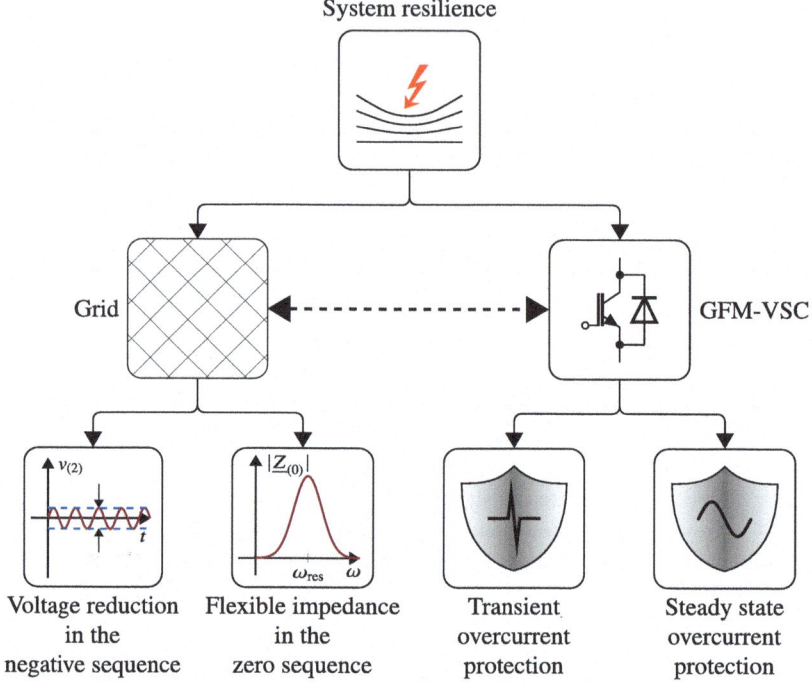

Fig. 6.1 Composition of system resilience by resilience of the grid operation and resilience of the VSC

The objective of the resilience of the grid operation is to ensure a symmetrical grid operation and the fault arc suppression. Addressing these different aims, a separation regarding the symmetrical components (see Sect. 3.1.2.2) is made. The voltage amplitude in the negative sequence is reduced by a series resonant circuit in the negative sequence, see Sect. 5.2.3.2. The limiting of the amplitudes of voltages and currents in the zero sequence is reached by a flexible absolute value of the impedance of the zero sequence realized by a parallel resonant circuit, see Sect. 6.2.3.1.

Supplementary, a different approach is mentioned here: The activation of a reserve wire for the redundant operation after a single-phase-to-ground fault is presented in [154].

The objective of the resilience of the VSC is to stay connected to the grid and to avoid damages of the power semiconductor devices caused by overcurrents in the transient or steady state. The limiting of currents in the transient and the steady state requires different approaches, see Sect. 6.3.1 and Sect. 6.3.2.

6.2 Resilience of the Grid Operation

In 75–85% of all electrical power grid faults, ground is considered to be involved [105]. The single-phase-to-ground fault represents the most common type. To manage these fault types and to continue the grid operation, the fault current must be reduced or at least controlled within a certain range. The reduction of the amplitude of the positive sequence voltage of a VSC is not the favoured option, since parallel subgrids have to be supplied with the nominal voltage amplitude in the positive sequence.

To solve this problem, a novel method is developed. Instead of adjusting the fault current by the voltage source in the positive sequence, it is adjusted by the impedance of the fault current path in the zero sequence. A virtual inductance is realized by the VSC in the zero sequence. Taking advantage of parasitic line capacities and the properties of a parallel resonance circuit, a flexible absolute value of the impedance of the zero sequence is reached (see also Fig. 6.1). Consequently, the fault current becomes controllable indirectly. Optionally, the reduction of the negative sequence voltage (see Sect. 5.2.3.2 and Sect. 5.2.3.3) can be added (see also Fig. 6.1). The control of the characteristics of the VSC in the zero sequence represents a novel approach for neutral point treatment, that is especially relevant for 100% IBPS.

The operation of the negative sequence as a parallel resonant circuit is not an option, because VSCs, that do not control the negative sequence, represent a short circuit. A single VSC representing a short circuit in the negative sequence ruins the concept of a parallel resonance circuit with a high absolute value of the negative sequence impedance. This principle is not possible in practical scenarios.

Conventional three-phase three-wire VSCs do not allow zero sequence currents due to Kirchhoff's current law and represent an open circuit in the zero sequence. These VSCs do not ruin the concept of a parallel resonance circuit in the zero sequence with a high absolute value of the zero sequence impedance. To make zero sequence currents possible and to control the zero sequence characteristics, the hardware and the software configuration of the conventional VSC (see Fig. 2.3) have to be extended.

6.2.1 Hardware Configuration

The extension of the hardware configuration of the conventional VSC leads to an extended VSC. In this subsection, the grid integration of the extended VSC is discussed with focus on the characteristics of a three-phase four-wire system. Subsequently, the extended VSC is integrated into the grid as direct-connected VSC without transformer.

6.2.1.1 Voltage Balancer and Extended VSC

The zero sequence current equals the scaled sum of the three phase currents, see also Eq. (3.14). As a consequence, a fourth wire is necessary to make zero sequence currents possible.

There are different options to connect a fourth wire to the VSC to provide a neutral line [106]. The first option is to connect the fourth wire to the midpoint of the DC-link. The problem is, that the parts of the DC-link voltage become unbalanced and the basic operating principle of the VSC (see Sect. 2.1.1) is threatened. The second option is to use a fourth half-bridge and to connect the fourth wire to its midpoint. Problems concerning the electromagnetic compatibility arise due to high voltages at high frequencies [106].

Combining these two options and using the fourth half-bridge as a voltage balancer (VB) of the DC-link voltage leads to the independently controlled neutral leg [106]. This concept is applied subsequently. The VB is connected via a choke (represented by L_{vb} and R_{vb}) to the DC-link midpoint. The VB can be realized as a two-level half-bridge or a three-level half-bridge [107, 108]. The equivalent circuit of the three-level VB is shown in Fig. 6.2.

The following state differential equations appear based on Kirchhoff's current law and the V/I-characteristics of the choke assuming $C_{d1} = C_{d2} = C_d$, see Eq. (6.1).

$$\frac{d}{dt}\begin{pmatrix}(v_{d1} - v_{d2})\\ i_{vb}\end{pmatrix} = \begin{pmatrix}0 & -\frac{1}{C_d}\\ 0 & -\frac{R_{vb}}{L_{vb}}\end{pmatrix}\begin{pmatrix}(v_{d1} - v_{d2})\\ i_{vb}\end{pmatrix} + \begin{pmatrix}0\\ \frac{1}{L_{vb}}\end{pmatrix} v_{vb} + \begin{pmatrix}-\frac{1}{C_d}\\ 0\end{pmatrix} 3i_{c0}$$

(6.1)

6.2 Resilience of the Grid Operation

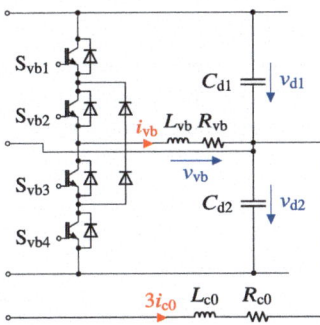

Fig. 6.2 Equivalent circuit of the three-level voltage balancer

Based on these differential equations, different control principles of the VB can be designed. The voltage v_{vb} depends on the switching state of the VB and thus represents the actuating variable. The DC-link voltages v_{d1} and v_{d2} are assumed to be greater than zero.

The control principle of the three-level VB is explained in Table 6.1 and Table 6.2 based on two different switching states and the state differential equations in Eq. (6.1). The resistance of the choke R_{vb} is neglected in this explanation.

Table 6.1 Influence of the switching state of the voltage balancer on the time derivative of the choke current i_{vb} based on Fig. 6.2 and the second line of Eq. (6.1)

S_{vb1}	S_{vb2}	S_{vb3}	S_{vb4}	v_{vb}	$\frac{d}{dt} i_{vb}$
on	on	off	off	v_{d1}	> 0
off	off	on	on	$-v_{d2}$	< 0

Table 6.2 Influence of the currents i_{vb} and i_{c0} on the time derivative of the DC-link voltage unbalance ($v_{d1} - v_{d2}$) based on the first line of Eq. (6.1)

$i_{vb} + 3i_{c0}$	$\frac{d}{dt}(v_{d1} - v_{d2})$
> 0	< 0
< 0	> 0

The combination of the three-level VSC and the three-level VB leads to the extended VSC, which allows zero sequence currents while balancing the DC-link voltages. A freewheeling path for the zero sequence current i_{c0} is available in each switching state of the VB. The equivalent circuit of the extended VSC based on Fig. 2.3 is shown in Fig. 6.3.

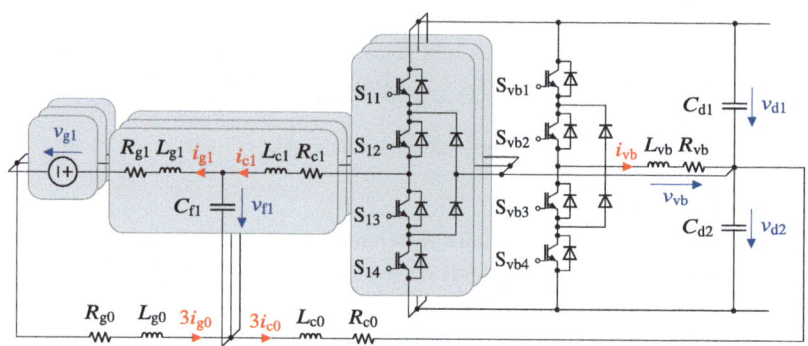

Fig. 6.3 Equivalent circuit of the extended three-level VSC with LCL-filter

The control algorithm of the VB is separated from the control algorithm of the conventional part of the VSC. Both control algorithms are designed independently of each other. The conventional part of the VSC controls the positive, the negative and the zero sequence. The VB realizes balanced DC-link voltages.

6.2.1.2 Direct-connected VSCs

Today, VSCs are often connected as three-phase three-wire systems via a transformer to the grid [6]. The transformer is essentially used to ensure the voltage compatibility between the VSC and the medium or the high voltage grid. Typically, a delta-wye transformer is used. The star point of the transformer on the grid side is often used to realize the desired type of neutral point treatment (NPT). On the VSC side, this type of transformer represents an open circuit in the zero sequence and consequently no zero sequence current can flow. Alternatively to this concept, other NPT devices can be used to realize the desired type of NPT.

6.2 Resilience of the Grid Operation

State-of-the-art VSCs with transformer are inflexible concerning NPT. Hardware modifications or the installation of new equipment are needed to change of the type of NPT. Regarding grid expansion associated with the integration of renewable energy sources may e.g. require the change from a resonant grounding system (RGS) to a solid grounding system (SGS) due to a higher capacitance in the zero sequence. Moreover, the grid operation reliability of state-of-the-art solutions for RGS are improvable due to the limited adjustment speed of arc suppression coils. Disconnections of subgrids lead to detuning of arc suppression coils and may cause further grid outages. Since no or only high-impedance freewheeling current paths are available, high transient overvoltages, especially in ungrounded systems (UGS) and RGS occur.

Modern VSCs are designed as switch-based or cell-based multilevel topologies to reduce the voltage rating per semiconductor power device in medium or high voltage applications [21, 109]. Besides, wide-bandgap (WBG) semiconductors, such as silicon carbide (SiC), allow power devices with higher voltage ratings [110]. Both aspects reduce the relevance of the transformer concerning its task to ensure the voltage compatibility between VSCs and medium or high voltage grids.

Several steps towards the integration of direct-connected (or transformerless) VSCs into the grid have been made. Transformerless VSCs are used for medium voltage large electric drives today [111]. Grid-connected (GC) transformerless VSCs have been investigated in laboratory tests [112, 113]. Beyond that, additional DC/DC-converters have been used to adjust the DC-link voltage for the proper operation of VSCs in medium voltage grids [114]. The application of the dual active bridge (DAB) (see [28, 29]) as a DC/DC-converter as shown in [114] ensures the galvanic isolation between the medium voltage grid and the low voltage side by a high frequency transformer. The modular integration of energy storage devices into a hybrid energy storage system is investigated in [146].

Regarding these aspects, the hardware configuration of the extended VSC (see Fig. 6.3) can make the transformer unnecessary. By using the extended VSC, the bulky transformer is replaced by an extension of the VSC representing an existing component. Direct-connected VSC provide higher flexibility for NPT regarding grid expansion because the NPT is realized by software, see Sect. 6.2.2.2. Higher grid operation reliability after disconnections of subgrids is reached through high adjustment speed. Besides, there is lower stress on equipment due to the absence of high transient overvoltages.

For the realization of direct-connected VSC during grid faults, the ratings of the semiconductor power devices and the DC-link have to be designed appropriately concerning maximum currents on medium voltage level, see [155]. Furthermore, the absence of the transformer leads to the investigation of the mean values of the phase current. They are not blocked inherently.

Non-zero mean values of the phase currents are induced by VSCs if the DC-link voltage is unbalanced. This effect is reduced by the VB, see Sect. 6.2.1.1. Another cause is the locking time of semiconductor power devices [92].

If non-zero mean values of the phase currents appear, inductive components with magnetic core can reach the saturation state. Moreover, the temperature of components can increase due to higher root-mean-square (RMS) values.

To avoid non-zero mean values of the phase currents, a novel control concept is developed. First, models of the LCL-filter (see Sect. 3.2.2.3) in the stationary reference frame (see Eq. (6.2)) and the zero sequence (see Eq. (3.43)) are derived.

$$\frac{d}{dt}\begin{pmatrix} i_{c\alpha} \\ i_{c\beta} \\ i_{g\alpha} \\ i_{g\beta} \\ v_{f\alpha} \\ v_{f\beta} \end{pmatrix} = \begin{pmatrix} -\frac{R_c}{L_c} & 0 & 0 & 0 & -\frac{1}{L_c} & 0 \\ 0 & -\frac{R_c}{L_c} & 0 & 0 & 0 & -\frac{1}{L_c} \\ 0 & 0 & -\frac{R_g}{L_g} & 0 & \frac{1}{L_g} & 0 \\ 0 & 0 & 0 & -\frac{R_g}{L_g} & 0 & \frac{1}{L_g} \\ \frac{1}{C_f} & 0 & -\frac{1}{C_f} & 0 & 0 & 0 \\ 0 & \frac{1}{C_f} & 0 & -\frac{1}{C_f} & 0 & 0 \end{pmatrix} \begin{pmatrix} i_{c\alpha} \\ i_{c\beta} \\ i_{g\alpha} \\ i_{g\beta} \\ v_{f\alpha} \\ v_{f\beta} \end{pmatrix} + \begin{pmatrix} \frac{1}{L_c} & 0 \\ 0 & \frac{1}{L_c} \\ 0 & 0 \\ 0 & 0 \\ 0 & 0 \\ 0 & 0 \end{pmatrix} \begin{pmatrix} v_{c\alpha} \\ v_{c\beta} \end{pmatrix}$$

(6.2)

$$+ \begin{pmatrix} 0 & 0 \\ 0 & 0 \\ -\frac{1}{L_g} & 0 \\ 0 & -\frac{1}{L_g} \\ 0 & 0 \\ 0 & 0 \end{pmatrix} \begin{pmatrix} v_{g\alpha} \\ v_{g\beta} \end{pmatrix}$$

6.2 Resilience of the Grid Operation

Second, the mean values of the variables are calculated. From Eq. (6.2) and Eq. (3.43), the state-space models according to Eq. (6.3) and Eq. (6.4) reveal.

$$\frac{d}{dt}\begin{pmatrix}\overline{i}_{c\alpha}\\ \overline{i}_{c\beta}\\ \overline{i}_{g\alpha}\\ \overline{i}_{g\beta}\\ \overline{v}_{f\alpha}\\ \overline{v}_{f\beta}\end{pmatrix} = \begin{pmatrix}-\frac{R_c}{L_c} & 0 & 0 & 0 & -\frac{1}{L_c} & 0\\ 0 & -\frac{R_c}{L_c} & 0 & 0 & 0 & -\frac{1}{L_c}\\ 0 & 0 & -\frac{R_g}{L_g} & 0 & \frac{1}{L_g} & 0\\ 0 & 0 & 0 & -\frac{R_g}{L_g} & 0 & \frac{1}{L_g}\\ \frac{1}{C_f} & 0 & -\frac{1}{C_f} & 0 & 0 & 0\\ 0 & \frac{1}{C_f} & 0 & -\frac{1}{C_f} & 0 & 0\end{pmatrix} \begin{pmatrix}\overline{i}_{c\alpha}\\ \overline{i}_{c\beta}\\ \overline{i}_{g\alpha}\\ \overline{i}_{g\beta}\\ \overline{v}_{f\alpha}\\ \overline{v}_{f\beta}\end{pmatrix} + \begin{pmatrix}\frac{1}{L_c} & 0\\ 0 & \frac{1}{L_c}\\ 0 & 0\\ 0 & 0\\ 0 & 0\\ 0 & 0\end{pmatrix}\begin{pmatrix}\overline{v}_{c\alpha}\\ \overline{v}_{c\beta}\end{pmatrix}$$

$$+ \begin{pmatrix}0 & 0\\ 0 & 0\\ -\frac{1}{L_g} & 0\\ 0 & -\frac{1}{L_g}\\ 0 & 0\\ 0 & 0\end{pmatrix}\begin{pmatrix}\overline{v}_{g\alpha}\\ \overline{v}_{g\beta}\end{pmatrix} \quad (6.3)$$

$$\frac{d}{dt}\begin{pmatrix}\overline{i}_{c0}\\ \overline{i}_{g0}\\ \overline{v}_{f0}\end{pmatrix} = \begin{pmatrix}-\frac{R_c^*}{L_c^*} & 0 & -\frac{1}{L_c^*}\\ 0 & -\frac{R_g^*}{L_g^*} & \frac{1}{L_g^*}\\ \frac{1}{C_f} & -\frac{1}{C_f} & 0\end{pmatrix}\begin{pmatrix}\overline{i}_{c0}\\ \overline{i}_{g0}\\ \overline{v}_{f0}\end{pmatrix} + \begin{pmatrix}\frac{1}{L_c^*}\\ 0\\ 0\end{pmatrix}\overline{v}_{c0} + \begin{pmatrix}0\\ -\frac{1}{L_g^*}\\ 0\end{pmatrix}\overline{v}_{g0} \quad (6.4)$$

Third, a linear—quadratic (LQ) controller with integral characteristics (see Sect. 5.3.5) is designed based on these models to control the mean values of the phase currents i_{gx} to zero.

This concept does not affect the positive, the negative and the zero sequence of any other (especially the fundamental) frequency (see Sect. 3.1.3) and is also adaptable for L- and LC-filters.

6.2.2 Software Configuration in the Zero Sequence

Positive, negative and zero sequence components are controlled independently of each other. Positive and negative sequence components are controlled using double synchronous reference frame (DSRF) components, see Sect. 3.1.2.5. The zero sequence components are controlled using synchronous reference frame (SRF) components in addition, see Sect. 3.1.2.4. Consequently, three rotating reference frames are used for the fundamental frequency.

6.2.2.1 Signal Transformations and Modelling

The objective of this section is to derive a model using DC quantities of zero sequence components according to the concept of a SRF. The starting point is Fig. 3.6 and Eq. (3.43) using the abbreviations from Eq. (3.38), Eq. (3.39), Eq. (3.44) and Eq. (3.45).

To reach SRF-components from single-phase AC-components in the zero sequence, two orthogonal components per variable are required. The original AC-component is used as the α-component according to [115, 143]. The β-component corresponds to the original component delayed by one fourth of the period duration. The α-component and the β-component show the same angular frequency ω.

DC-quantities are obtained by transforming these $\alpha\beta$-components into dq-components using the SRF with the angular frequency ω. These components are labelled by the indices 0d and 0q. The transformation matrix from Eq. (3.18) is used to transform the $\alpha\beta$-components of each variable. The transformation angle of the zero sequence γ_{T0} is set to the transformation angle of the positive sequence γ_T^+, see Sect. 4.3.3 and Sect. 4.3.4.1.

The state differential equations according to Eq. (6.5) are derived. These equations complement Eq. (3.42) by the transformed zero sequence components.

6.2 Resilience of the Grid Operation

$$\frac{d}{dt}\begin{pmatrix} i_{c0d} \\ i_{c0q} \\ i_{g0d} \\ i_{g0q} \\ v_{f0d} \\ v_{f0q} \end{pmatrix} = \begin{pmatrix} -\frac{R_c^*}{L_c^*} & \omega & 0 & 0 & -\frac{1}{L_c^*} & 0 \\ -\omega & -\frac{R_c^*}{L_c^*} & 0 & 0 & 0 & -\frac{1}{L_c^*} \\ 0 & 0 & -\frac{R_g^*}{L_g^*} & \omega & \frac{1}{L_g^*} & 0 \\ 0 & 0 & -\omega & -\frac{R_g^*}{L_g^*} & 0 & \frac{1}{L_g^*} \\ \frac{1}{C_f} & 0 & -\frac{1}{C_f} & 0 & 0 & \omega \\ 0 & \frac{1}{C_f} & 0 & -\frac{1}{C_f} & -\omega & 0 \end{pmatrix} \begin{pmatrix} i_{c0d} \\ i_{c0q} \\ i_{g0d} \\ i_{g0q} \\ v_{f0d} \\ v_{f0q} \end{pmatrix} + \begin{pmatrix} \frac{1}{L_c^*} & 0 \\ 0 & \frac{1}{L_c^*} \\ 0 & 0 \\ 0 & 0 \\ 0 & 0 \\ 0 & 0 \end{pmatrix} \begin{pmatrix} v_{c0d} \\ v_{c0q} \end{pmatrix}$$

$$+ \begin{pmatrix} 0 & 0 \\ 0 & 0 \\ -\frac{1}{L_g^*} & 0 \\ 0 & -\frac{1}{L_g^*} \\ 0 & 0 \\ 0 & 0 \end{pmatrix} \begin{pmatrix} v_{g0d} \\ v_{g0q} \end{pmatrix} \qquad (6.5)$$

Based on Eq. (6.5), a LQ-controller (see Sect. 5.3.4 and optionally Sect. 5.3.5) is designed. The pair of actuating variables v_{c0d}/v_{c0q} is transformed back into αβ-components and $v_{c0\alpha}$ is added to the actuating variables in the natural reference frame as a zero sequence component.

6.2.2.2 Options of Grid Characteristics and Operation Modes

The extended hardware configuration using the extended VSC (see Sect. 6.2.1.1) allows zero sequence currents due to Kirchhoff's current law. From the fundamentals of VSCs connecting different DC-link potentials to the AC-side (see Sect. 2.1), the monofrequency symmetrical components equivalent circuit extending Fig. 5.3 of a three-phase four-wire VSC is derived, see Fig. 6.4.

New options of grid-side characteristics of a GC VSC reveal [153]. Based on Fig. 5.5, the extended generalized steady state symmetrical components equivalent circuit of the GC VSC appears.

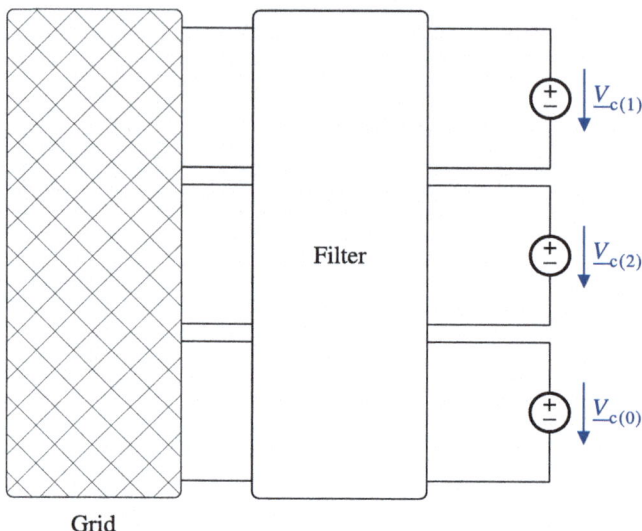

Fig. 6.4 Steady state monofrequency symmetrical components equivalent circuit of the three-phase four-wire VSC representing a voltage source in the positive sequence (1), a voltage source in the negative sequence (2) and a voltage source in the zero sequence (0)

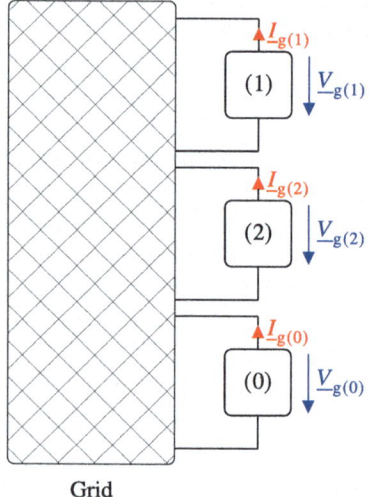

Fig. 6.5 Generalized steady state symmetrical components equivalent circuit of the GC VSC regarding the positive sequence (1), the negative sequence (2) and the zero sequence (0) from the grid perspective

The options of steady state grid characteristics in the zero sequence are now analogue to the ones in the positive and the negative sequence, see Fig. 6.6 and also Fig. 5.6 and Fig. 5.7.

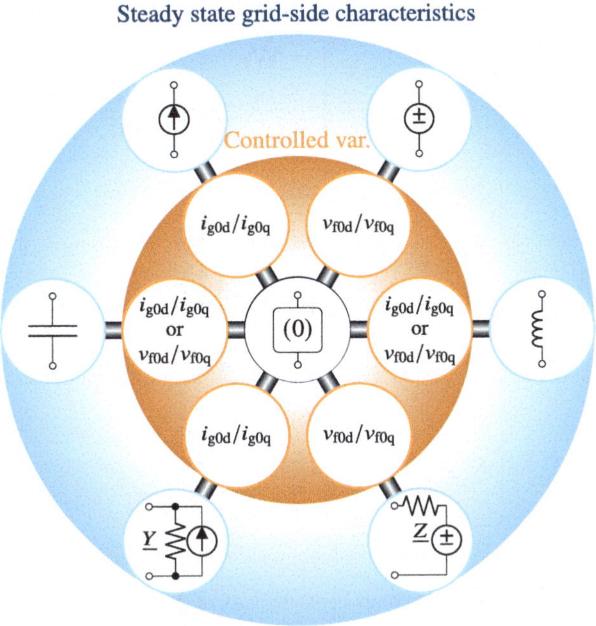

Fig. 6.6 Options of grid characteristics of a GC VSC in the zero sequence (0) in the steady state and the according controlled variables

Different control modes of the zero sequence reveal. They differ from each other depending on the choice of the pair of the controlled variables.

An ungrounded system (UGS) is realized by controlling the pair i_{g0d}/i_{g0q} to zero. The steady state symmetrical components network according to Fig. 6.7 appears. The problem of the UGS-mode during a single-phase-to-ground fault is, that the amplitude of the voltages becomes very high and the insulation limits of equipment can be exceeded [153]. In addition, only the zero sequence current at the controlling

VSC is zero. That means that the absolute value of the impedance of the zero sequence network as a parallel circuit to the positive sequence circuit is not adjusted in general. Consequently, fault currents of other VSCs in the grid are not affected.

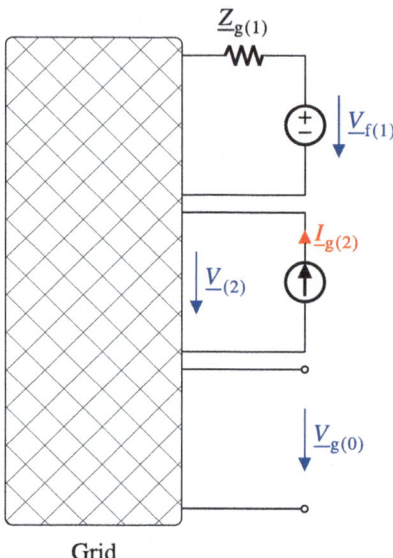

Fig. 6.7 Steady state symmetrical components network of the UGS-mode

A solid grounded system (SGS) is realized by controlling the pair v_{f0d}/v_{f0q} to zero. The following steady state symmetrical components network according to Fig. 6.8 reveals. The problem of the SGS-mode during a single-phase-to-ground fault is, that the amplitude of the phase currents becomes very high [153]. Since the ability of VSCs to allow overcurrents is very limited, this control mode is not suitable for practical applications.

Furthermore, a resonant grounding system (RGS) is realized. This control mode represents a compromise between the UGS-mode and the SGS-mode concerning the stress on the equipment by high amplitudes of voltages or currents.

6.2 Resilience of the Grid Operation

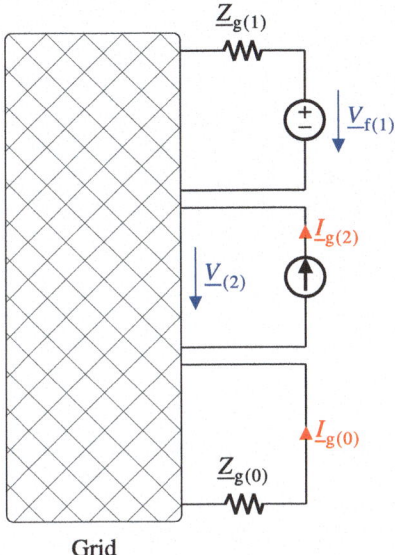

Fig. 6.8 Steady state symmetrical components network of the SGS-mode

6.2.3 Resonant Grounding System Mode

6.2.3.1 Parallel Resonant Circuit in the Zero Sequence

The RGS-mode of the VSC is presented assuming a single-phase-to-ground fault, see also [155]. The fault current is caused by a voltage source in the positive sequence, see Sect. 4.4.1.2 and Sect. 4.4.2. Optionally, reactive current is injected in the negative sequence, see Sect. 5.2.3.2.

The fault location has hardly no effect on the characteristics of the zero sequence from the VSC perspective. The reason is, that the series impedances in the zero sequences are significantly smaller than the impedances of the zero sequence capacitances (see [116]) and the realized virtual impedance. Neglecting the series impedances, the zero sequence is interpreted as a parallel circuit consisting of parallel line capacitances and the realized virtual impedance. This simplification leads to a non-compensated part of the fault current in the range of a few percent using the RGS-mode.

The objective is to ensure a symmetrical grid operation, the reduction of the fault current and the arc suppression. In addition, the fault currents of other VSCs also have to be reduced. Concerning the stress on the equipment by high amplitudes of voltages or currents, a compromise between the UGS-mode and the SGS-mode is reached (see Sect. 6.2.2.2). To realize these objectives and characteristics, the absolute value of the impedance of the zero sequence as a parallel circuit to the positive sequence circuit has to be adjustable and represent a high value in general. This fact is reached by operating the zero sequence circuit as a parallel resonant circuit near the resonance frequency [156, 157]. The advantage over the UGS-mode is, that not only the fault current of one VSC is reduced.

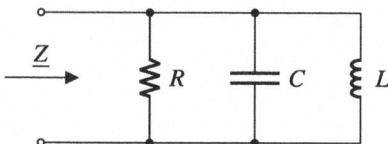

Fig. 6.9 Generalized equivalent circuit of a parallel resonant circuit

The general equivalent circuit of a parallel resonant circuit is shown in Fig. 6.9. The absolute value of the impedance of the parallel resonant circuit is dependent on the frequency and calculated according to Eq. (6.6).

$$|\underline{Z}| = \frac{1}{|\underline{Y}|} \tag{6.6a}$$

$$= \frac{1}{\sqrt{\frac{1}{R^2} + \left(\omega C - \frac{1}{\omega L}\right)^2}} \tag{6.6b}$$

The maximum of the absolute value of the impedance appears at the resonance angle frequency ω_{res}, see Eq. (6.7). The argument of the impedance is calculated as in accordance with Eq. (6.8).

$$\omega_{res} = \frac{1}{\sqrt{LC}} \tag{6.7}$$

$$\arg\{\underline{Z}\} = -\arg\{\underline{Y}\} \tag{6.8a}$$

$$= -\mathrm{atan2}\left(\omega C - \frac{1}{\omega L}, \frac{1}{R}\right) \tag{6.8b}$$

To use Eq. (6.6b) to adjust the absolute value of the zero sequence impedance $|\underline{Z}_{(0)}|$, the variables in this equation have to be evaluated. The operating angle frequency is given by the voltage source in the positive sequence. The parasitic capacitances of overhead lines are dominant in the zero sequence. The value of the zero sequence

6.2 Resilience of the Grid Operation

capacitance is dependent on the grid dimension. The value of the parallel resistance of the zero sequence is often very high and is thus to be neglected for the calculation of the impedance. Consequently, the value of the inductance of the zero sequence represents the parameter to be adjusted to control $|\underline{Z}_{(0)}|$.

The inductance of the zero sequence is realized by a virtual inductance $L_{(0),\text{VI}}$. The starting point of the subsequent derivation is the SRF-model of the zero sequence in Eq. (6.5). Analogue to Sect. 5.2.4.1, the Eq. (6.9) is derived. It leads to the equivalent circuit in the zero sequence in the steady state shown in Fig. 6.10.

$$\begin{pmatrix} i_{g0d} \\ i_{g0q} \end{pmatrix} = \begin{pmatrix} 0 & -\frac{1}{\omega L_{(0),\text{VI}}} \\ \frac{1}{\omega L_{(0),\text{VI}}} & 0 \end{pmatrix} \begin{pmatrix} v_{g0d} \\ v_{g0q} \end{pmatrix} \tag{6.9}$$

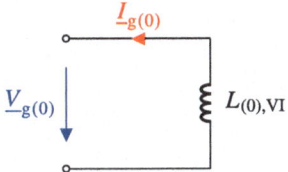

Fig. 6.10 Equivalent circuit of the zero sequence (0) of a GC VSC in the steady state applying a virtual inductance $L_{(0),\text{VI}}$

Implementing the previously described characteristics of the positive, the negative and the zero sequence into Fig. 6.5, the following generalized steady state symmetrical components equivalent circuit of the GC VSC in the RGS-mode appears, see Fig. 6.11.

To realize the virtual inductance in the zero sequence, a LQ-controller (see Sect. 5.3.4 and optionally Sect. 5.3.5) is designed. Only the zero sequence voltage at the point of interconnection of the VSC has to be measured to calculate the reference variables. The absolute value of the $|\underline{Z}_{(0)}|$ is adjustable by the value of $L_{(0),\text{VI}}$. The parallel resonant circuit does not necessarily have to be exactly operated at the point of resonance to realize high impedance.

The presented concept also works for phase-to-phase-to-ground faults, since zero sequence currents are avoided. The phase-to-phase-to-ground fault develops to a phase-to-phase fault, which is handled by reactive current injection in the negative sequence, see Sect. 5.2.3.3.

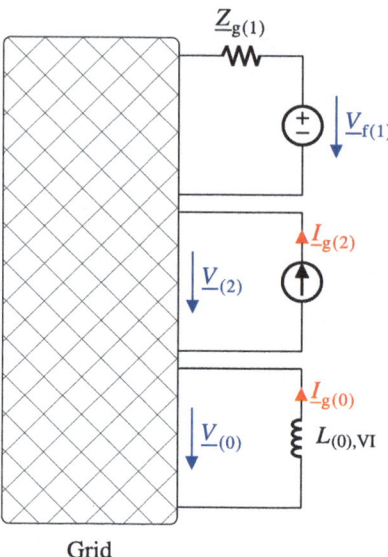

Fig. 6.11 Steady state symmetrical components network of the RGS-mode

6.2.3.2 Scenario: Evaluation of the RGS-mode

In the following scenario, the operating principle of the RGS-mode using the virtual inductance in the zero sequence is demonstrated via simulations in MATLAB/Simulink. The objective of the RGS-mode is to ensure symmetrical phase currents, the reduction of the fault current and the arc suppression. Table 6.3 gives an overview of the relevant figures of this section.

The investigated scenario is similar to the scenario from Sect. 5.2.3.3 in general. In contrast to Sect. 5.2.3.3, a single-phase-to-ground fault in phase 1 with the fault resistance $R_F = 0\,\Omega$ is applied at $t = 0.1$ s. The relative fault distance m equals 0.8 seen from busbar A. The scenario is shown in Fig. 6.12.

The reference value of the virtual inductance of the zero sequence $L_{(0),\text{VI}}$ after the fault equals 0.94 H. This value is chosen according to the zero sequence capacitance to reach the resonance frequency approximately at the operating frequency of 50 Hz, see also Eq. (6.7). The resonance frequency is chosen to be greater than the operating frequency to avoid crossing the resonance frequency during the transient state and thus to avoid high amplitudes of the voltages.

The VSC is connected as a three-phase four-wire system with LCL-filter, see Fig. 6.3. The symmetrical components networks are shown in Fig. 6.11 and Fig.

6.2 Resilience of the Grid Operation

Table 6.3 Overview of the relevant figures of Sect. 6.2.3.2 containing the evaluation of the RGS-mode

Content	Figure
Scenario under investigation	Fig. 6.12
Symmetrical components network of the VSC	Fig. 6.11
Symmetrical components network of the VSC and the fault	Fig. 6.13
Phase currents i_{gx} at the point of interconnection	Fig. 6.14
Voltages v_{gx} at the point of interconnection	Fig. 6.15
Zero sequence voltage v_{g0} and zero sequence current i_{g0}	Fig. 6.16
Virtual inductance $L_{(0),VI}$ in the zero sequence	Fig. 6.17
Absolute value of the zero sequence impedance $|\underline{Z}_{(0)}|$	Fig. 6.18
Current i_F through the fault resistance R_F	Fig. 6.19

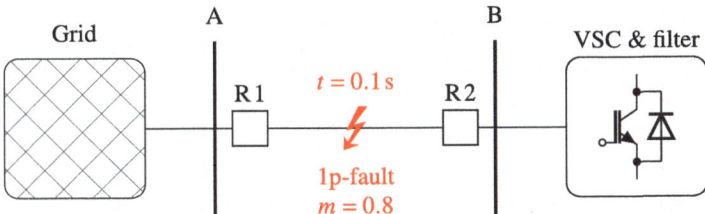

Fig. 6.12 Scenario for evaluating the fault characteristics of a GC VSC during a phase-to-ground fault in phase 1

6.13. In Fig. 6.13, the parallel resonant circuit in the zero sequence similar to Fig. 6.9 reveals. The voltage source in the positive sequence and the current source in the negative sequence are realized via a LQ-controller according to Sect. 5.3.4. Moreover, the mean values of the phase currents i_{gx} are controlled to zero according to Sect. 6.2.1.2.

In the transient state, the amplitudes of the phase currents i_{gx} (see Fig. 6.14) are increased approximately by a factor up to 2. In the steady state, symmetrical currents close to normal operation conditions are reached.

The amplitude of the voltages v_{gx} at the point of interconnection (see Fig. 6.15) of the healthy phases are increased approx. by the factor $\sqrt{3}$ in the steady state. The DC-link voltage has to designed appropriately due to the basic operating principle of the VSC (see Sect. 2.1.1). For example, a DC/DC-converter inside the DC-link can be used to adjust the DC-link voltage. Moreover, wide-bandgap semiconductors or

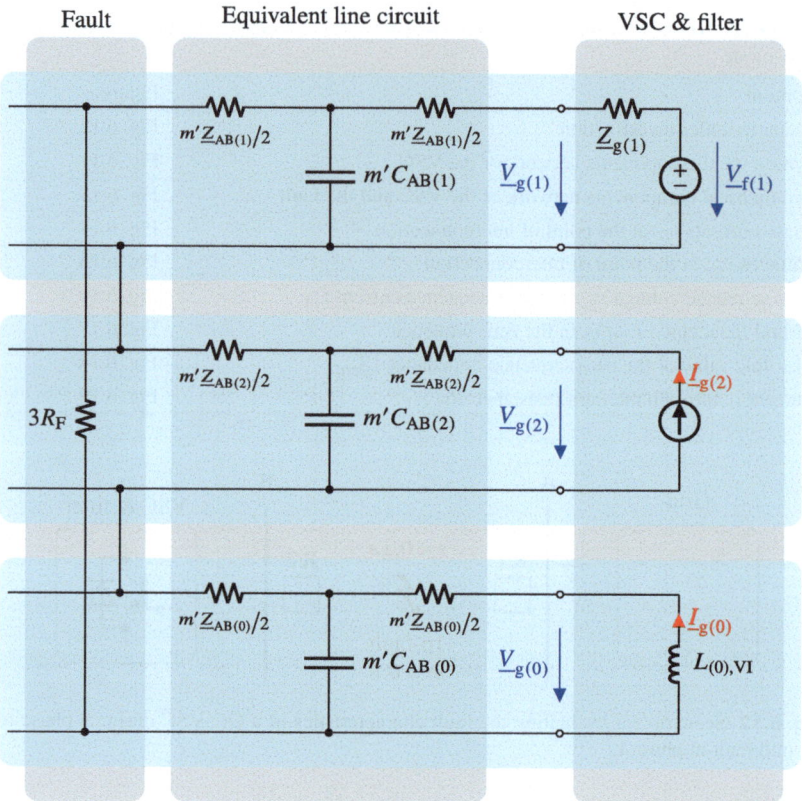

Fig. 6.13 Steady state symmetrical components equivalent circuit of the three-phase four-wire VSC applying a voltage source in the positive sequence (1), a current source in the negative sequence (2) and a virtual inductance in the zero sequence (0) during a single-phase-to-ground fault in phase 1. The relative fault distance $m' = (1 - m)$ seen from the point of interconnection at busbar B is used

multilevel topologies can avoid the excess of the voltage rating per semiconductor power device. Due to the increased voltages in the healthy phases, which is not a VSC-induced effect and also appears in conventional grids, this type of neutral point treatment is often not applicable in high voltage grids due to limited electrical field strength of components.

From Fig. 6.16 it is seen that the zero sequence current i_{g0} shows an amplitude of up to 800 A in the transient state. In the steady state, i_{g0} leads the zero sequence

6.2 Resilience of the Grid Operation

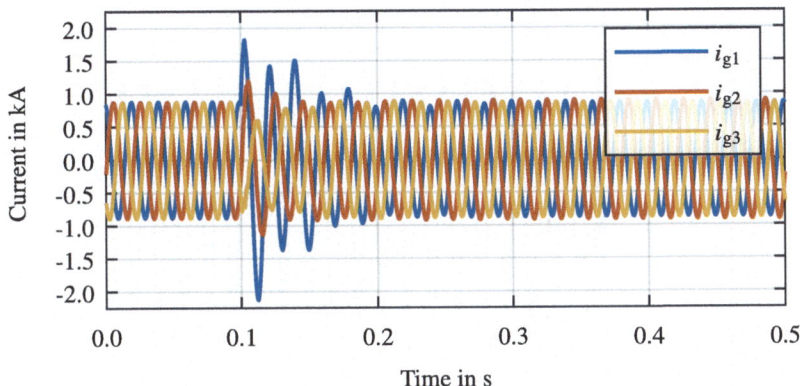

Fig. 6.14 Phase currents i_{gx} at the point of interconnection at busbar B before and after a single-phase-to-ground fault in phase 1 at $t = 0.1$ s. The VSC is operated in the RGS-mode

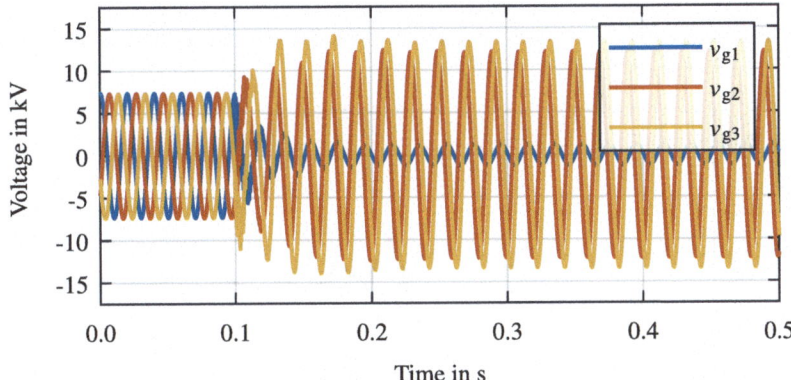

Fig. 6.15 Voltages v_{gx} at the point of interconnection at busbar B before and after a single-phase-to-ground fault in phase 1 at $t = 0.1$ s. The VSC is operated in the RGS-mode

voltage v_{g0} by $\pi/2$. Regarding the active sign convention in the zero sequence (see Fig. 6.5), it is demonstrated that a virtual inductance is realized.

Fig. 6.17 shows, that the reference value of the virtual inductance $L_{(0),VI}$ of 0.94 H is realized within a few period durations after the single-phase-to-ground fault is applied at $t = 0.1$ s.

The curves of Fig. 6.18 are calculated from $L_{(0),VI}$ at $t = 0.15$ s, $t = 0.2$ s and $t = 0.4$ s and the zero sequence capacitance and resistance from Eq. (6.6b). Since

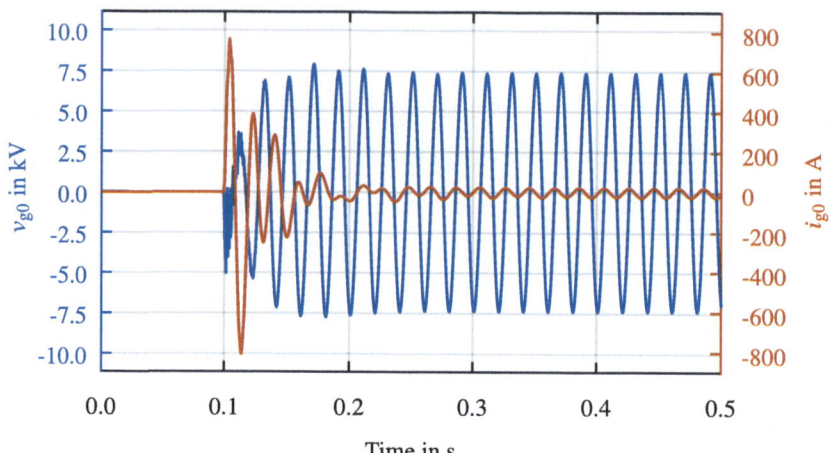

Fig. 6.16 Zero sequences of the voltages v_{gx} and the phase currents i_{gx} at the point of interconnection at busbar B before and after a single-phase-to-ground fault in phase 1 at $t = 0.1$ s. The VSC is operated in the RGS-mode

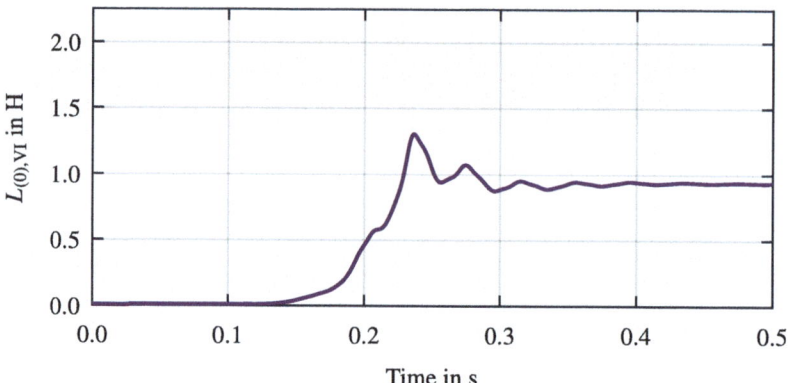

Fig. 6.17 Virtual inductance $L_{(0),VI}$ in the zero sequence at the point of interconnection at busbar B before and after a single-phase-to-ground fault in phase 1 at $t = 0.1$ s. The VSC is operated in the RGS-mode

6.2 Resilience of the Grid Operation

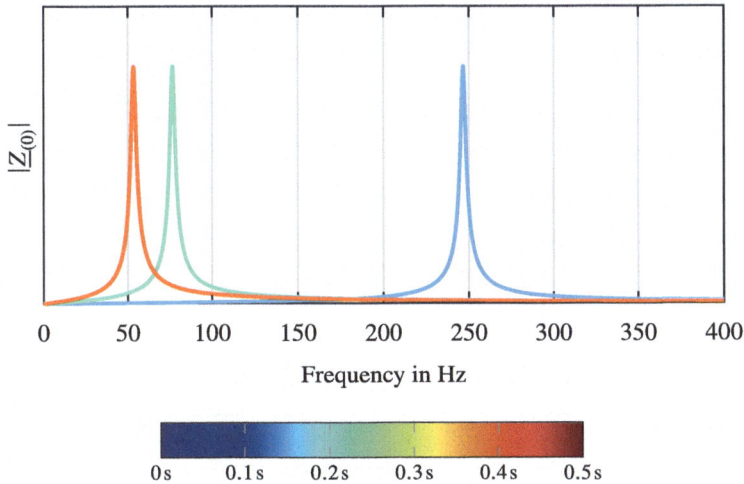

Fig. 6.18 Qualitative presentation of the absolute value of the impedance of the zero sequence $|\underline{Z}_{(0)}|$ at $t = 0.15$ s, $t = 0.2$ s and $t = 0.4$ s over frequency after a single-phase-to-ground fault in phase 1 at $t = 0.1$ s. The VSC is operated in the RGS-mode and the operation frequency is 50 Hz

Eq. (6.6b) is only valid regarding a time interval (see Sect. 3.1.1), this method is not thoroughly correct and consequently no values for the absolute value of the zero sequence impedance $|\underline{Z}_{(0)}|$ are given. However, Fig. 6.18 clearly demonstrates the operating principle of the RGS-mode and its dynamic behaviour. The curve of $|\underline{Z}_{(0)}|$ appears from the right side (equals high resonance frequency) by continuously increasing $L_{(0),VI}$, see Fig. 6.17. In this way, the resonance angle frequency ω_{res} is reduced (see Eq. (6.7)) and approaches the operating frequency of 50 Hz. The value of $|\underline{Z}_{(0)}|$ consequently increases. Depending on the breakdown field strength of the equipment in practical scenarios, it has to be decided, if the maximum of $|\underline{Z}_{(0)}|$ should be reached or the overcompensated interval left of the maximum should be set. In practical applications, often the overcompensated operation is preferred, since disconnections of grid areas evoking reducing the value of the zero sequence capacitance do not lead to the voltage maximum.

The Fig. 6.19 demonstrates the operation of the RGS-mode, the reduction of the fault current from both sides and the potential for the arc suppression. In the transient state, the amplitude of the current i_F through the fault resistance R_F (see Fig. 6.13) exceeds the value of 2 kA in this scenario. The amplitude is decreased due to the rising value of $|\underline{Z}_{(0)}|$. In the steady state, the amplitude of i_F is very low, see also Fig. 6.24.

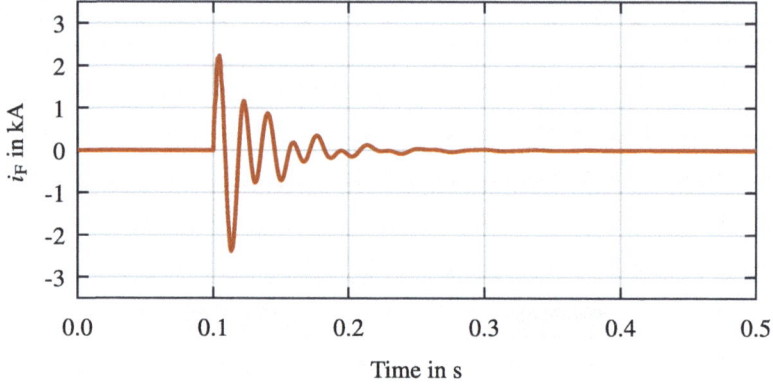

Fig. 6.19 Current i_F flowing through the fault resistance R_F before and after a single-phase-to-ground fault in phase 1 at $t = 0.1$ s. The VSC is operated in the RGS-mode

It is demonstrated that a VSC realizes a virtual inductance in the zero sequence applying the RGS-mode. In this way, a parallel resonant circuit in the zero sequence appears and the absolute value of its impedance is adjusted. After a single-phase-to-ground fault, symmetrical phase currents are reached. The amplitude of the fault current is reduced and the arc is suppressed. The resilience of the grid operation is increased. Moreover, the mean values of the phase currents are controlled to zero and VSCs are directly connected to the grid.

6.2.3.3 Scenario: Changing Zero Sequence Capacitance

In the following scenario, the advantages of the RGS-mode under changing zero sequence capacitance are presented via simulations in MATLAB/Simulink. Therefore, the scenario from Sect. 6.2.3.2 is extended after $t = 0.5$ s. Table 6.4 gives an overview of the relevant figures of this section.

6.2 Resilience of the Grid Operation

Table 6.4 Overview of the relevant figures of Sect. 6.2.3.3 containing the evaluation of the RGS-mode under changing zero sequence capacitance

Content	Figure
Scenario under investigation	Fig. 6.20
Symmetrical components network of the VSC	Fig. 6.11
Symmetrical components network of the VSC and the fault	Fig. 6.13
Phase currents i_{gx} at the point of interconnection	Fig. 6.21
Voltages v_{gx} at the point of interconnection	Fig. 6.22
Virtual inductance $L_{(0),VI}$ in the zero sequence	Fig. 6.23
Vertical zoom of the current i_F through the fault resistance R_F	Fig. 6.24

A single-phase-to-ground fault in phase 1 with the fault resistance $R_F = 0\,\Omega$ is applied at $t = 0.1\,\text{s}$. The disconnection of a subgrid at $t = 0.5\,\text{s}$ leads to the reduction of the zero sequence capacitance by 40%. The scenario is shown in Fig. 6.20.

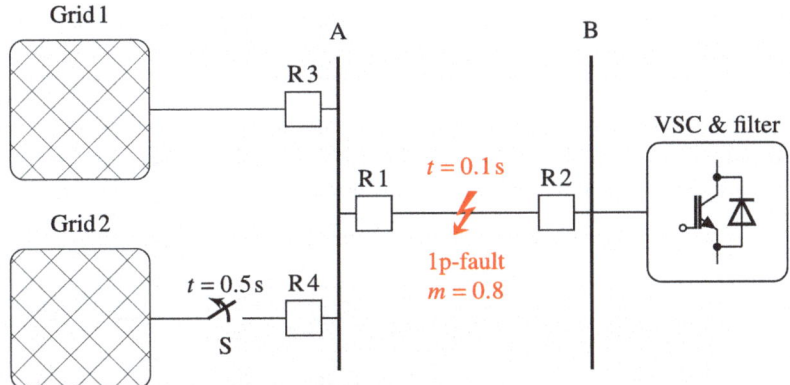

Fig. 6.20 Scenario for evaluating the fault characteristics of a VSC in the RGS-mode during a single-phase-to-ground fault in phase 1 at $t = 0.1\,\text{s}$. At $t = 0.5\,\text{s}$ a subgrid is disconnected by the switch S and the zero sequence capacitance is reduced

The parameters are identical to the ones in Sect. 6.2.3.2 and Fig. 6.11 and Fig. 6.13 are valid. The reference value of the virtual inductance of the zero sequence $L_{(0),VI}$ before $t = 0.5\,\text{s}$ equals $0.94\,\text{H}$. The disconnection of the subgrid leads to the detuning of the resonant circuit in the zero sequence and the reduction of the absolute value of the zero sequence impedance $|\underline{Z}_{(0)}|$. At $t = 0.7\,\text{s}$, the reference value of $L_{(0),\mathrm{VI}}$ is adjusted to 1.56 H and consequently $|\underline{Z}_{(0)}|$ rises again. The time interval between the disconnection and the adjustment of the reference value is chosen freely and can be smaller in practical scenarios. Furthermore, the actual zero sequence capacitance must—at least approximately—be known.

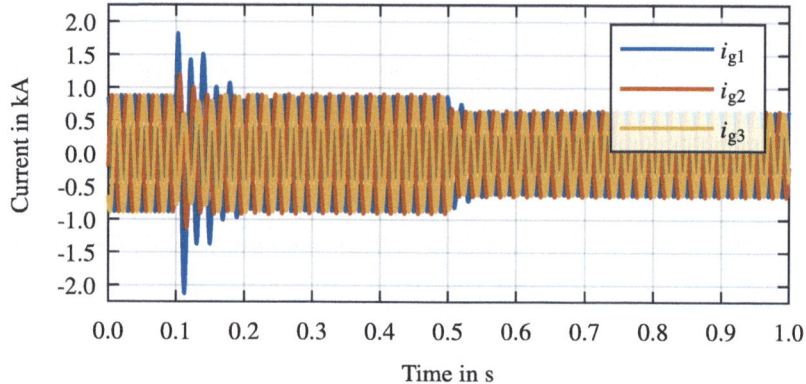

Fig. 6.21 Phase currents i_{gx} at the point of interconnection before and after a single-phase-to-ground fault in phase 1 at $t = 0.1$ s. At $t = 0.5$ s the zero sequence capacitance is reduced by 40% and at $t = 0.7$ s the reference value of the virtual inductance $L_{(0),\mathrm{VI}}$ is adjusted

Fig. 6.21 shows, that the disconnection does not increase the amplitude of the phase currents i_{gx} significantly. In the steady state after the disconnection, the load current is reduced due to the reduction of the power consumption.

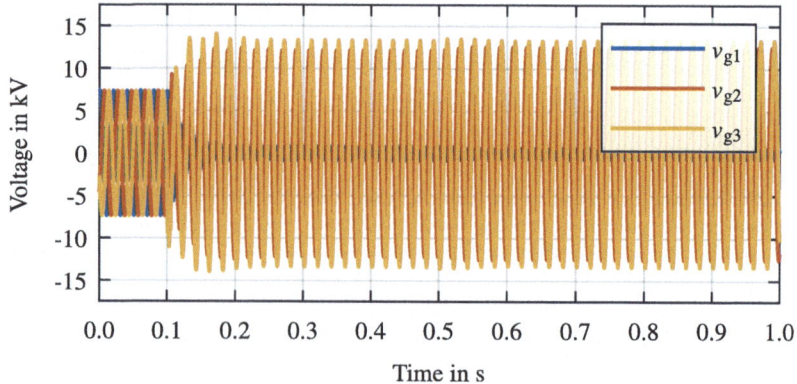

Fig. 6.22 Voltages v_{gx} at the point of interconnection before and after a single-phase-to-ground fault in phase 1 at $t = 0.1$ s. At $t = 0.5$ s the zero sequence capacitance is reduced by 40% and at $t = 0.7$ s the reference value of the virtual inductance $L_{(0),\mathrm{VI}}$ is adjusted

6.2 Resilience of the Grid Operation

It is presented in Fig. 6.22, that the amplitude of the voltages v_{gx} at the interconnection point is increased approx. by the factor $\sqrt{3}$ in the steady state.

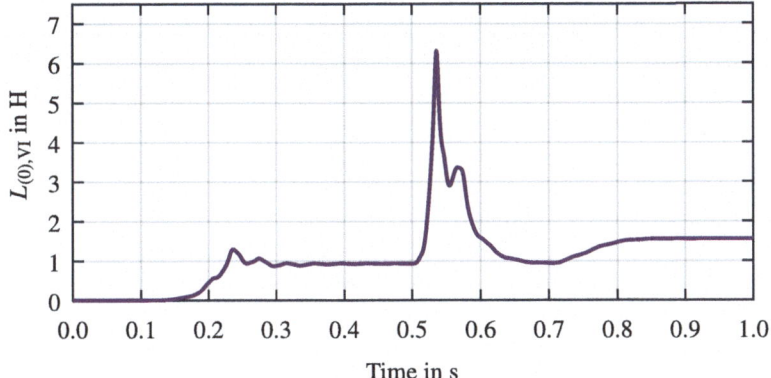

Fig. 6.23 Virtual inductance $L_{(0),VI}$ in the zero sequence at the point of interconnection before and after a single-phase-to-ground fault in phase 1 at $t = 0.1$ s. At $t = 0.5$ s the zero sequence capacitance is reduced by 40% and at $t = 0.7$ s the reference value of the virtual inductance $L_{(0),VI}$ is adjusted

The disconnection at $t = 0.5$ s (see Fig. 6.23) represents a disturbance to the virtual inductance $L_{(0),VI}$. The initial reference value of 0.94 H is realized within $t = 150$ ms after the disconnection. The new reference value of 1.56 H is reached within $t = 100$ ms after the adjustment of the reference value.

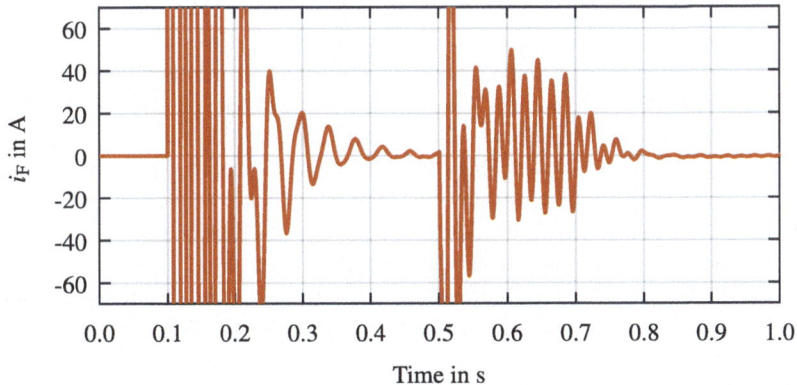

Fig. 6.24 Current i_F flowing through the fault resistance R_F before and after a single-phase-to-ground fault in phase 1 at $t = 0.1$ s. At $t = 0.5$ s the zero sequence capacitance is reduced by 40% and at $t = 0.7$ s the reference value of the virtual inductance $L_{(0),VI}$ is adjusted

From Fig. 6.24 it is seen, that at $t = 0.3$ s (200 ms after the fault), the amplitude of the current i_F through the fault resistance R_F (see Fig. 6.13) is below a typical value to get self-extinguished. This value is exceeded after the detuning of the resonant circuit. After the adjustment of the reference value of $L_{(0),VI}$, the amplitude decreases again.

Extending the conclusions of Sect. 6.2.3.2 it is demonstrated, that the value of the virtual inductance in zero sequence is adjusted at high speed. In this way, the parallel resonant circuit is adjusted to a changing zero sequence capacitance very fast. Even after the disconnection of a subgrid after a single-phase-to-ground fault, the grid operation can be continued.

6.2.3.4 Comparison to Conventional Methods and Perspective

Conventional methods for NPT via transformers or other NPT devices represent static methods due to their limited adjustment speed. Short-term low impedance grounding is initiated by discrete switching operations. In contrast, the NPT via VSCs shows very high adjustment speed (see Sect. 6.2.3.2—especially Fig. 6.23) and transitions from the different operating modes are realized continuously. The operation modes of the VSC (see Sect. 6.2.2.2) refer to the steady state.

The voltage and current characteristics of conventional methods differ significantly to NPT via VSCs. Before the fault occurs, the VSC approximately represents a short circuit, see Fig. 6.17. The VSC changes its characteristics in the zero sequence, if the amplitude of the measured zero sequence voltage v_{g0} is non-zero. In this way, the VSC shows intrinsic characteristics immediately after the fault, which are similar to reverse short-term low-impedance grounding, see Fig. 6.14. These characteristics appear in the UGS-mode, the RGS-mode and the SGS-mode [153]. Transient overvoltages are avoided in this low-impedance time interval.

The duration of the low-impedance time interval can for example be increased, if conventional distance protection algorithms are used for the fault localization. Consequently, highly dynamic coordination between the VSC NPT and the ground fault protection is conceivable. This coordination is not investigated in this thesis. Instead, a new protection algorithm (see Chap. 8), which is universally applicable for any fault characteristics, is developed. In this way, the fault characteristics of the VSC are optimized independently of the protection algorithm.

Some starting points of further research of the RGS-mode are given subsequently. The VSC allows the continuous adjustment of the zero sequence inductance $L_{(0),VI}$ at high speed as a so-called "inductance ride". This property could be used for grids, in which the zero sequence capacitance is not or only inaccurately known. Furthermore, a perturb-and-observe algorithm similar to photovoltaic plants for maximum power point tracking (see [117, 118]) could be developed, if the zero sequence capacitance is unknown. The distance from the operating point to the impedance maximum is adjusted (see also Fig. 6.18) and it can be determined, if the operating frequency is lower or higher than the resonance frequency. Additional ideas are discussed in Sect. 11.2.

6.3 Resilience of the VSC

The objective of the resilience of the VSC (see Sect. 6.1) is to stay connected to the grid and to avoid damages of the power semiconductor devices caused by overcurrents in the transient or steady state. In contrast to synchronous machines, the ability to allow overcurrents is very limited. The current limiting is essential for grid-forming (GFM) VSCs, since the components of the currents are not selected as controlled variables. In contrast to grid-following (GFL) VSCs, their currents are not limited inherently, see also Sect. 4.2.

To ensure a holistic current limiting concept, different states and consequently time intervals are considered. Different approaches are required for the transient state (see Sect. 6.3.1) and the steady state (see Sect. 6.3.2).

6.3.1 Transient State Current Limiting

6.3.1.1 Integration of the Cost Layer

State-of-the-art LQ-controllers are designed to minimize the control error and the actuating energy, see Eq. (5.51) in Sect. 5.3.4.2. Additional control objectives are not considered. The modulator calculates the gate signals from the actuating variables, see Fig. 5.1 and Fig. 2.2.

A novel method is developed, which allows adding further control objectives. A similar concept is presented in [158]. These additional control objectives can compete with the existing control objectives. The developed method allows prioritising different control objectives via cost functions. The hierarchical control concept from Fig. 5.1 is extended by the calculation of virtual costs for each switching state of the VSC. Based on these costs, the best possible switching state of the VSC minimizing the costs is chosen and the according gate signals are sent to power semiconductor devices, see Fig. 6.25.

The switching state **s** of the VSC is defined by the entirety of the switching functions s_x of the three half-bridges with $x \in \{1, 2, 3\}$, see Eq. (6.10).

$$\mathbf{s} = (s_1, s_2, s_3)^\mathrm{T} \tag{6.10}$$

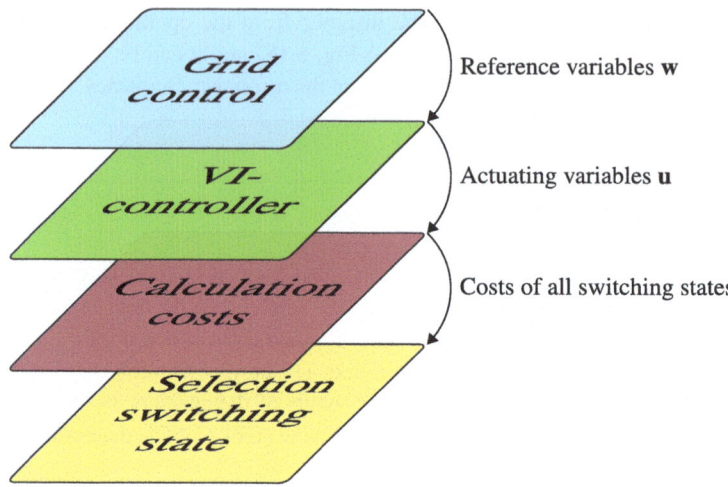

Fig. 6.25 Cost layer as a hierarchical layer between the V/I-controller and the selection of the switching state

Assuming a three-level VSC (see Fig. 6.3), the switching functions can take the following values: $s_x \in \{1, 0, -1\}$ and $3^3 = 27$ different switching states are obtained. The actuating variables in the natural reference frame are calculated assuming a constant and balanced DC-link voltage v_{DC}, see Eq. (6.11).

$$\mathbf{u}_{c123} = \mathbf{s}\frac{v_{DC}}{2} \qquad (6.11)$$

Each switching state **s** of the three-level VSC leads to a discrete voltage space vector $\underline{v}_c$ in the complex plane applying Eq. (3.13), see Fig. 6.26. Redundancies regarding $\underline{v}_c$ appear and 18 different voltage space vectors can be realized.

The method described above and in Fig. 6.25 works independently of the type and design of the V/I-controller. A LQ-controller (see Sect. 5.3.4.3 and Sect. 5.3.5) is applied. This concept allows formulating additional control objectives via cost functions for variables in the natural reference frame, the stationary reference frame or rotating reference frames. Consequently, the tracking of controlled variables representing DC-quantities in rotating reference frames and the limiting of state variables as arbitrary quantities in the natural reference frame in the transient state is possible.

6.3 Resilience of the VSC

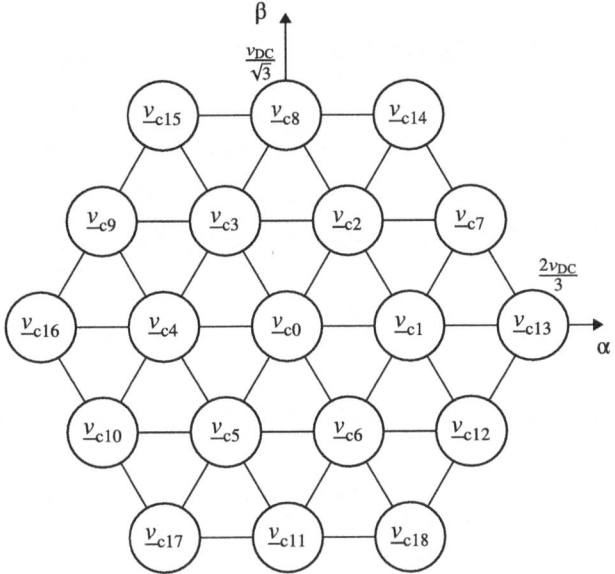

Fig. 6.26 Realizable discrete voltage space vectors $\underline{v}_c$ by a three-level VSC in the complex plane depending on the switching state **s**

The total cost function is designed from parts representing the reference tracking and the current limiting. The reference tracking is realized by the tracking of the actuating variables calculated by the V/I-controller. In this way, the V/I-controller is not adjusted. The Fig. 6.27 shows the cost functions for actuator tracking (AT) and current limiting (CL).

The point of intersection of these functions allows defining separate sections depending on the input parameters, in which AT or CL is predominant. This point of intersection is defined by different characteristics and slopes of the cost functions for AT and CL. The design of the cost function of AT is described in Sect. 6.3.1.2 and the cost function of CL is described in Sect. 6.3.1.3. The total cost function J_{VSC} is calculated with $x \in \{1, 2, 3\}$ for each switching state **s** according to Eq. (6.12).

$$J_{\text{VSC}} = J_{\text{AT}} + \sum_x J_{\text{CL},x} + \sum_x J_{\text{CLmin},x} \qquad (6.12)$$

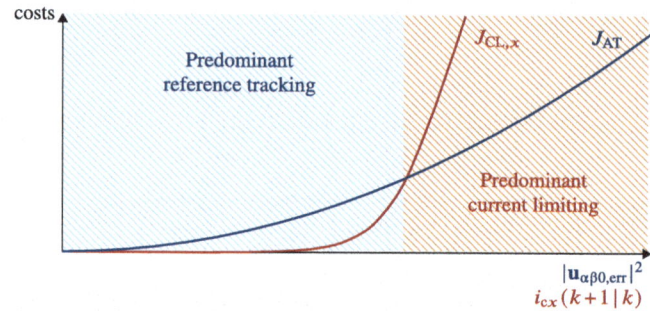

Fig. 6.27 Cost functions for actuator tracking (AT) and current limiting (CL)

Further research work can include the investigation of the total costs for the grid fault identification. The total costs can take high values due to high costs of CL. Alternatively, the temperature of power semiconductor devices are used to monitor the state of these devices via an additional cost function.

6.3.1.2 Cost Function for Actuator Tracking

The principle of the actuator tracking (AT) is to follow the calculated actuating variables by the V/I-controller. These variables are called the optimal actuating variables. The optimal actuating vector in the stationary reference frame $\mathbf{u}_{\alpha\beta 0,\text{opt}}$ contains the space vector components of the positive sequence ($v_{c\alpha}^+$ and $v_{c\beta}^+$), the space vector components of the negative sequence ($v_{c\alpha}^-$ and $v_{c\beta}^-$) and the zero sequence component $v_{c0\alpha}$ (see Sect. 6.2.2.1), see Eq. (6.13).

$$\mathbf{u}_{\alpha\beta 0,\text{opt}} = \begin{pmatrix} v_{c\alpha}^+ + v_{c\alpha}^- \\ v_{c\beta}^+ + v_{c\beta}^- \\ v_{c0\alpha} \end{pmatrix} \tag{6.13}$$

The positive sequence components are calculated according to Eq. (3.18) using the transformation matrix Eq. (3.19) according to Eq. (6.14). The negative sequence components are calculated from Eq. (6.15). The transformation angle γ_T^+ is calculated by the PLL-controller (see Sect. 4.3.1) using the decoupling cell from Sect. 4.3.3 or Sect. 4.3.4.1.

6.3 Resilience of the VSC

$$\begin{pmatrix} v_{c\alpha}^+ \\ v_{c\beta}^+ \end{pmatrix} = \mathbf{T}(-\gamma_T^+) \begin{pmatrix} v_{cd}^+ \\ v_{cq}^+ \end{pmatrix} \tag{6.14}$$

$$\begin{pmatrix} v_{c\alpha}^- \\ v_{c\beta}^- \end{pmatrix} = \mathbf{T}(\gamma_T^+) \begin{pmatrix} v_{cd}^- \\ v_{cq}^- \end{pmatrix} \tag{6.15}$$

The zero sequence components are calculated according to Eq. (6.16), see also Sect. 6.2.2.1. The actuating variable of the zero sequence equals the α-component $v_{c0\alpha}$.

$$\begin{pmatrix} v_{c0\alpha} \\ v_{c0\beta} \end{pmatrix} = \mathbf{T}(-\gamma_{T0}) \begin{pmatrix} v_{c0d} \\ v_{c0q} \end{pmatrix} \tag{6.16}$$

In general, it is not possible to realize the optimal actuating vector $\mathbf{u}_{\alpha\beta 0,\text{opt}}$ calculated by the V/I-controller due to the discrete switching states of the VSC. The difference between the optimal actuating vector and the real actuating vector $\mathbf{u}_{\alpha\beta 0,\text{err}}$ is calculated in accordance with Eq. (6.17) using Eq. (6.11) and the inverse transformation matrix from Eq. (3.12).

$$\mathbf{u}_{\alpha\beta 0,\text{err}} = \mathbf{u}_{\alpha\beta 0,\text{opt}} - \mathbf{T}_{\alpha\beta 0 \to 123}^{-1} \mathbf{s} \frac{v_{DC}}{2} \tag{6.17}$$

The cost function for the AT (see also Eq. (6.12) and Fig. 6.27) is designed to penalize $\mathbf{u}_{\alpha\beta 0,\text{err}}$ via the sum of its square elements, see Eq. (6.18). The factor k_{AT} is a design parameter.

$$J_{AT} = k_{AT} \left(\mathbf{u}_{\alpha\beta 0,\text{err}}^T \mathbf{u}_{\alpha\beta 0,\text{err}} \right) \tag{6.18}$$

6.3.1.3 Cost Function for Current Limiting

The principle of the current limiting (CL) is to predict the values of the phase currents i_{cx} with $x \in \{1, 2, 3\}$ in the natural reference frame and to penalize predicted values, whose absolute value exceed an upper limit [158]. For the prediction, the basic principle of the finite control set model predictive control (FCS-MPC, see [104]) is used. Consequently, the term finite control set model prediction (FCS-MP) is used subsequently.

The FCS-MP fits to the principle of the VSC realizing discrete switching states in so far, as it precalculates the system states from discrete-time state difference equations based on discrete actuating vectors. The models in the stationary reference frame (see Eq. (6.2)) and the zero sequence (see Eq. (3.43)) are used. Each actuating vector according to Eq. (6.19) is applied.

$$\mathbf{u}_{\alpha\beta 0}(k) = \mathbf{T}^{-1}_{\alpha\beta 0 \to 123}\mathbf{s}(k)\frac{v_{\text{DC}}(k)}{2} \tag{6.19}$$

The prediction at $t = k \cdot T_S$ with the sampling time T_S is presented subsequently. The predicted state variables at the next sampling point $k+1$ yield Eq. (6.20) (see also Eq. (3.46)). At $k+2$, the state variables are predicted in accordance with Eq. (6.21). In general, the law according to Eq. (6.22) is obtained.

$$\mathbf{x}(k+1 \mid k) = \mathbf{A}_\text{d}\mathbf{x}(k) + \mathbf{B}_\text{d}\mathbf{u}(k) + \mathbf{E}_\text{d}\mathbf{z}(k) \tag{6.20}$$

$$\mathbf{x}(k+2 \mid k) = \mathbf{A}_\text{d}\mathbf{x}(k+1 \mid k) + \mathbf{B}_\text{d}\mathbf{u}(k+1 \mid k) + \mathbf{E}_\text{d}\mathbf{z}(k+1 \mid k) \tag{6.21a}$$
$$= \mathbf{A}_\text{d}^2\mathbf{x}(k) + \mathbf{A}_\text{d}\mathbf{B}_\text{d}\mathbf{u}(k) + \mathbf{B}_\text{d}\mathbf{u}(k+1 \mid k) + \mathbf{A}_\text{d}\mathbf{E}_\text{d}\mathbf{z}(k) + \mathbf{E}_\text{d}\mathbf{z}(k+1 \mid k) \tag{6.21b}$$

$$\mathbf{x}(k+i \mid k) = \mathbf{A}_\text{d}^i\mathbf{x}(k) + \sum_{j=0}^{i-1}\mathbf{A}_\text{d}^{i-j-1}\mathbf{B}_\text{d}\mathbf{u}(k+j \mid k) + \sum_{j=0}^{i-1}\mathbf{A}_\text{d}^{i-j-1}\mathbf{E}_\text{d}\mathbf{z}(k+j \mid k) \tag{6.22}$$

This concept is visualized in Fig. 6.28. The prediction horizon n_p equals the number of predicted sampling points in the future and thus the length of the predicted time interval. The control horizon n_c equals the number of sampling points, in which

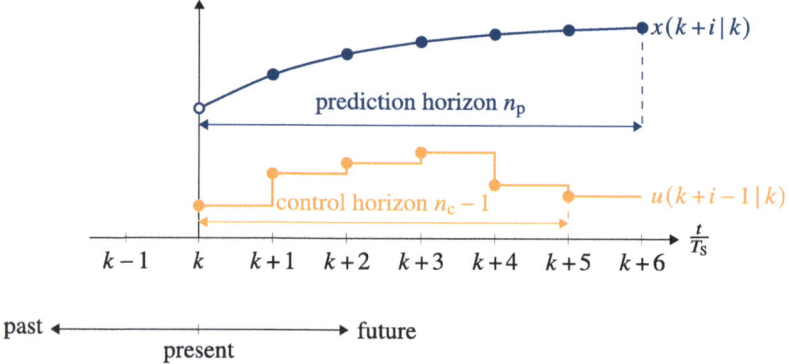

Fig. 6.28 Principle of the model predictive control (MPC) visualizing the prediction horizon n_p and the control horizon n_c

6.3 Resilience of the VSC

the actuating vector is varied to change the state variables in the respective next sampling point. Consequently, the control horizon is always less than or equal to the prediction horizon. In Fig. 6.28, $n_p = n_c = 6$ is chosen exemplary.

The predicted state variables contain the components of the predicted phase currents $\mathbf{i}_{c\alpha\beta 0}$, see also Fig. 6.3. These components are transformed into the natural reference frame using Eq. (3.12) and i_{cx} are obtained. Subsequently, the prediction horizon $n_p = 1$ is chosen.

The cost function $J_{\text{CL},x}$ (see also Eq. (6.12) with $x \in \{1, 2, 3\}$ and Fig. 6.27) is used to penalize high values of i_{cx}. The variables $i_{cx}(k+1 \mid k)$ represent the precalculated currents at $t = k \cdot T_S$, see Eq. (6.23).

$$J_{\text{CL},x} = \frac{k_{1,\text{CL}}}{1 + \exp\left(-k_{3,\text{CL}}(i_{cx}(k+1 \mid k) - k_{2,\text{CL}})\right)} \tag{6.23}$$

The costs of currents of infinite high values are described by the parameter $k_{1,\text{CL}}$, see Eq. (6.24). The parameter $k_{2,\text{CL}}$ determines the value of the currents with half of the maximum costs, see Eq. (6.25). The slope of the sigmoid curve is characterized by the parameter $k_{3,\text{CL}}$, see Eq. (6.26).

$$k_{1,\text{CL}} = J_{\text{CL},x}\left(i_{cx}(k+1 \mid k) \to \infty\right) \tag{6.24}$$

$$J_{\text{CL},x}\left(i_{cx}(k+1 \mid k) = k_{2,\text{CL}}\right) = \frac{k_{1,\text{CL}}}{2} \tag{6.25}$$

$$J_{\text{CL},x}\left(i_{cx}(k+1 \mid k) = i_{cx,\text{set}}\right)$$
$$= \frac{k_{1,\text{CL}}}{1 + \exp(-k_{3,\text{CL}}(i_{cx,\text{set}} - k_{2,\text{CL}}))} \tag{6.26a}$$
$$\stackrel{!}{=} p k_{1,\text{CL}} \tag{6.26b}$$

The parameter p determines the costs $J_{\text{CL},x} = p k_{1,\text{CL}}$ at a predefined value $i_{cx,\text{set}}$. The Eq. (6.27) to calculate $k_{3,\text{CL}}$ is obtained.

$$k_{3,\text{CL}} = -\frac{\ln\left(\frac{1}{p} - 1\right)}{i_{cx,\text{set}} - k_{2,\text{CL}}} \tag{6.27}$$

The cost function for maximum currents with a negative sign $J_{\text{CLmin},x}$ (see also Eq. (6.12)) and the tuning parameters are defined in an analogous way.

6.3.1.4 Space Vector Modulation for Constant Switching Frequency

Constant switching frequency of each power semiconductor device within one fundamental grid period is crucial to gain even thermal stress on these switching devices. In addition, constant switching frequency leads to a clearer frequency spectrum of the phase currents and capacitor voltages [92]. This property helps to design the parameters of the LCL-filter [143]. Using the principle of Sect. 6.3.1.1, Sect. 6.3.1.2 and Sect. 6.3.1.3, the switching frequency per power semiconductor device is not constant.

In this section, a method is described to realize constant switching frequency of each power semiconductor device. The idea is to select an optimal switching sector in the complex plane instead of selecting an optimal switching state. The duty cycles of the surrounding space vectors at the corners of this switching sector are calculated from the corresponding costs of according switching states. The general idea of combining the FCS-MPC and the space vector modulation (SVM) is presented in [119, 120] and the term modulated MPC is used. The Fig. 6.29 extends Fig. 6.26 by

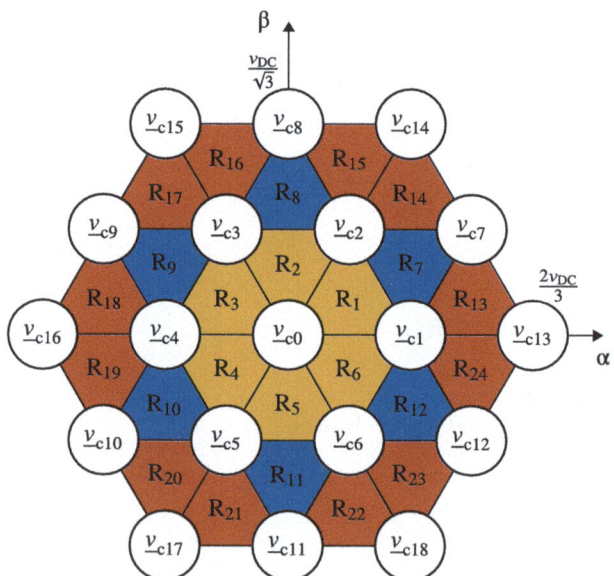

Fig. 6.29 Realizable discrete voltage space vectors $\underline{v}_c$ by a three-level VSC in the complex plane and switching sectors R_j

6.3 Resilience of the VSC

introducing the 24 switching sectors R_j. These switching sectors are divided into three groups, which are highlighted by different colours according to the absolute value of the surrounding voltage space vectors.

Adopting this concept to the control strategies of this thesis, the selection layer in Fig. 6.25 is modified. However, the principles of Sect. 6.3.1.1, Sect. 6.3.1.2 and Sect. 6.3.1.3 are not affected.

An optimization problem is solved to calculate the optimal switching sector, see Eq. (6.28) [119]. In this context, the variable j equals the number of the switching sector and i equals the number of the voltage space vector at the corners of the sector j. The costs of the switching states g_{ij} and the duty cycles d_{ij} are used.

$$\min \ G_j = \sum_{i \in R_j} g_{ij} d_{ij}^2 \tag{6.28}$$

The optimization problem Eq. (6.28) is solved considering the constraints in Eq. (6.29) and Eq. (6.30) [119].

$$\sum_{i \in R_j} d_{ij} = 1 \tag{6.29}$$

$$0 \le d_{ij} \le 1 \tag{6.30}$$

The duty cycles d_{ij} are calculated in accordance with Eq. (6.31) [119]. In Eq. (6.31), the variable Q_j is used, see Eq. (6.32) [119].

$$d_{ij} = \frac{Q_j}{g_{ij}} \tag{6.31}$$

$$Q_j = \frac{1}{\sum_{i \in R_j} g_{ij}^{-1}} \tag{6.32}$$

After solving the optimization problem given in Eq. (6.28), the optimal switching sector is chosen. The duty cycles of the surrounding space vectors at the corners of this switching sector are used to generate the gate signals of the semiconductor power devices. The redundancy of the voltage space vector $\underline{v}_{c0}$ is used to realize the actuating variable $v_{c0\alpha}$ of the zero sequence.

Optionally, this concept allows arranging the switching states of one switching period to reach an optimized pulse pattern without affecting neither the duty cycles nor the switching frequency within one grid period. For example, the harmonic performance can be increased, or the switching losses can be reduced [121].

6.3.1.5 Scenario: Transient Current Limiting

In the following scenario, the operating principle of the RGS-mode using the virtual inductance (see Sect. 6.2.3) in the zero sequence with transient current limiting and avoiding non-zero mean values of the phase currents (see Sect. 6.2.1.2) is demonstrated via simulations in MATLAB/Simulink. The investigated scenario equals the scenario from Sect. 6.2.3.2 in general. In contrast to Sect. 6.2.3.2, another operating point is chosen and the constant switching frequency according to Sect. 6.3.1.1, Sect. 6.3.1.2, Sect. 6.3.1.3 and Sect. 6.3.1.4 is activated. In Sect. 6.2.3.2 the phase currents i_{gx} are increased approximately by a factor up to 2 in the transient state immediately after the fault occurs, see Fig. 6.14. The objective of the RGS-mode with transient current limiting is to stay connected to the grid and to avoid damages of the power semiconductor devices by limiting the currents in the transient state.

The investigated scenario is shown in Fig. 6.12. A VSC is connected via an overhead line between the busbars A and B to a 10 kV-grid. The overhead line has a length of 10 km. A single-phase-to-ground fault in phase 1 with the fault resistance $R_F = 0\,\Omega$ is applied at $t = 0.1$ s. The electrical distance between the VSC and the fault equals 2 km. The symmetrical components network is shown in Fig. 6.13.

The parameters of the three-phase four-wire VSC are equal to the parameters in Sect. 6.2.3.2 and the mean values of the phase currents i_{gx} are controlled to zero according to Sect. 6.2.1.2. The current limit of the phase currents is set to ± 600 A via the parameter $k_{2,\mathrm{CL}}$, see Eq. (6.25). This limit can be adjusted and should be in accordance with the allowed current of the power semiconductor devices.

The phase currents i_{gx} are presented in Fig. 6.30. The predefined current limits are met and overcurrents are avoided. The transient current limiting is activated near the predefined current limits if overshooting is predicted. During these short time intervals, the sinusoidal characteristics are left and harmonic components appear. Switching states of the VSC, that would further increase the current of the concerning phase, are avoided.

The Fig. 6.31 shows the voltages v_{gx} at the point of interconnection. Similar to Fig. 6.15, the amplitudes of the healthy phases are increased approximately by the factor $\sqrt{3}$ in the steady state.

Extending the conclusions of Sect. 6.2.3.2 and Sect. 6.2.3.3 it is demonstrated, that the transient current limiting of a VSC works reliably after a single-phase-to-ground fault and overcurrents are avoided. In this way, the resilience of VSCs is increased. VSCs stay connected to the grid and damages of the power semiconductor devices are avoided.

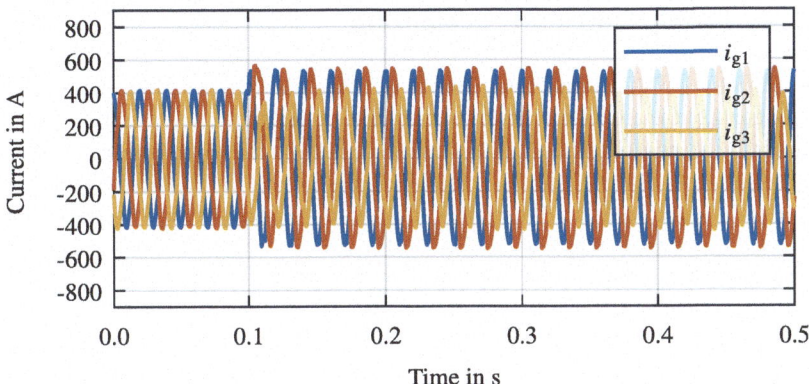

Fig. 6.30 Phase currents i_{gx} at the point of interconnection at busbar B before and after a single-phase-to-ground fault in phase 1 at $t = 0.1$ s. The VSC is operated in the RGS-mode with transient current limiting

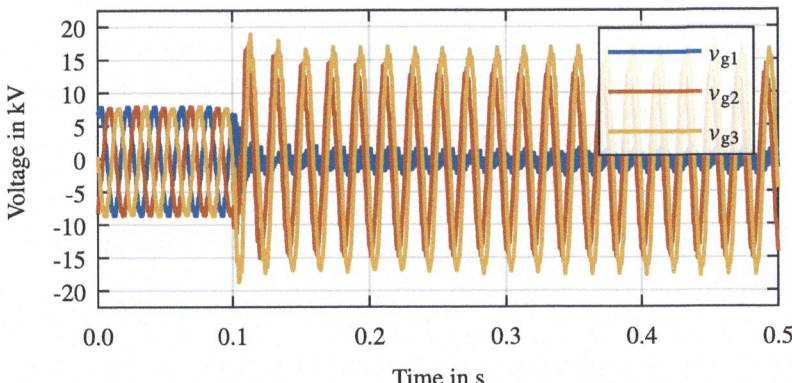

Fig. 6.31 Voltages v_{gx} at the point of interconnection at busbar B before and after a single-phase-to-ground fault in phase 1 at $t = 0.1$ s. The VSC is operated in the RGS-mode with transient current limiting

6.3.2 Steady State Current Limiting

The objective of the steady state current limiting is to check before sending the gate signals, if the aimed operating point consisting of characteristics in the positive sequence, the negative sequence and the zero sequence can be reached without exceeding current limits. The presented concept is to precalculate the phase currents i_{gx} in the steady state in a three-phase four-wire system (see also Fig. 6.3). The currents i_{gx} approximately equal the currents i_{cx} in the steady state at the operating frequency. The calculation is based on the description of space vectors because their components are defined as controlled variables, see Sect. 5.3. In the steady state at constant frequency, the space vectors are described by phasors, see Eq. (3.17) and Eq. (3.24). Therefore, the frequency in the subsequent calculations has to be adjusted to the actual frequency f_{vg} calculated by a PLL-controller, see Sect. 4.3.1. An additional advantage of this concept is, that the DC-link voltage can be designed based on the calculations of $\underline{\hat{V}}_{cx}$ and the operating principle of VSCs.

The prerequisites of this concept are the measurement of v_{gx} (see Fig. 6.3) and the calculation the voltage phasors of the positive sequence, the negative sequence and the zero sequence. The principles of Sect. 4.3.3 and Sect. 4.3.4.1 are used to calculate the absolute value and the phase angle of the phasors.

Some alternatives to the presented concept for symmetrical disturbances are discussed in [122]. Here, current limiters, voltage limiters and virtual impedances (VIs) in the positive sequence are investigated. The latter principle is based on the concept presented in Sect. 5.2.4.2, preferably without affecting the X/R-ratio (see Sect. 4.1.1.3). This principle can be interpreted as the increase of the electrical distance between the VSC and the fault location, or the reduction of the amplitude of the actuating voltage. The problem of VIs is, that the fault location is unknown and thus the design of the virtual parameters is difficult.

A similar concept to the presented concept in this thesis is given in [77]. Since the consideration of the zero sequence is emphasized in this thesis, the principles of [77] are extended and the zero sequence is included in the steady state current limiting concept presented in the following sections.

6.3.2.1 Calculation of the Time Signals from the Space Vector and the Zero Sequence Component

The basic idea of the steady state current limiting is the calculation of the time signals from the current space vector and the zero sequence current. The information of the positive sequence, the negative sequence and the zero sequence are combined.

The space vector regarding positive and negative sequence components in the steady state (see also Eq. (3.24)) represents the starting point of the following considerations, see Eq. (6.33). The zero sequence current is described in accordance with Eq. (6.34).

$$\underline{i} = \hat{I}_{(1)}e^{j(\omega t + \varphi_{i(1)})} + \hat{I}_{(2)}e^{-j(\omega t + \varphi_{i(2)})} \tag{6.33}$$

$$i_0 = \hat{I}_{(0)}\cos(\omega t + \varphi_{i(0)}) \tag{6.34}$$

The phase currents are calculated in Eq. (6.35), see also [76] with $x \in \{1, 2, 3\}$ and the complex number $\underline{a}$ in Eq. (3.8).

$$i_x = \text{Re}\{\underline{i}(\underline{a}^{(x-1)})^*\} + i_0 \tag{6.35}$$

For reasons of clarity, the following considerations are limited to phase 1. Analogously, phase 2 and phase 3 are investigated. The phase current of phase 1 yield Eq. (6.36).

$$i_1 = \text{Re}\{\underline{i}\} + i_0 \tag{6.36}$$

The real part of the space vector is calculated according to Eq. (6.37) [75].

$$\text{Re}\{\underline{i}\} = \hat{I}_{(1)}\cos(\omega t + \varphi_{i(1)}) + \hat{I}_{(2)}\cos(-\omega t - \varphi_{i(2)}) \tag{6.37a}$$

$$= \hat{I}_{(1)}\cos(\omega t + \varphi_{i(1)}) + \hat{I}_{(2)}\cos(\omega t + \varphi_{i(2)}) \tag{6.37b}$$

$$= \hat{I}_{(1)}\sin\left(\omega t + \varphi_{i(1)} + \frac{\pi}{2}\right) + \hat{I}_{(2)}\sin\left(\omega t + \varphi_{i(2)} + \frac{\pi}{2}\right) \tag{6.37c}$$

$$= \sqrt{\hat{I}_{(1)}^2 + \hat{I}_{(2)}^2 + 2\hat{I}_{(1)}\hat{I}_{(2)}\cos(\varphi_{i(1)} - \varphi_{i(2)})}\sin\left(\omega t + \widetilde{\varphi}_{i1} + \frac{\pi}{2}\right) \tag{6.37d}$$

$$= \widetilde{I}_1\cos(\omega t + \widetilde{\varphi}_{i1}) \tag{6.37e}$$

The amplitude $\tilde{I}_1$ appears in Eq. (6.38). The phase angle $\tilde{\varphi}_{i1}$ is given in Eq. (6.39).

$$\tilde{I}_1 = \sqrt{\hat{I}_{(1)}^2 + \hat{I}_{(2)}^2 + 2\hat{I}_{(1)}\hat{I}_{(2)}\cos(\varphi_{i(1)} - \varphi_{i(2)})} \quad (6.38)$$

$$\tilde{\varphi}_{i1} = \operatorname{atan2}\left(\left[\hat{I}_{(1)}\sin(\varphi_{i(1)}) + \hat{I}_{(2)}\sin(\varphi_{i(2)})\right], \left[\hat{I}_{(1)}\cos(\varphi_{i(1)}) + \hat{I}_{(2)}\cos(\varphi_{i(2)})\right]\right) \quad (6.39)$$

The addition of the zero sequence current leads to the phase current i_1, see Eq. (6.40).

$$\begin{aligned}
i_1 &= \tilde{I}_1\cos(\omega t + \tilde{\varphi}_{i1}) + \hat{I}_{(0)}\cos(\omega t + \varphi_{i(0)}) & (6.40\text{a}) \\
&= \tilde{I}_1\sin\left(\omega t + \tilde{\varphi}_{i1} + \frac{\pi}{2}\right) + \hat{I}_{(0)}\sin\left(\omega t + \varphi_{i(0)} + \frac{\pi}{2}\right) & (6.40\text{b}) \\
&= \sqrt{\tilde{I}_1^2 + \hat{I}_{(0)}^2 + 2\tilde{I}_1\hat{I}_{(0)}\cos(\tilde{\varphi}_{i1} - \varphi_{i(0)})}\sin\left(\omega t + \varphi_{i1} + \frac{\pi}{2}\right) & (6.40\text{c}) \\
&= \hat{I}_1\cos(\omega t + \varphi_{i1}) & (6.40\text{d})
\end{aligned}$$

The amplitude of i_1 is calculated in Eq. (6.41). The phase angle of i_1 is given in Eq. (6.42).

$$\hat{I}_1 = \sqrt{\tilde{I}_1^2 + \hat{I}_{(0)}^2 + 2\tilde{I}_1\hat{I}_{(0)}\cos(\tilde{\varphi}_{i1} - \varphi_{i(0)})} \quad (6.41)$$

$$\varphi_{i1} = \operatorname{atan2}\left(\left[\tilde{I}_1\sin(\tilde{\varphi}_{i1}) + \hat{I}_{(0)}\sin(\varphi_{i(0)})\right], \left[\tilde{I}_1\cos(\tilde{\varphi}_{i1}) + \hat{I}_{(0)}\cos(\varphi_{i(0)})\right]\right) \quad (6.42)$$

Using Eq. (6.38), Eq. (6.39) and Eq. (6.34), the maximum value of i_1 is calculated using Eq. (6.41).

6.3 Resilience of the VSC

The characteristics of the space vector in Eq. (6.33) depend on the control characteristics of the positive sequence and the negative sequence. The direct power control (see Sect. 6.3.2.2) and the indirect power control (see Sect. 6.3.2.3) have to be distinguished. Combinations are possible, too. The zero sequence current in Eq. (6.34) using the RGS-mode (see Sect. 6.2.3.1) is calculated from Eq. (6.9) and applying the concepts of Sect. 4.3.3 to gain the amplitude and the phase angle.

6.3.2.2 Current Space Vector During Direct Power Control

Direct power control characteristics appear, if the active and the reactive power are controlled in the positive sequence or the negative sequence. For example, the hybrid synchronization (see Sect. 4.5.1) and the P/Q-control (see Sect. 5.2.3.1) lead to the direct power control. Especially, the reactive current injection in the negative sequence (see Sect. 5.2.3.2) leads to direct power control characteristics.

According to Eq. (5.9), Eq. (5.10) and Eq. (5.11), the Eq. (6.43) reveals.

$$\underline{i} = \frac{2}{3}\frac{P_{(1)}}{|\underline{v}^+|^2}\underline{v}^+ - j\frac{2}{3}\frac{Q_{(1)}}{|\underline{v}^+|^2}\underline{v}^+ + \frac{2}{3}\frac{P_{(2)}}{|\underline{v}^-|^2}\underline{v}^- - j\frac{2}{3}\frac{Q_{(2)}}{|\underline{v}^-|^2}\underline{v}^- \quad (6.43a)$$

$$= (C_1 - jC_3)\underline{v}^+ + (C_2 - jC_4)\underline{v}^- \quad (6.43b)$$

The following factors according to [77] are used, see Eq. (6.44), Eq. (6.45), Eq. (6.46) and Eq. (6.47).

$$C_1 = \frac{2}{3}\frac{P_{(1)}}{|\underline{v}^+|^2} \quad (6.44)$$

$$C_2 = \frac{2}{3}\frac{P_{(2)}}{|\underline{v}^-|^2} \quad (6.45)$$

$$C_3 = \frac{2}{3}\frac{Q_{(1)}}{|\underline{v}^+|^2} \quad (6.46)$$

$$C_4 = \frac{2}{3}\frac{Q_{(2)}}{|\underline{v}^-|^2} \quad (6.47)$$

It becomes obvious, that these factors are similar to the active currents and reactive currents given in Eq. (5.12), Eq. (5.13), Eq. (5.14) and Eq. (5.15).

Translating Eq. (6.43b) to Eq. (6.33), the components of the positive sequence space vector reveal, see Eq. (6.48) and Eq. (6.49). Likewise, the components of the negative sequence space vector appear, see Eq. (6.50) and Eq. (6.51).

$$\hat{I}_{(1)} = \hat{V}_{(1)}\sqrt{C_1^2 + C_3^2} \qquad (6.48)$$

$$\varphi_{i(1)} = \varphi_{v(1)} - \operatorname{atan2}(C_3, C_1) \qquad (6.49)$$

$$\hat{I}_{(2)} = \hat{V}_{(2)}\sqrt{C_2^2 + C_4^2} \qquad (6.50)$$

$$\varphi_{i(2)} = \varphi_{v(2)} - \operatorname{atan2}(C_4, C_2) \qquad (6.51)$$

Using the pair Eq. (6.48)/Eq. (6.49) and the pair Eq. (6.50)/Eq. (6.51), the maximum values of the phase currents are calculated according to Sect. 6.3.2.1.

6.3.2.3 Current Space Vector During Indirect Power Control

Indirect power control characteristics appear, if the active and the reactive power are controlled indirectly in the positive sequence or the negative sequence, e.g. by voltage control (see Sect. 4.4). Especially, the configurable natural droop controller (see Sect. 4.4.1.2) leads to indirect power control characteristics.

The Eq. (6.33) shows the correlation of the space vector and the sequence phasor in the steady state. If only one frequency component is regarded in the steady state, the calculation of symmetrical components equivalent circuits using phasors is sufficient. Based on the symmetrical components, phasors of $\underline{\hat{V}}_{gx}$ and the symmetrical components phasors of $\underline{\hat{V}}_{fx}$ representing the reference values of the controlled variables (see also Eq. (4.54b) and Eq. (4.71)), the symmetrical components phasors of $\underline{\hat{I}}_{gx}, \underline{\hat{I}}_{cx}$ and $\underline{\hat{V}}_{cx}$ are calculated. Therefore, the Fig. 6.32 shows the steady state symmetrical components equivalent circuit of a three-phase four-wire LCL-structure based on Fig. 3.6. The abbreviations from Eq. (3.38), Eq. (3.39), Eq. (3.44) and Eq. (3.45) are used.

From the symmetrical components phasors of $\underline{\hat{I}}_{gx}$, the components of the positive sequence space vector reveal and the maximum values of the phase currents are calculated according to Sect. 6.3.2.1.

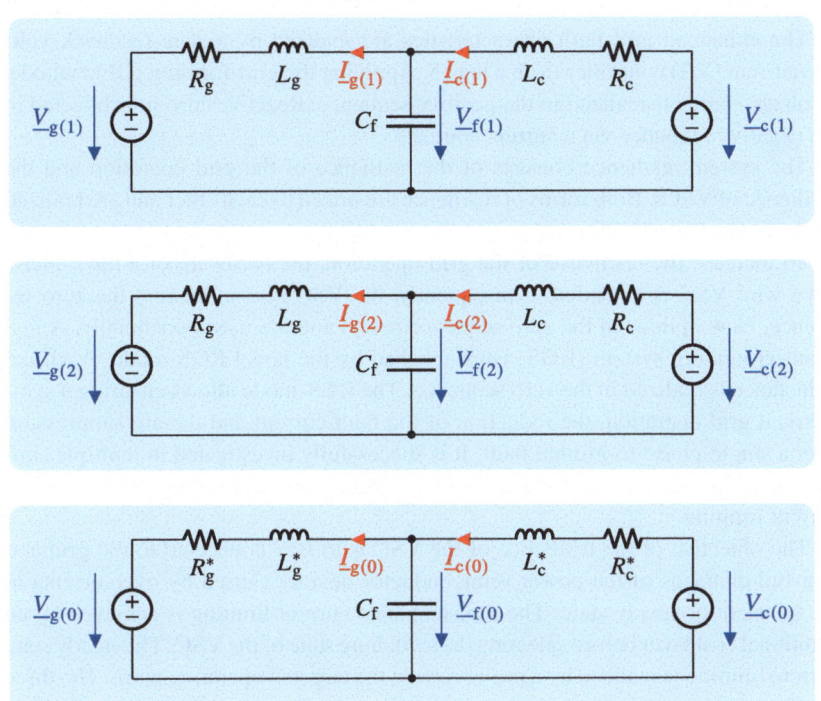

Fig. 6.32 Steady state symmetrical components equivalent circuit of a three-phase four-wire LCL-structure connecting two voltage sources

6.4 Summary

In this chapter, the realization of enhanced grid fault characteristics of voltage source converters (VSCs) is presented. These fault characteristics lead to a resilient power system and are a prerequisite for the operation of a 100% inverter-based power system. A reliable supply of energy is ensured.

The enhanced grid fault characteristics are realized by a state-feedback voltage/current (V/I) controller from Chap. 5. Applying the grid-forming (GFM) mode, a voltage source is realized in the positive sequence. Reactive current is injected in the negative sequence via a current source.

The system resilience consists of the resilience of the grid operation and the resilience of VSCs. Both forms of resilience are linked to each other and are realized by the controller of VSCs.

To increase the resilience of the grid operation, the hardware of a three-phase three-wire VSC is extended. Consequently, the VSC also influences the zero sequence. New options in the zero sequence reveal and are used beneficially. A resonant grounding system (RGS) is established by the novel RGS-mode. A virtual inductance is realized in the zero sequence. The RGS-mode allows ensuring a symmetrical grid operation, the reduction of the fault current and the arc suppression after a single-phase-to-ground fault. It is successfully investigated in multiple simulation scenarios, e.g. applying changing zero sequence capacitance and transient current limiting.

The objective of the resilience of the VSC is to stay connected to the grid and to avoid damages of the power semiconductor devices caused by overcurrents in the transient or steady state. The transient state current limiting is achieved by an additional cost layer before selecting the switching state of the VSC. The steady state current limiting is realized by a pre-review of the targeted operating point. The three steady state symmetrical components equivalent circuits are used to precalculate the current amplitudes.

The achieved V/I fault characteristics differ from conventional fault characteristics of synchronous machines. The compatibility of these fault characteristics and existing protection algorithms is not guaranteed. Therefore, a novel universal protection algorithm is presented in Chap. 8 based on state-space models derived in Chap. 7. To follow the key aspects of the central theme of this thesis on a shortcut: The conclusion of the next chapter is provided in Sect. 7.5.

6.4 Summary

Open Access This chapter is licensed under the terms of the Creative Commons Attribution 4.0 International License (http://creativecommons.org/licenses/by/4.0/), which permits use, sharing, adaptation, distribution and reproduction in any medium or format, as long as you give appropriate credit to the original author(s) and the source, provide a link to the Creative Commons license and indicate if changes were made.

The images or other third party material in this chapter are included in the chapter's Creative Commons license, unless indicated otherwise in a credit line to the material. If material is not included in the chapter's Creative Commons license and your intended use is not permitted by statutory regulation or exceeds the permitted use, you will need to obtain permission directly from the copyright holder.

7 Grid Fault Modelling

This chapter presents the grid fault modelling as the basic prerequisite of the model-based protection algorithm. In Sec. 7.1, the model-based fault analysis is defined. The basic concept of the model-based protection algorithm is introduced and compared to existing protection schemes. State-space models of selected grid topologies are derived in Sec. 7.2. Rules for the formulation of state-space models of more complex grid topologies are identified. The modelling of transverse faults and longitudinal faults is discussed in Sec. 7.3. A methodical approach shows different fault resistance configurations. Sec. 7.4 contains the modelling of lines. State-space models of a healthy line and faulty lines are developed.

7.1 Model-Based Fault Analysis

If a grid fault appears in an electrical power system, the characteristics of voltages and currents can differ from their characteristics during normal operation conditions. The characteristics of voltages and currents are analysed and decisions for the tripping of circuit breakers to protect living beings and hardware equipment are taken. The fault analysis is for example based on the comparison of measured values with predefined limits or calculations using measured values aided by models reflecting physical laws. A binary decision for the tripping of a circuit breaker is made from this comparison or this calculation. The fault clearing is executed during the zero crossing of the respective phase current.

The fault analysis and the derived decision is called the protection scheme. Subsequently, three different conventional protection schemes are categorized. The overcurrent protection implies, that the tripping of the circuit breaker is initiated if a current limit is exceeded.

© The Author(s) 2025
F. Mahr, *Control and Protection of 100% Inverter-based Power Systems*,
https://doi.org/10.1007/978-3-658-47217-7_7

The distance protection calculates the impedances of current loops based on Kirchhoff's voltage law (KVL), see [41, 143]. Consequently, it is based on a simple electrical model. The Fig. 7.1 shows a generalized model representing a fault current loop of a single-phase-to-ground fault in phase x with $x \in \{1, 2, 3\}$ used by the distance protection scheme, see also Fig. 7.12 assuming the fault resistance R_F. In this example, the loop impedance is calculated from the measured variables $\underline{V}_{Ax}$ and $\underline{I}_{ax}$.

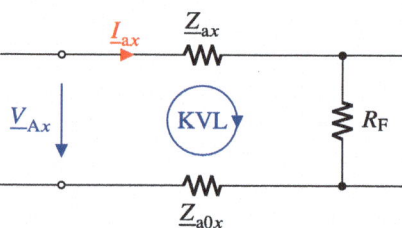

Fig. 7.1 Generalized model representing a fault current loop of a single-phase-to-ground fault in phase x with $x \in \{1, 2, 3\}$ used by the distance protection scheme, see also Fig. 7.12. Kirchhoff's voltage law (KVL) is applied

The differential protection evaluates Kirchhoff's current law (KCL) [123]. It is therefore based on a simple electrical model of the protection zone to be protected. Often, an individual component, e.g. a transformer or an overhead line, is protected and external measurements are used. The Fig. 7.2 shows a generalized model representing a protection zone in phase x with $x \in \{1, 2, 3\}$ used by the differential protection scheme, see also Fig. 7.12. In this example, KCL is evaluated from the measured variables $\underline{I}_{ax}$ and $\underline{I}_{bx}$.

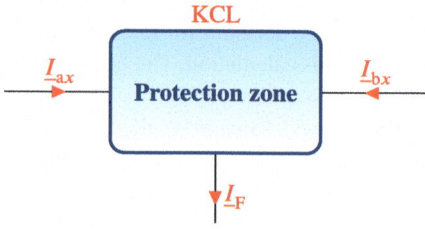

Fig. 7.2 Generalized model representing a protection zone in phase x with $x \in \{1, 2, 3\}$ used by the differential protection scheme, see also Fig. 7.12. Kirchhoff's current law (KCL) is applied

7.1 Model-Based Fault Analysis

Extending the differential protection, the "Setting-Less Protection (SLP)" also known as the "Dynamic State Estimation (DSE) Based Protection" is presented in [124–127]. The basic idea is to apply "all physical laws" including also thermodynamic laws (describing e.g. the temperature T) and mechanical laws (describing e.g. the mechanical torque M_m and the rotor shaft velocity ω_m) in addition to electrical laws. External and additional internal measurements are used. The Fig. 7.3 represents a conceptual illustration of the DSE-based protection as an evolution of the differential protection.

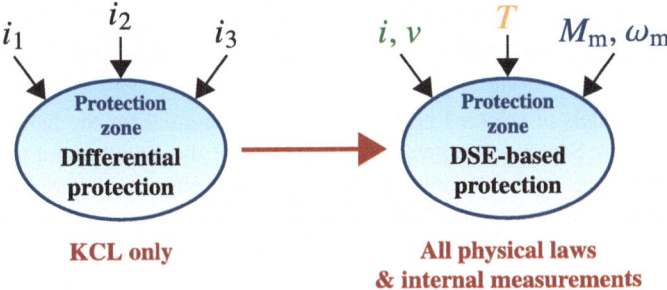

Fig. 7.3 Conceptual illustration of the dynamic state estimation (DSE) based protection applying all physical laws instead of using only Kirchhoff's current law (KCL), adapted from [126]

The fault analysis of the model-based protection scheme developed in this thesis (see Chap. 8) is based on state-space models (see Sec. 3.2.1). Measured variables are compared with calculated variables from models. Afterwards, it is investigated whether they match or not. On this basis, information about the fault appearance, the fault configuration and the fault localization is gained. A comparison of the developed model-based protection scheme in this thesis (see Chap. 8) and the DSE-based protection scheme is given in Sec. 8.4.5.2.

Subsequently, models based on space-vector components (see Sec. 3.1.2.3) in the time domain are used. Using state-space components in the stationary reference frame ($\alpha\beta$-components) leads to the decoupling of differential equations representing healthy systems and compact and clear models. The usage of direct current (DC) quantities is not essential, since instantaneous values are processed and no integration is necessary.

According to the concept of the model-based protection scheme (see Sec. 8.4.1) using three protection layers (see Fig. 8.6), state-space models of grid areas and state-space models of lines are needed. Models of selected grid topologies (see Sec. 7.2) and lines (see Sec. 7.4) are derived. These state-space models are the basic prerequisite of the model-based protection algorithm.

It is concluded, that the distance protection scheme, the differential protection scheme, the DSE-based protection scheme and the novel model-based protection scheme use models. The essence of these models differ from each other. Furthermore, it is identified, that the design of protection schemes as well as the design of controllers of voltage source converters (VSC) (see also Sec. 5.1 and Sec. 5.3) is often based on models emulating real physical characteristics.

7.2 State-Space Models of Selected Grid Topologies

State-space models of selected grid topologies are essential for the fault identification layer (see Sec. 8.4.2) representing one layer of the model-based protection algorithm, see Sec. 8.4.1. The objective of the fault identification layer is to decide, if normal operation conditions inside the monitored grid area are present or not.

Modelling the zero sequence of a meshed grid area is not a trivial task because the parameters are often unknown and may change due to environmental conditions. Consequently, models of selected healthy grid topologies are used and zero sequence components are not considered. The input and output variables vary by model. The selection of input and output variables is crucial to ensure the observability, see Sec. 3.3.4.

7.2.1 Single Π-Section Line

A single line is subsequently described as a Π-section equivalent circuit with concentrated elements [71]. The capacitance of the line is shared equally upon both sides of the line. A Π-section equivalent circuit is an appropriate option to describe overhead lines and cables. Overhead lines and cables are distinguished by different parameter values. The Fig. 7.4 shows the equivalent circuit of a healthy line (named line 1, index: L1) based on a CLC-structure between the grid nodes A and B.

The equivalent circuit in Fig. 7.4 includes serial resistances, serial inductances, phase-to-ground capacitances, ground impedances and mutual coupling inductances between the phases to describe electrical characteristics and effects. The mutual coupling inductances $M_{\text{L1}xy}$ with $x \in \{1, 2, 3\}$, $y \in \{1, 2, 3\}$ and $x \neq y$ are described by the following KVLs according to Fig. 7.4, see Eq. (7.1), Eq. (7.2) and Eq. (7.3).

7.2 State-Space Models of Selected Grid Topologies

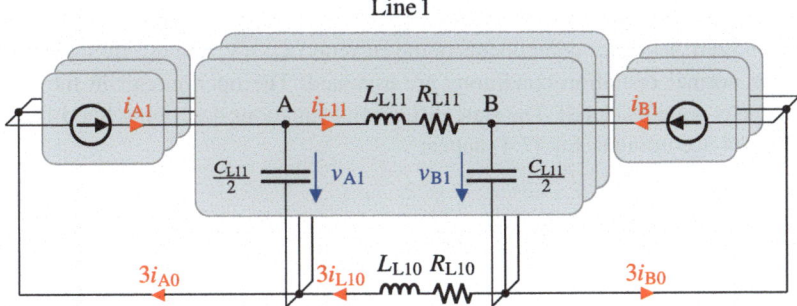

Fig. 7.4 Equivalent circuit of a healthy line based on a CLC-structure between the grid nodes A and B

$$v_{A1} - v_{B1} - 3R_{L10}i_{L10} - 3L_{L10}\frac{d}{dt}i_{L10} = \quad (7.1)$$
$$R_{L11}i_{L11} + L_{L11}\frac{d}{dt}i_{L11} + M_{L112}\frac{d}{dt}i_{L12} + M_{L113}\frac{d}{dt}i_{L13}$$

$$v_{A2} - v_{B2} - 3R_{L10}i_{L10} - 3L_{L10}\frac{d}{dt}i_{L10} = \quad (7.2)$$
$$R_{L12}i_{L12} + L_{L12}\frac{d}{dt}i_{L12} + M_{L121}\frac{d}{dt}i_{L11} + M_{L123}\frac{d}{dt}i_{L13}$$

$$v_{A3} - v_{B3} - 3R_{L10}i_{L10} - 3L_{L10}\frac{d}{dt}i_{L10} = \quad (7.3)$$
$$R_{L13}i_{L13} + L_{L13}\frac{d}{dt}i_{L13} + M_{L131}\frac{d}{dt}i_{L11} + M_{L132}\frac{d}{dt}i_{L12}$$

In Fig. 7.4, the conductances between different phases and conductances between the phases and ground are neglected. The capacitances between different phases are also neglected. Essentially, the previously mentioned assumptions of describing a single line according to the equivalent circuit in Fig. 7.4 are equal to the common assumptions made for distance protection algorithms [143].

Subsequently, it is assumed, that corresponding resistances, inductances and capacitances in each phase are equal. Furthermore, all mutual coupling inductances are regarded to be equal: $M_{L112} = M_{L113} = M_{L121} = M_{L123} = M_{L131} = M_{L132} = M_{L1}$. The zero sequence components of the currents i_{Ax}, i_{L1x} and i_{Bx} are set to zero because normal operation conditions are assumed. The open circuit in the zero sequence is not considered. The state differential equations describing line 1 based on Fig. 7.4 according to Eq. (7.4) appear.

$$\frac{d}{dt}\begin{pmatrix} i_{L1\alpha} \\ i_{L1\beta} \\ v_{A\alpha} \\ v_{A\beta} \\ v_{B\alpha} \\ v_{B\beta} \end{pmatrix} = \begin{pmatrix} -\frac{R_{L1}}{L_{L1}-M_{L1}} & 0 & \frac{1}{L_{L1}-M_{L1}} & 0 & -\frac{1}{L_{L1}-M_{L1}} & 0 \\ 0 & -\frac{R_{L1}}{L_{L1}-M_{L1}} & 0 & \frac{1}{L_{L1}-M_{L1}} & 0 & -\frac{1}{L_{L1}-M_{L1}} \\ -\frac{2}{C_{L1}} & 0 & 0 & 0 & 0 & 0 \\ 0 & -\frac{2}{C_{L1}} & 0 & 0 & 0 & 0 \\ \frac{2}{C_{L1}} & 0 & 0 & 0 & 0 & 0 \\ 0 & \frac{2}{C_{L1}} & 0 & 0 & 0 & 0 \end{pmatrix} \begin{pmatrix} i_{L1\alpha} \\ i_{L1\beta} \\ v_{A\alpha} \\ v_{A\beta} \\ v_{B\alpha} \\ v_{B\beta} \end{pmatrix}$$

$$+ \begin{pmatrix} 0 & 0 & 0 & 0 \\ 0 & 0 & 0 & 0 \\ \frac{2}{C_{L1}} & 0 & 0 & 0 \\ 0 & \frac{2}{C_{L1}} & 0 & 0 \\ 0 & 0 & \frac{2}{C_{L1}} & 0 \\ 0 & 0 & 0 & \frac{2}{C_{L1}} \end{pmatrix} \begin{pmatrix} i_{A\alpha} \\ i_{A\beta} \\ i_{B\alpha} \\ i_{B\beta} \end{pmatrix} \qquad (7.4)$$

Separating the state differential equations of line currents and voltages at node A and B from Eq. (7.4) leads to Eq. (7.5), Eq. (7.6) and Eq. (7.7).

$$\frac{d}{dt}\mathbf{i}_{L1\alpha\beta} = -\frac{R_{L1}}{L_{L1}-M_{L1}}\mathbf{i}_{L1\alpha\beta} + \frac{1}{L_{L1}-M_{L1}}(\mathbf{v}_{A\alpha\beta} - \mathbf{v}_{B\alpha\beta}) \qquad (7.5)$$

$$\frac{d}{dt}\mathbf{v}_{A\alpha\beta} = \frac{2}{C_{L1}}(\mathbf{i}_{A\alpha\beta} - \mathbf{i}_{L1\alpha\beta}) \qquad (7.6)$$

$$\frac{d}{dt}\mathbf{v}_{B\alpha\beta} = \frac{2}{C_{L1}}(\mathbf{i}_{B\alpha\beta} + \mathbf{i}_{L1\alpha\beta}) \qquad (7.7)$$

Based on Eq. (7.5), Eq. (7.6) and Eq. (7.7), rules for the formulation of state differential equations of more complex grid topologies are derived. Basically, Eq. (7.5) represents KVL and Eq. (7.6) and Eq. (7.7) represent KCLs describing three-phase systems using αβ-components. The identified rules for the formulation are used to derive the state differential equations presented in Sec. 7.2.2. For reasons of clarity,

7.2 State-Space Models of Selected Grid Topologies

the equivalent circuit of line 1 according to Fig. 7.4 is represented by the single-line diagram according to Fig. 7.5 subsequently.

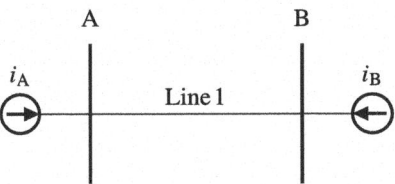

Fig. 7.5 Single-line diagram of one line connecting the grid nodes A and B

7.2.2 Combination of Multiple Π-Section Lines

7.2.2.1 Parallel Lines

To increase the electrical energy transport capacity between two grid nodes A and B, two lines are operated in parallel. Based on Fig. 7.5, the Fig. 7.6 shows two parallel lines in a single-line diagram.

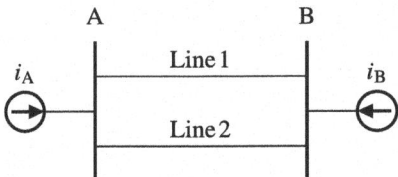

Fig. 7.6 Single-line diagram of two parallel lines connecting the grid nodes A and B

Analogue to Fig. 7.4, the index L2 is used for line 2. The assumptions from Sec. 7.2.1 are adopted. In addition to Eq. (7.5), the following state differential equations to describe two parallel lines reveal, see Eq. (7.8), Eq. (7.9) and Eq. (7.10).

$$\frac{d}{dt}\mathbf{i}_{L2\alpha\beta} = -\frac{R_{L2}}{L_{L2} - M_{L2}}\mathbf{i}_{L2\alpha\beta} + \frac{1}{L_{L2} - M_{L2}}(\mathbf{v}_{A\alpha\beta} - \mathbf{v}_{B\alpha\beta}) \quad (7.8)$$

$$\frac{d}{dt}\mathbf{v}_{A\alpha\beta} = \frac{2}{C_{L1} + C_{L2}}(\mathbf{i}_{A\alpha\beta} - \mathbf{i}_{L1\alpha\beta} - \mathbf{i}_{L2\alpha\beta}) \quad (7.9)$$

$$\frac{\mathrm{d}}{\mathrm{d}t}\mathbf{v}_{B\alpha\beta} = \frac{2}{C_{L1} + C_{L2}}(\mathbf{i}_{B\alpha\beta} + \mathbf{i}_{L1\alpha\beta} + \mathbf{i}_{L2\alpha\beta}) \tag{7.10}$$

7.2.2.2 Serial Lines

To transport electrical energy over long distances, two lines are operated in series. Grid node A is connected to node C via node B. Based on Fig. 7.5, the Fig. 7.7 shows two serial lines in a single-line diagram.

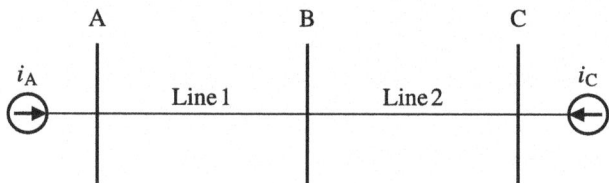

Fig. 7.7 Single-line diagram of two serial lines connecting the grid nodes A, B and C

The assumptions from Sec. 7.2.1 are adopted. In addition to Eq. (7.5), the following state differential equations representing KVLs to describe two serial lines reveal, see Eq. (7.11).

$$\frac{\mathrm{d}}{\mathrm{d}t}\mathbf{i}_{L2\alpha\beta} = -\frac{R_{L2}}{L_{L2} - M_{L2}}\mathbf{i}_{L2\alpha\beta} + \frac{1}{L_{L2} - M_{L2}}(\mathbf{v}_{B\alpha\beta} - \mathbf{v}_{C\alpha\beta}) \tag{7.11}$$

Moreover, the following KCLs in addition to Eq. (7.6) appear, see Eq. (7.12) and Eq. (7.13).

$$\frac{\mathrm{d}}{\mathrm{d}t}\mathbf{v}_{B\alpha\beta} = \frac{2}{C_{L1} + C_{L2}}(\mathbf{i}_{L1\alpha\beta} - \mathbf{i}_{L2\alpha\beta}) \tag{7.12}$$

$$\frac{\mathrm{d}}{\mathrm{d}t}\mathbf{v}_{C\alpha\beta} = \frac{2}{C_{L2}}(\mathbf{i}_{C\alpha\beta} + \mathbf{i}_{L2\alpha\beta}) \tag{7.13}$$

7.2.2.3 Intermediate Infeed

The integration of renewable energy sources (RES) into the electrical power grid often leads to intermediate infeed because RES are usually locally distributed. Extending Fig. 7.7, the Fig. 7.8 shows the according grid topology with three lines in a single-line diagram.

Using the assumptions from Sec. 7.2.1, the KVLs from Eq. (7.5), Eq. (7.11) and Eq. (7.14) yield.

7.3 Modelling of Faults

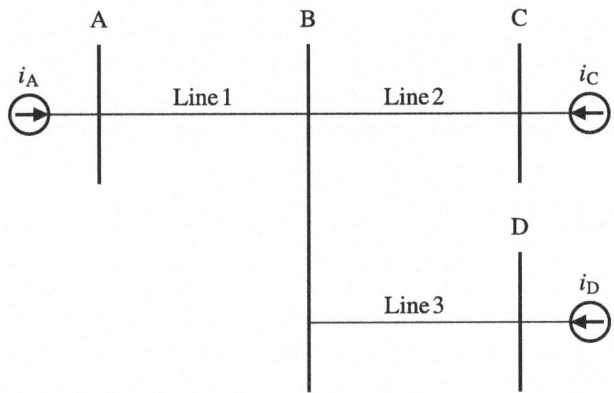

Fig. 7.8 Single-line diagram of two serial lines connecting the grid nodes A, B and C with an intermediate infeed at grid node B from grid node D via a third line

$$\frac{d}{dt}\mathbf{i}_{L3\alpha\beta} = -\frac{R_{L3}}{L_{L3} - M_{L3}}\mathbf{i}_{L3\alpha\beta} + \frac{1}{L_{L3} - M_{L3}}(\mathbf{v}_{C\alpha\beta} - \mathbf{v}_{D\alpha\beta}) \quad (7.14)$$

The KCLs from Eq. (7.6), Eq. (7.13), Eq. (7.15) and Eq. (7.16) appear.

$$\frac{d}{dt}\mathbf{v}_{B\alpha\beta} = \frac{2}{C_{L1} + C_{L2} + C_{L3}}(\mathbf{i}_{L1\alpha\beta} - \mathbf{i}_{L2\alpha\beta} - \mathbf{i}_{L3\alpha\beta}) \quad (7.15)$$

$$\frac{d}{dt}\mathbf{v}_{D\alpha\beta} = \frac{2}{C_{L3}}(\mathbf{i}_{D\alpha\beta} + \mathbf{i}_{L3\alpha\beta}) \quad (7.16)$$

7.3 Modelling of Faults

Grid faults are categorized into transverse and longitudinal faults [71]. In general, both fault categories differ from each other in terms of the transformation of the line topology at the fault location. Transverse fault are often of greater practical importance than longitudinal faults. In Sec. 7.3.1, a generic method for modelling transverse faults is presented. Sec. 7.3.2 shows a generic method for modelling longitudinal faults. Moreover, combinations of transverse faults and longitudinal faults can occur, e.g. due to the disconnection of one phase by a circuit breaker after a single-phase-to-ground fault.

During transverse faults and longitudinal faults, fault impedances appear. These impedances characterize an arc, for example. Subsequently, the fault impedances are modelled by resistors for simplification. Furthermore, the values of the fault resistances are assumed to be constant.

7.3.1 Transverse Faults

Transverse faults appear between one phase and ground or between different phases with or without ground connection. Transverse faults occur, for example if a tree hits an overhead line or a cable is destroyed by an excavator. All types of transverse faults (index: t) are modelled by four resistors R_{t1}, R_{t2}, R_{t3} and R_{t0}. The values of the resistances can differ from each other. The Fig. 7.4 shows the general configuration of fault resistances modelling a transverse fault, see [41, 143].

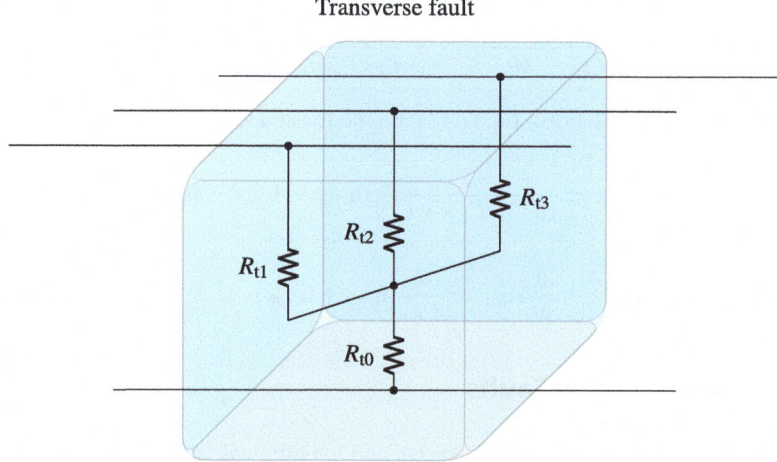

Fig. 7.9 General configuration of fault resistances modelling a transverse fault

7.3 Modelling of Faults

The fault resistance configuration of transverse faults is described in Tab. 7.1.

Table 7.1 Fault resistance configuration of transverse faults

		fault resistance configuration			
		R_{t1}	R_{t2}	R_{t3}	R_{t0}
1	single-phase-to-ground fault in phase 1	R_{t1}	∞	∞	R_{t0}
2	single-phase-to-ground fault in phase 2	∞	R_{t2}	∞	R_{t0}
3	single-phase-to-ground fault in phase 3	∞	∞	R_{t3}	R_{t0}
4	phase-to-phase fault between phase 1 and 2	R_{t1}	R_{t2}	∞	∞
5	phase-to-phase fault between phase 2 and 3	∞	R_{t2}	R_{t3}	∞
6	phase-to-phase fault between phase 1 and 3	R_{t1}	∞	R_{t3}	∞
7	phase-to-phase-to-ground fault between phase 1 and 2	R_{t1}	R_{t2}	∞	R_{t0}
8	phase-to-phase-to-ground fault between phase 2 and 3	∞	R_{t2}	R_{t3}	R_{t0}
9	phase-to-phase-to-ground fault between phase 1 and 3	R_{t1}	∞	R_{t3}	R_{t0}
10	three-phase fault	R_{t1}	R_{t2}	R_{t3}	∞
11	three-phase-to-ground fault	R_{t1}	R_{t2}	R_{t3}	R_{t0}

7.3.2 Longitudinal Faults

Longitudinal faults appear along one phase or multiple phases. Ground is not affected. Longitudinal faults occur, for example, if an overhead line or a cable breaks or a circuit breaker is damaged. All types of longitudinal faults (index: l) are modelled by three resistors R_{l1}, R_{l2} and R_{l3}. The values of the resistances can differ from each other and are often very high (range of kΩ). The general configuration of fault resistances modelling a longitudinal fault is presented in Fig. 7.10 [71].

The Tab. 7.2 describes the fault resistance configuration of transverse faults.

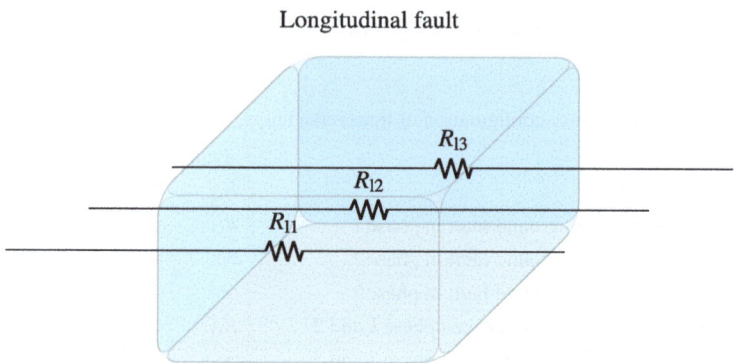

Fig. 7.10 General configuration of fault resistances modelling a longitudinal fault

Table 7.2 Fault resistance configuration of longitudinal faults

		fault resistance configuration		
		R_{11}	R_{12}	R_{13}
1	single-phase break in phase 1	R_{11}	0	0
2	single-phase break in phase 2	0	R_{12}	0
3	single-phase break in phase 3	0	0	R_{13}
4	double-phase break in phase 1 and 2	R_{11}	R_{12}	0
5	double-phase break in phase 2 and 3	0	R_{12}	R_{13}
6	double-phase break in phase 1 and 3	R_{11}	0	R_{13}
7	three-phase break	R_{11}	R_{12}	R_{13}

7.4 State-Space Models of Lines

State-space models of lines are fundamental for the fault characterization layer (see Sec. 8.4.3) and the fault localization layer (see Sec. 8.4.4) representing two layers of the model-based protection algorithm, see Sec. 8.4.1. In the fault characterization layer each line is monitored. In the first step it is examined, if the monitored line is healthy or not. The identification of healthy lines is essential to reach selectivity by proper tripping actions. Afterwards, the faulty line or multiple faulty lines are investigated regarding fault characterization. The fault characterization includes the determination of the fault resistance configuration and the values of the appearing fault resistance. Each faulty line is further monitored in detail in the fault local-

7.4 State-Space Models of Lines

ization layer to specify the fault location based on the identified fault resistance configuration.

Consequently, a model of a healthy line (see Sec. 7.4.1) and models of selected faulty lines (see Sec. 7.4.2, Sec. 7.4.3 and Sec. 7.4.4) are presented. The input and output variables are equal for each line model, see Sec. 8.4.3.

A grid fault on one line leads to the split of the model at the fault location based on the model of the healthy line. The structural change of the line model at the fault location depends on the fault resistance configuration. Fault-characteristic model structures of the line reveal, see e.g. Sec. 7.4.3 and Sec. 7.4.4. Beyond the line resistance, the line inductance, the mutual line coupling inductance, the derived fault-characteristic models are parametrized regarding the fault resistances and the fault location. The parametrization regarding the fault resistances (see Fig. 7.9 and Fig. 7.10) allows estimating the real values of the involved fault resistances. The parametrization regarding the fault location allows locating the fault, see Sec. 8.4.4. Fault location is not for possible longitudinal faults in general.

The simplifying assumption from Sec. 7.2.1 are considered. Furthermore, the phase-to-ground capacitances are neglected because lines show dominant inductive characteristics during a fault [143].

7.4.1 Model of a Healthy Line

The equivalent circuit of a healthy line based on a L-structure is shown in Fig. 7.11. To prepare the modelling of faulty lines and to ensure compatibility, the healthy line in Fig. 7.11 is split into two parts. The currents i_{ax} with $x \in \{1, 2, 3\}$ and represented by the index a appear between the fault location and the grid node A. The currents i_{bx} represented by the index b appear between the fault location and the grid node B. The share of the line resistances, the line inductances and the mutual coupling inductances depend on the relative fault distance m regarding the actual length of the line.

Based on Fig. 7.11, the state differential equations are derived. The mutual coupling inductances are described analogue to Eq. (7.1), Eq. (7.2) and Eq. (7.3). Furthermore, the corresponding line parameters are assumed to be equal in each phase. The state differential equations of the zero sequence are explicitly considered increasing the quality of the model to distinguish between a healthy and a faulty line. The state differential equations in accordance with Eq. (7.17), Eq. (7.18), Eq. (7.19) and Eq. (7.20) appear.

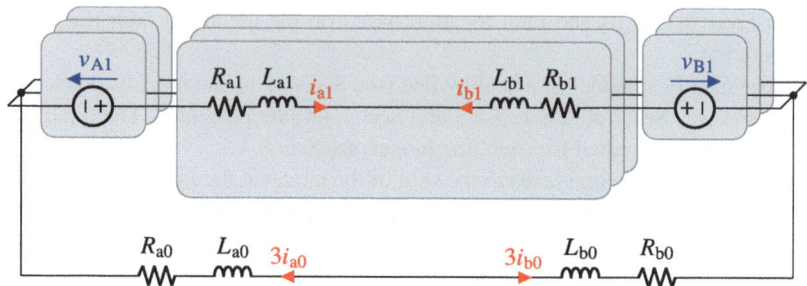

Fig. 7.11 Equivalent circuit of a healthy line based on a L-structure

$$\frac{d}{dt}\mathbf{i}_{\alpha\beta} = -\frac{R_a + R_b}{L_a + L_b - M_a - M_b}\mathbf{i}_{\alpha\beta} + \frac{1}{L_a + L_b - M_a - M_b}(\mathbf{v}_{A\alpha\beta} - \mathbf{v}_{B\alpha\beta}) \tag{7.17}$$

$$\frac{d}{dt}i_{a0} = -\frac{R_a + R_b + 3R_{a0} + 3R_{b0}}{L_a + L_b + 2M_a + 2M_b + 3L_{a0} + 3L_{b0}}i_{a0}$$
$$+ \frac{1}{L_a + L_b + 2M_a + 2M_b + 3L_{a0} + 3L_{b0}}(v_{A0} - v_{B0}) \tag{7.18}$$

$$\frac{d}{dt}\mathbf{i}_{b\alpha\beta} = -\frac{R_a + R_b}{L_a + L_b - M_a - M_b}\mathbf{i}_{b\alpha\beta} + \frac{1}{L_a + L_b - M_a - M_b}(\mathbf{v}_{B\alpha\beta} - \mathbf{v}_{A\alpha\beta}) \tag{7.19}$$

$$\frac{d}{dt}i_{b0} = -\frac{R_a + R_b + 3R_{a0} + 3R_{b0}}{L_a + L_b + 2M_a + 2M_b + 3L_{a0} + 3L_{b0}}i_{b0}$$
$$+ \frac{1}{L_a + L_b + 2M_a + 2M_b + 3L_{a0} + 3L_{b0}}(v_{B0} - v_{A0}) \tag{7.20}$$

7.4.2 Universal Representation of Faulty Lines

Based on the equivalent circuit of a healthy line in Fig. 7.11, the general configuration of fault resistances modelling a transverse fault in Fig. 7.9 and Tab. 7.1, a universal representation of faulty lines with a transverse fault is found, see Fig. 7.12. Models for all transverse fault configurations according to Tab. 7.1 are derived.

7.4 State-Space Models of Lines

They are used by the fault characterization layer and the fault localization layer of the model-based protection algorithm, see Sec. 8.4.3 and Sec. 8.4.4. All models are parametrized regarding the line resistance, the line inductance, the mutual line coupling inductance, the fault resistances and the fault location.

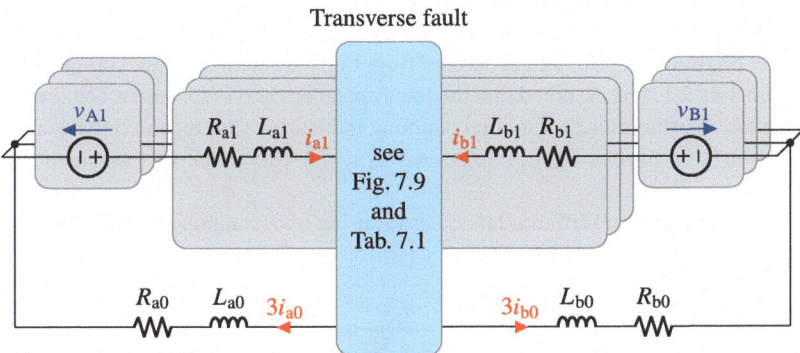

Fig. 7.12 Universal equivalent circuit of a line with a transverse fault

Analogously, the general configuration of fault resistances modelling a longitudinal fault (see Fig. 7.10) and Tab. 7.2, a universal representation of faulty lines with a longitudinal fault is found, see Fig. 7.13. All models are parametrized regarding the line resistance, the line inductance, the mutual line coupling inductance and the fault resistances.

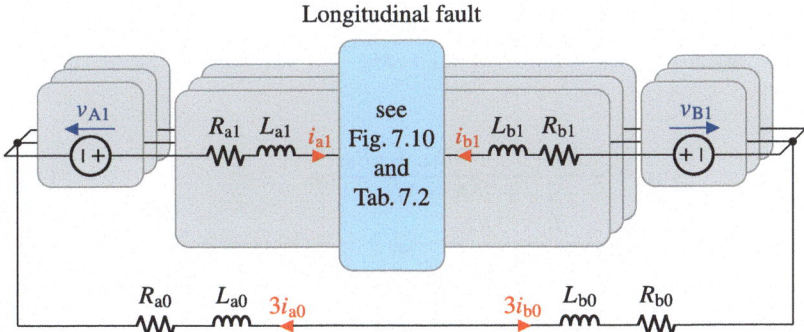

Fig. 7.13 Universal equivalent circuit of a line with a longitudinal fault

Selected models of faulty lines are presented by their state differential equations using state-space components in the stationary reference frame (αβ-components, see Sec. 3.1.2.3) and the zero sequence component in Sec. 7.4.3 and Sec. 7.4.4.

7.4.3 Three-Phase-to-Ground Fault

A three-phase-to-ground fault is derived from Fig. 7.12 with the fault configuration 11 from Tab. 7.1. It is assumed that the line resistances and inductances are equal in each phase. Furthermore, all mutual coupling inductances are regarded to be equal: $M_{12} = M_{13} = M_{21} = M_{23} = M_{31} = M_{32} = M$. Moreover, it is assumed, that $R_{t1} = R_{t2} = R_{t3} = R_t$.

The following state-differential equations using αβ0-components at grid node A appear, see Eq. (7.21) and Eq. (7.22).

$$\frac{d}{dt}\mathbf{i}_{a\alpha\beta} = -\frac{R_a - R_t}{L_a - M_a}\mathbf{i}_{a\alpha\beta} + \frac{R_t}{L_a - M_a}\mathbf{i}_{b\alpha\beta} + \frac{1}{L_a - M_a}\mathbf{v}_{A\alpha\beta} \quad (7.21)$$

$$\frac{d}{dt}i_{a0} = -\frac{R_a - R_t + 3R_{a0} + 3R_{t0}}{L_a + 2M_a + 3L_{a0}}i_{a0} \\ + \frac{R_t - 3R_{t0}}{L_a + 2M_a + 3L_{a0}}i_{b0} + \frac{1}{L_a + 2M_a + 3L_{a0}}v_{A0} \quad (7.22)$$

Analogously, the state-differential equations using αβ0-components at grid node B appear, see Eq. (7.23) and Eq. (7.24).

$$\frac{d}{dt}\mathbf{i}_{b\alpha\beta} = -\frac{R_b - R_t}{L_b - M_b}\mathbf{i}_{b\alpha\beta} + \frac{R_t}{L_b - M_b}\mathbf{i}_{a\alpha\beta} + \frac{1}{L_b - M_b}\mathbf{v}_{B\alpha\beta} \quad (7.23)$$

$$\frac{d}{dt}i_{b0} = -\frac{R_b - R_t + 3R_{b0} + 3R_{t0}}{L_b + 2M_b + 3L_{b0}}i_{b0} \\ + \frac{R_t - 3R_{t0}}{L_b + 2M_b + 3L_{b0}}i_{a0} + \frac{1}{L_b + 2M_b + 3L_{b0}}v_{B0} \quad (7.24)$$

The Eq. (7.21), Eq. (7.22), Eq. (7.23) and Eq. (7.24) are aggregated into a holistic state-space model.

7.4.4 Three-Phase Break

A three-phase break is derived from Fig. 7.13 with the fault configuration 7 from Tab. 7.2. The line resistances and inductances are assumed to be equal in all phases. Besides, all mutual coupling inductances are regarded to be equal: $M_{12} = M_{13} = M_{21} = M_{23} = M_{31} = M_{32} = M$. Moreover, it is assumed, that $R_{11} = R_{12} = R_{13} = R_1$.

At grid node A, the following state-differential equations using $\alpha\beta 0$-components appear, see Eq. (7.25) and Eq. (7.26).

$$\frac{d}{dt}\mathbf{i}_{a\alpha\beta} = -\frac{R_a + R_b + R_1}{L_a + L_b - M_a - M_b}\mathbf{i}_{a\alpha\beta} + \frac{1}{L_a + L_b - M_a - M_b}(\mathbf{v}_{A\alpha\beta} - \mathbf{v}_{B\alpha\beta}) \tag{7.25}$$

$$\frac{d}{dt}i_{a0} = -\frac{R_a + R_b + 3R_{a0} + 3R_{b0} + R_1}{L_a + L_b + 2M_a + 2M_b + 3L_{a0} + 3L_{b0}}i_{a0} \tag{7.26}$$
$$+ \frac{1}{L_a + L_b + 2M_a + 2M_b + 3L_{a0} + 3L_{b0}}(v_{A0} - v_{B0})$$

Analogously, the state-differential equations using $\alpha\beta 0$-components at grid node B appear in accordance with Eq. (7.27) and Eq. (7.28).

$$\frac{d}{dt}\mathbf{i}_{b\alpha\beta} = -\frac{R_a + R_b + R_1}{L_a + L_b - M_a - M_b}\mathbf{i}_{b\alpha\beta} + \frac{1}{L_a + L_b - M_a - M_b}(\mathbf{v}_{B\alpha\beta} - \mathbf{v}_{A\alpha\beta}) \tag{7.27}$$

$$\frac{d}{dt}i_{b0} = -\frac{R_a + R_b + 3R_{a0} + 3R_{b0} + R_1}{L_a + L_b + 2M_a + 2M_b + 3L_{a0} + 3L_{b0}}i_{b0} \tag{7.28}$$
$$+ \frac{1}{L_a + L_b + 2M_a + 2M_b + 3L_{a0} + 3L_{b0}}(v_{B0} - v_{A0})$$

The Eq. (7.25), Eq. (7.26), Eq. (7.27) and Eq. (7.28) are aggregated into a holistic state-space model.

7.5 Summary

In this chapter, the grid fault modelling in the state-space is presented. The grid fault modelling is the basic prerequisite of the developed model-based protection algorithm.

The models of this chapter are based on the state-space representation using space vector components in the stationary reference frame presented in Chap. 3. Like the design of the state-feedback voltage/current (V/I) controller of voltage source converters (VSCs) (see Sec. 5.1 and Sec. 5.3), the design of the developed model-based protection algorithm is also based on models emulating real physical characteristics. Existing protection algorithms like the distance protection and the differential protection are also based on models. These models are of different nature and are less detailed compared to the state-space models in this chapter.

If a grid fault appears in an electrical power system, the characteristics of voltages and currents can differ from their characteristics during normal operation conditions. V/I-characteristics can be analysed in different ways. The analysis of the model-based protection algorithm is based on the comparison of measured variables with corresponding calculated variables from models. It is investigated, whether they match or not, and the goodness-of-fit is quantified. Decisions based on this analysis for the tripping of circuit breakers to protect living beings and hardware equipment are taken.

State-space models of selected grid topologies, which are especially relevant for 100 % inverter-based power system, are analytically derived. They are used inside the fault identification layer of the model-based protection algorithm. Here, a grid area is monitored. Rules for the formulation of state differential equations of more complex grid topologies are identified. Consequently, it is possible to extend the fault identification as a part of the fault analysis for more complex grid topologies.

Grid faults are categorized into transverse faults and longitudinal faults. Both fault categories are generalized. Different fault resistance configurations are distinguished.

State-space models of lines are derived. They are used inside the fault characterization and the fault localization, representing two layers of the model-based protection algorithm. In the fault characterization layer, each line of the grid area investigated by the fault identification layer is monitored. The line models are parametrized regarding the line parameters, the fault resistance configuration, the values of the fault resistances and the fault location.

The derived state-space models are essential for the analysis and the decision of the model-based protection algorithm. The model-based protection algorithm is

7.5 Summary

presented in Chap. 8. To follow the key aspects of the central theme of this thesis on a shortcut: The conclusion of the next chapter is provided in Sec. 8.5.

Open Access This chapter is licensed under the terms of the Creative Commons Attribution 4.0 International License (http://creativecommons.org/licenses/by/4.0/), which permits use, sharing, adaptation, distribution and reproduction in any medium or format, as long as you give appropriate credit to the original author(s) and the source, provide a link to the Creative Commons license and indicate if changes were made.

The images or other third party material in this chapter are included in the chapter's Creative Commons license, unless indicated otherwise in a credit line to the material. If material is not included in the chapter's Creative Commons license and your intended use is not permitted by statutory regulation or exceeds the permitted use, you will need to obtain permission directly from the copyright holder.

Model-Based Protection Algorithm 8

This chapter presents the model-based protection (MBP) algorithm. In Sect. 8.1, the fundamentals of the model-based fault analysis based on estimated values are explained. The principles of the state observer and the Kalman filter as state estimators are described in Sect. 8.2. Sect. 8.3 contains the derivation of an appropriate protection criterion based on hypothesis testing. The MBP scheme is presented in Sect. 8.4. It contains the basic structure and the discussion of the properties of the MBP. The three layers of the MBP (fault identification, fault characterization and fault localization) and their interactions are explained in detail. Furthermore, the MBP is compared to alternative protection principles. Distinctions between the MBP and state-of-the-art protection schemes and alternative model-based protection schemes are pointed out.

8.1 Model-Based Fault Analysis Based on Estimated Values

A novel grid protection algorithm named "model-based protection (MBP)" for 100 % inverter-based power systems (IBPS) is presented. The focus of this thesis is the model-based fault analysis containing the fault identification, the fault characterization and the fault localization. The dynamics and the stability after the fault and after the fault clearance are defined by the grid control strategy, see Sect. 2.2.3. The basic objective is to maintain the grid operation.

The MBP is inspired by the "Setting-Less Protection (SLP)" also known as the "Dynamic State Estimation (DSE) Based Protection" presented in [124–127]. Distinctions between this scheme and the presented MBP are discussed in Sect. 8.4.5.2.

The basic idea of the presented MBP is to evaluate, if measured values and calculated values based on models (see Chap. 7) of corresponding variables match or not. In this way, selected output variables of the regarded system are measured and calculated as well. The output variables are a subset of the state variables using the state-space representation, see Sect. 3.2.1.

The MBP turns out to be very suitable for 100 % IBPS based on renewable energy system (RES). It operates independently of signal characteristics. Differences between measured values and calculated values based on models are crucial using the MBP. Consequently, the flexibility options of voltage source converters (VSCs) (see especially Sect. 2.3 and Chap. 6) are not restricted. Furthermore, transient state characteristics are evaluated. Moreover, unidirectional power flow from higher voltage levels to lower voltage levels is not assumed using the MBP. The direction of power flows will not be predetermined any more regarding storage-based RES, see Sect. 2.3 and Sect. 1.1.4.

There are various possibilities to calculate the output variables of a system from its input variables. The direct calculation from a state-space model using Eq. (6.22) and Eq. (3.47) is disadvantageous, since the initial value of each state variable ($\mathbf{x}(k)$ in Eq. (6.22) with $k = 0$) is often unknown in practical applications. An alternative calculation principle is to use the dynamic state estimation principle, see Sect. 8.2. This approach is beneficial over the direct calculation because the initial state variables do not have to be known. State estimation in power systems is already discussed in early publications, see [128].

8.2 State Estimation

8.2.1 State Observer

State observers are state estimation algorithms, which are often applied in the context of state-feedback controllers (see also Sect. 5.3) [129–131]. They are used, if not each state variable is measured or measurable. The objective of a state observer is to estimate the state variables $\mathbf{x}$ aided by the measurement of the input variables and the output variables of the control plant. From this objective, the term "state observer" is derived.

Applying a state observer (subsequently abbreviated as observer) to the state-feedback V/I-controller in Fig. 5.18, the control structure according to Fig. 8.1 is derived [81]. The plant and the observer are influenced by the identical measured actuating variables $\mathbf{u}$ and disturbance variables $\mathbf{z}$ representing the input variables.

8.2 State Estimation

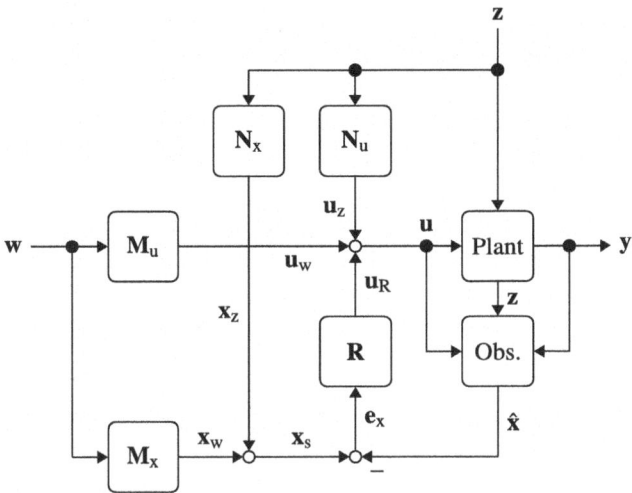

Fig. 8.1 Block diagram of the state-feedback V/I-controller using the feedback controller **R** and a state observer (Obs.) extending Fig. 5.18

The state variables of the control plant **x** and the estimated state variables by the observer $\hat{\mathbf{x}}$ can differ from each other. The reason is, that the model used for the design of the observer does not represent the real characteristics exactly.

A discrete-time state-space model of the control plant is the basic prerequisite for the design of the observer. From Eq. (3.46) and Eq. (3.47), the model according to Eq. (8.1b) and Eq. (8.2) appears. It is assumed, that the actuating variables **u** and the disturbance variables **z** are measured and aggregated into the vector $\tilde{\mathbf{u}}$ representing the input variables.

$$\mathbf{x}(k+1) = \mathbf{A}_d \mathbf{x}(k) + \mathbf{B}_d \mathbf{u}(k) + \mathbf{E}_d \mathbf{z}(k) \tag{8.1a}$$

$$= \mathbf{A}_d \mathbf{x}(k) + \widetilde{\mathbf{B}}_d \widetilde{\mathbf{u}}(k) \tag{8.1b}$$

$$\mathbf{y}(k) = \mathbf{C}\mathbf{x}(k) \tag{8.2}$$

The basic concept of the observer is to feed back the estimation error calculated from the measured output variables **y** and the estimated output variables by the observer $\hat{\mathbf{y}}$ via the observer matrix **L** [81]. In this way, the next estimation of the state variables $\hat{\mathbf{x}}(k+1)$ is updated. This concept based on the model in Eq. (8.1b) and Eq. (8.2) is represented by Eq. (8.3) and Eq. (8.4) [78].

$$\hat{\mathbf{x}}(k+1) = \mathbf{A}_d\hat{\mathbf{x}}(k) + \widetilde{\mathbf{B}}_d\widetilde{\mathbf{u}}(k) + \mathbf{L}[\mathbf{y}(k) - \hat{\mathbf{y}}(k)] \quad (8.3)$$

$$\hat{\mathbf{y}}(k) = \mathbf{C}\hat{\mathbf{x}}(k) \quad (8.4)$$

The estimation error of the state variables $\mathbf{e}_{\hat{x}}$ equals the difference of the real state variables $\mathbf{x}$ and the estimated state variables $\hat{\mathbf{x}}$ by the observer, see Eq. (8.5).

$$\mathbf{e}_{\hat{x}}(k) = \mathbf{x}(k) - \hat{\mathbf{x}}(k) \quad (8.5)$$

The feedback of the estimation error of the output variables (see the third addend in Eq. (8.3)) is performed in a way, that $\mathbf{e}_{\hat{x}}$ reaches zero [81]. It is assumed, that the output matrix $\mathbf{C}$ of the model describes the real output characteristics sufficiently. If $\mathbf{e}_{\hat{x}} = 0$ is reached, the estimated state variables equal the actual state variables and thus the real system state is exactly known.

The Fig. 8.2 shows the block diagram of the observer based on Eq. (8.3) and Eq. (8.4) [78]. The initial state variables $\mathbf{x}(0)$ are often partly or entirely unknown. The unknown initial state of the system represents a deterministic disturbance to the state observer. Consequently, the initial estimation of the state variables $\hat{\mathbf{x}}(0)$ is often set to zero.

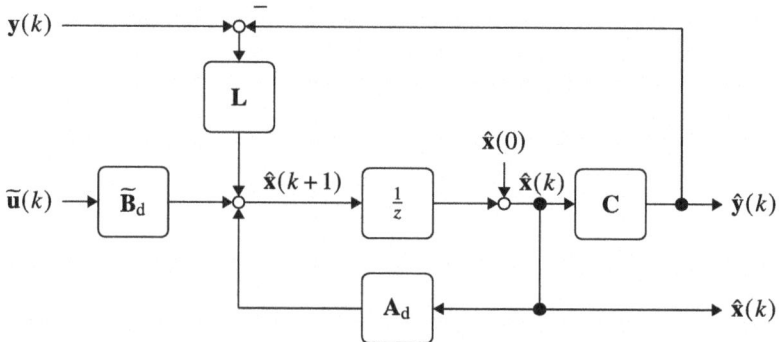

Fig. 8.2 Block diagram of the state observer based on a linear feedback of the difference of the measured output variables $\mathbf{y}$ and the estimated output variables $\hat{\mathbf{y}}$ using the observer matrix $\mathbf{L}$ and a discrete-time state-space model of the regarded system

8.2 State Estimation

The state difference equations of the closed control loop of the observer using Eq. (8.5), Eq. (8.1b), Eq. (8.3), Eq. (8.2) and Eq. (8.4) yield Eq. (8.6).

$$\mathbf{e}_{\hat{x}}(k+1) = \mathbf{x}(k+1) - \hat{\mathbf{x}}(k+1) \tag{8.6a}$$
$$= \mathbf{A}_d\mathbf{x}(k) + \widetilde{\mathbf{B}}_d\widetilde{\mathbf{u}}(k) - \mathbf{A}_d\hat{\mathbf{x}}(k) - \widetilde{\mathbf{B}}_d\widetilde{\mathbf{u}}(k) - \mathbf{L}\big[\mathbf{y}(k) - \hat{\mathbf{y}}(k)\big] \tag{8.6b}$$
$$= \mathbf{A}_d\big[\mathbf{x}(k) - \hat{\mathbf{x}}(k)\big] - \mathbf{LC}\big[\mathbf{x}(k) - \hat{\mathbf{x}}(k)\big] \tag{8.6c}$$
$$= \big[\mathbf{A}_d - \mathbf{LC}\big]\mathbf{e}_{\hat{x}}(k) \tag{8.6d}$$

Similar feedback concepts of the state-feedback V/I-controller in Fig. 5.18 and the observer in Fig. 8.2 are identified. The state difference equations of the closed control loop of the observer in Eq. (8.6d) are similar to the state difference equations of the closed control loop of the state-feedback V/I-controller in Eq. (5.31b). The observer controls the estimation error of the output variables (see the third addend in Eq. (8.3)) to zero. The state-feedback V/I-controller controls the state error of the control plant $\mathbf{e}_x$ (see Eq. (5.50) and Eq. (5.48)) to zero.

Analogue to the system matrix $(\mathbf{A}_d - \mathbf{B}_d\mathbf{R})$ of the state-feedback V/I-controller in Eq. (5.31b), the asymptotic stability (see Sect. 3.3.2) of the observer is determined by the system matrix $(\mathbf{A}_d - \mathbf{LC})$ in Eq. (8.6d).

The controllability of the control plant (see Sect. 3.3.3) is the prerequisite for the design of the state-feedback V/I-controller [78]. The observability of the control plant (see Sect. 3.3.4) is the prerequisite for the design of the observer [78]. The duality of controllability and observability (see Sect. 3.3.5) underlines the analogies of the state-feedback V/I-controller and the observer. Furthermore, the design principles of the state-feedback V/I-controller controller matrix $\mathbf{R}$ can be applied to design the observer matrix $\mathbf{L}$ [78, 81]. Various options to calculate the observer matrix $\mathbf{L}$ appear, e.g. using pole placement or the linear-quadratic regulator (LQR) algorithm, see Sect. 5.3.4.2 and Sect. 5.3.4.3.

8.2.2 Kalman Filter

Kalman filters (KFs) are state estimation algorithms, which differ from state observers (see Sect. 8.2.1) regarding the nature of considered disturbances to the state estimator [132]. Initial estimation errors of the state variables represent the deterministic disturbance of state observers. KFs take stochastic errors into account. However, the basic objective of state observers and KFs to estimate unknown state variables is the same.

Model uncertainties (also termed model inaccuracies or process noise) are included in the state difference equations of the model extending Eq. (8.1b). These equations are named "state extrapolation equations", since they are interpreted as the extrapolation or prediction of the state variables in the future. Measurement errors and measurement noise are included in the output equations extending Eq. (8.2). These equations are named "measurement equations" because the output variables are measured.

Process noise and measurement noise are stochastic disturbances, which are represented by random variables. It is assumed that these random variables follow a normal distribution. The state estimation error and the measured output variables are also regarded as random variables. Consequently, statistic methods are used for the state estimation. To clearly distinguish KFs from state observers, the time-variant feedback matrix of KFs is named $\mathbf{K}(k)$ instead of $\mathbf{L}$. The state estimation error is named $\epsilon_{\hat{x}}$ instead of $\mathbf{e}_{\hat{x}}$.

From this perspective, an interim conclusion is drawn: KFs represent state estimation algorithms based on statistical methods. The objective is to reach a sufficient estimation of non-measured variables based on an inaccurate model and noisy measurements of output variables. Taking these non-ideal boundary conditions into account makes KFs particularly suitable for various practical areas of application. If $\epsilon_{\hat{x}} = 0$ is reached, the estimated state variables equal the real state variables and thus the actual system state is exactly known.

KFs are designed based on the linear state extrapolation equations in Eq. (8.7) and the linear measurement equations in Eq. (8.8). The matrices $\mathbf{A}_d$, $\widetilde{\mathbf{B}}_d$, $\mathbf{C}$ and $\mathbf{K}$ can also be time-variant.

$$\mathbf{x}(k+1) = \mathbf{A}_d \mathbf{x}(k) + \widetilde{\mathbf{B}}_d \widetilde{\mathbf{u}}(k) + \mathbf{G}_d \boldsymbol{\mu}(k) \tag{8.7}$$

$$\mathbf{y}(k) = \mathbf{C}\mathbf{x}(k) + \boldsymbol{v}(k) \tag{8.8}$$

The state estimation error is calculated according to Eq. (8.9).

$$\epsilon_{\hat{x}}(k) = \mathbf{x}(k) - \hat{\mathbf{x}}(k) \tag{8.9}$$

Analogue to Eq. (8.6a), the closed control loop of the KF yields Eq. (8.10).

$$\epsilon_{\hat{x}}(k+1) = \mathbf{x}(k+1) - \hat{\mathbf{x}}(k+1) \tag{8.10a}$$
$$= \left[\mathbf{A}_d - \mathbf{K}(k)\mathbf{C}\right]\epsilon_{\hat{x}}(k) + \boldsymbol{\mu}(k) - \mathbf{K}(k)\boldsymbol{v}(k) \tag{8.10b}$$

8.2 State Estimation

To reach a sufficient estimation, an optimization problem is solved. The cost function of the KF J_{KF} is defined as the expected value E of the sum of the squares of the estimation error of each state variable, see Eq. (8.11).

$$J_{\mathrm{KF}} = \mathrm{E}\big[\boldsymbol{\epsilon}_{\hat{\mathbf{x}}}^{\mathrm{T}}(k)\boldsymbol{\epsilon}_{\hat{\mathbf{x}}}(k)\big] \tag{8.11}$$

In contrast to the cost function of the LQR-algorithm (see Eq. (5.51)), the cost function J_{KF} does not include any weighting matrix [103].

The idea of the optimization problem is to calculate $\hat{\mathbf{x}}$ to minimize J_{KF} and thus to reach $\boldsymbol{\epsilon}_{\hat{\mathbf{x}}} = 0$. The estimated state variables $\hat{\mathbf{x}}$ are linked to the feedback matrix $\mathbf{K}(k)$ of the KF (see also Eq. (8.19) and Eq. (8.13)). The matrix $\mathbf{K}(k)$ is time-variant and corresponds to a continuous correction to minimize the estimation error $\boldsymbol{\epsilon}_{\hat{\mathbf{x}}}$. All measurements up to the current sample point are included.

The following assumptions for the design of the KF are made [103]:

- Process noise and measurement noise are zero-mean
- Process noise and measurement noise are pairwise uncorrelated
- Process noise and measurement noise are not cross-correlated
- Process noise and measurement noise are uncorrelated to state variables

Minimizing the cost function J_{KF} according to Eq. (8.11) leads to the five equations of the KF, see Eq. (8.12), Eq. (8.13), Eq. (8.14), Eq. (8.15) and Eq. (8.16) [103].

$$\mathbf{K}(k) = \mathbf{\Pi}(k\,|\,k-1)\mathbf{C}^{\mathrm{T}}\big[\mathbf{C}\mathbf{\Pi}(k\,|\,k-1)\mathbf{C}^{\mathrm{T}} + \mathrm{Var}[\boldsymbol{\nu}(k)]\big]^{-1} \tag{8.12}$$

$$\hat{\mathbf{x}}(k\,|\,k) = \hat{\mathbf{x}}(k\,|\,k-1) + \mathbf{K}(k)\big[\mathbf{y}(k) - \mathbf{C}\hat{\mathbf{x}}(k\,|\,k-1)\big] \tag{8.13}$$

$$\mathbf{\Pi}(k\,|\,k) = \big(\mathbf{I} - \mathbf{K}(k)\mathbf{C}\big)\mathbf{\Pi}(k\,|\,k-1) \tag{8.14}$$

$$\hat{\mathbf{x}}(k+1\,|\,k) = \mathbf{A}_{\mathrm{d}}\hat{\mathbf{x}}(k\,|\,k) + \widetilde{\mathbf{B}}_{\mathrm{d}}\widetilde{\mathbf{u}}(k) \tag{8.15}$$

$$\mathbf{\Pi}(k+1\,|\,k) = \mathbf{A}_{\mathrm{d}}\mathbf{\Pi}(k\,|\,k)\mathbf{A}_{\mathrm{d}}^{\mathrm{T}} + \mathbf{G}_{\mathrm{d}}\mathrm{Var}[\boldsymbol{\mu}(k)]\mathbf{G}_{\mathrm{d}}^{\mathrm{T}} \tag{8.16}$$

The variances (Var) of the measurement noise appear in Eq. (8.12). The variances of the process noise appear in Eq. (8.16). These variances describe the dispersions of the according variables around their expected values.

The covariance matrix of the estimation error $\epsilon_{\hat{x}}$ is named $\Pi(k)$ [103]. The covariance is a measure of the joint variability of two random variables. It equals the product of the deviations of two random variables from their expected value [103].

The Kalman filter algorithm (see Eq. (8.12), Eq. (8.13), Eq. (8.14), Eq. (8.15) and Eq. (8.16)) represents a recursive and iterative algorithm. The Fig. 8.3 visualizes the two steps of the KF at one sampling point k as a time sequence, see also [133].

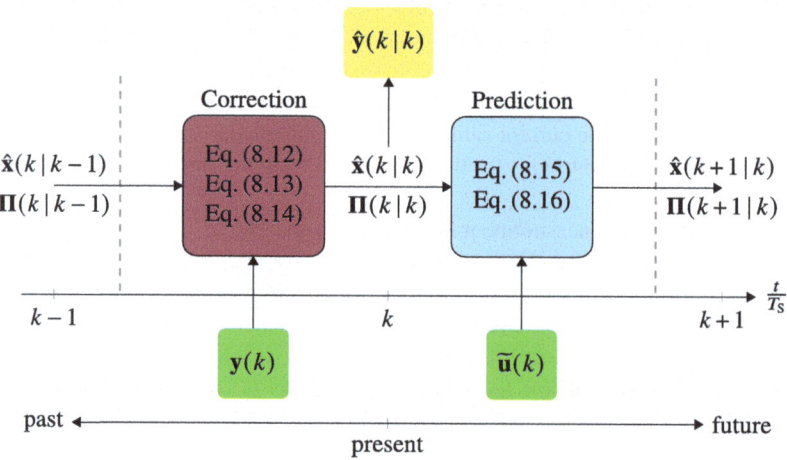

Fig. 8.3 Time sequence of the Kalman filter at the sampling point k consisting of the correction step and the prediction step

The first step in Fig. 8.3 is the correction step using the measurement equations (see Eq. (8.8)) to improve the estimation by updating the predicted state variables $\hat{x}(k \mid k-1)$ and updating the covariance matrix of the estimation error $\Pi(k \mid k-1)$ from the previous sampling point $(k-1)$ (see Eq. (8.13) and Eq. (8.14)). The correction is realized using the current measurement variables $y(k)$ (see Eq. (8.13)). The Kalman matrix $K(k)$ defines the impact of the correction using the current measurement variables $y(k)$. It is updated in each sampling point (see Eq. (8.12)). The correction step is interpreted as a a-posteriori system monitoring step [103].

The second step in Fig. 8.3 is the predicting step using the state extrapolation equations according to Eq. (8.7) neglecting the process noise to predict the state variables $\hat{x}(k+1 \mid k)$ and the covariance matrix of the estimation error $\Pi(k+1 \mid k)$ of the next sampling point $(k+1)$ (see Eq. (8.15) and Eq. (8.16)). The prediction

8.2 State Estimation

is realized using the current input variables $\widetilde{\mathbf{u}}(k)$ (see Eq. (8.15)). The predicted variables are used in the next sampling point. The predicted state variables $\hat{\mathbf{x}}(k+1 \mid k)$ are used to update the predicted state variables in the next sampling point, see Eq. (8.13). The predicted covariance matrix is used to update itself and to calculate the matrix $\mathbf{K}(k)$ in the next sampling point, see Eq. (8.12) and Eq. (8.14). The prediction step is interpreted as a-priori step to ensure real-time capability [103].

A connection between the KF and a low-pass filter (LPF) is identified. The prediction step in the discrete-time domain is equivalent to a time derivative in the continuous-time domain, see Sect. 3.2.3 and compare Eq. (3.46) and Eq. (3.29). Calculating time derivatives of noisy measurement signals according to Eq. (3.29) leads to high magnitudes. A LPF is often used to attenuate these signals. In contrast to the KF, a LPF is not real-time capable due to its group delay describing a time shift.

The estimated state variables $\hat{\mathbf{x}}(k)$ and the covariance matrix of the estimation error $\mathbf{\Pi}(k)$ have to be initialized. Often, no information is available. Consequently, both variables are initialized by zero.

The updated estimated state variables $\hat{\mathbf{x}}(k \mid k)$ and the updated estimated measurement variables $\hat{\mathbf{y}}(k \mid k)$ using Eq. (8.4) are extracted in each sampling point, see Fig. 8.3.

In practical applications using a one-step prediction, the five KF equations (see Eq. (8.12), Eq. (8.13), Eq. (8.14), Eq. (8.15) and Eq. (8.16)) are simplified, see Eq. (8.17), Eq. (8.18) and Eq. (8.19) [103].

$$\mathbf{K}(k) = \mathbf{A}_d \mathbf{\Pi}(k) \mathbf{C}^T \left[\mathbf{C} \mathbf{\Pi}(k) \mathbf{C}^T + \text{Var}[\mathbf{v}(k)] \right]^{-1} \tag{8.17}$$

$$\mathbf{\Pi}(k+1 \mid k) = \left[\mathbf{A}_d - \mathbf{K}(k) \mathbf{C} \right] \mathbf{\Pi}(k) \mathbf{A}_d^T + \mathbf{G}_d \text{Var}[\boldsymbol{\mu}(k)] \mathbf{G}_d^T \tag{8.18a}$$

$$= \mathbf{A}_d \mathbf{\Pi}(k) \mathbf{A}_d^T + \mathbf{G}_d \text{Var}[\boldsymbol{\mu}(k)] \mathbf{G}_d^T - \mathbf{K}(k) \mathbf{C} \mathbf{\Pi}(k) \mathbf{A}_d^T \tag{8.18b}$$

$$\hat{\mathbf{x}}(k+1 \mid k) = \mathbf{A}_d \hat{\mathbf{x}}(k \mid k-1) + \widetilde{\mathbf{B}}_d \widetilde{\mathbf{u}}(k) + \mathbf{K}(k) \left[\mathbf{y}(k) - \mathbf{C} \hat{\mathbf{x}}(k \mid k-1) \right] \tag{8.19}$$

The simplified covariance update equation in Eq. (8.18b) represents a dual representation of the discrete-time Riccati equation in Eq. (5.53) for V/I-controllers of VSCs. In contrast to Eq. (5.53), the Eq. (8.18b) is solved through integration forwards in time.

The basic idea of the model-based protection (MBP) algorithm is to evaluate, if measured values and estimated values based on models of corresponding variables match or not, see Sect. 8.1. In this way, the most suitable model to the actual system state is found. From the consideration of this section, the KF represents an appropriate state estimation algorithm for practical applications. The KF reduces estimation errors caused by noisy measurement signals. Consequently, estimation errors are allocated to false model configurations and the most suitable model is found and protection decisions are made. Furthermore, the KF shows real-time capability and high speed is a basic prerequisite of protection algorithms, see Fig. 8.6.

The presented KF is suitable for the models of linear systems in Chap. 7. However, there are extensions of the KF (see [132]), which could be applied to the MBP in the future. The extended Kalman filter (EKF) uses first order Taylor approximations of non-linear systems around one point [134]. The unscented Kalman filter (UKF) is based on a different principle of linear estimation approximating around multiple so-called sigma points [135]. Some relations between the EKF and the UKF are presented in [136].

8.3 Protection Criterion Based on Hypothesis Testing

The distinction between normal operation conditions and faulty operation conditions is based on a protection criterion. Today, the $VI\varphi$-starting mode is used for example [143]. A suitable protection criterion of the model-based protection (MBP) algorithm is derived from hypothesis testing in this section. The identified protection criterion is used subsequently in the fault identification layer to decide, if normal operation conditions inside the monitored grid area are present or not, see Sect. 8.4.2.

The updated estimated measurement variables from the Kalman filter (KF) are named $\hat{\mathbf{y}}(k \mid k)$ (see Sect. 8.2.2 and Fig. 8.3). The difference between the measured output variables $\mathbf{y}(k)$ and $\hat{\mathbf{y}}(k \mid k)$ is named the estimation error of the output variables $\epsilon_{\hat{y}}(k)$. It is calculated according to Eq. (8.20).

$$\epsilon_{\hat{y}}(k) = \mathbf{y}(k) - \hat{\mathbf{y}}(k \mid k) \tag{8.20}$$

The estimation error of the output variables $\epsilon_{\hat{y}}$ depends on the state extrapolation equations in Eq. (8.7) and the measurement equations in Eq. (8.8). Subsequently it is assumed that the KF operates in its steady state and the initial estimation error is compensated. It is concluded that the estimation error $\epsilon_{\hat{y}}$ is low, if Eq. (8.7) and Eq. (8.8) reflect the real characteristics of the monitored grid area sufficiently. The estimation error $\epsilon_{\hat{y}}$ is high, if Eq. (8.7) and Eq. (8.8) reflect the actual characteristics of the monitored grid area insufficiently.

8.3 Protection Criterion Based on Hypothesis Testing

A statistical hypothesis test (see [137]) allows to draw conclusions from $\epsilon_{\hat{y}}$ to evaluate the quality of a model given by Eq. (8.7) and Eq. (8.8). In this way, the goodness-of-fit of a model is quantified. The estimation error $\epsilon_{\hat{y}}$ represents a random variable. In this way, the confidence level (see [137]) of the model used inside the KF regarding the measured output variables $\mathbf{y}$ is quantified. The following null-hypothesis H_0 is formulated: The model given by Eq. (8.7) and Eq. (8.8) reflects the real characteristics of the monitored grid area. The null-hypothesis H_0 is tested using $\epsilon_{\hat{y}}$.

The chi-squared distribution test (abbreviated as χ^2-test) is a mathematical tool, which is often used in practical applications of hypothesis tests [137]. The random variable $\zeta(k)$ is introduced as the test variable of the χ^2-test. It is calculated according to Eq. (8.21), see [138]. It represents the weighted sum of the squared elements of $\epsilon_{\hat{y}}(k)$.

$$\zeta(k) = \epsilon_{\hat{y}}^{\mathrm{T}}(k) \mathbf{W} \epsilon_{\hat{y}}(k) \tag{8.21}$$

The weighting matrix $\mathbf{W}$ contains the standard deviation σ_i of each element of $\epsilon_{\hat{y}}$, see [138]. It is calculated according to Eq. (8.22). Here, n represents the number of elements in $\epsilon_{\hat{y}}$. The weighting matrix $\mathbf{W}$ emphasizes reliable elements of $\epsilon_{\hat{y}}$ with a low standard deviation. Less reliable elements of $\epsilon_{\hat{y}}$ with a high standard deviation have a lower impact on the random variable ζ. In this way, the protection criterion gets independent of the grid voltage level.

$$\mathbf{W} = \mathrm{diag}\left[1/\sigma_1^2,\ 1/\sigma_2^2,\ ...,\ 1/\sigma_n^2\right] \tag{8.22}$$

It is assumed that the elements of $\epsilon_{\hat{y}}(k)$ are normally distributed and thus ζ is χ^2-distributed [137]. The probability density function (PDF) $f_\nu(x)$ of a χ^2-distributed random variable depending on the degrees of freedom ν is described according to Eq. (8.23) [137].

$$f_\nu(x) = \frac{x^{\frac{1}{2}(\nu-2)} e^{-\frac{1}{2}x}}{2^{\frac{1}{2}\nu} \Gamma\left(\frac{1}{2}\nu\right)} \tag{8.23}$$

The Fig. 8.4 shows the PDF $f_\nu(x)$ of a χ^2-distributed random variable depending on the degrees of freedom ν according to Eq. (8.23). Regarding the MBP, ν equals the difference of the number of state variables $\mathbf{x}$ and the number of measured output variables $\mathbf{y}$ [124, 138]. In Fig. 8.4, the gamma function Γ (see [75]) is used.

To evaluate H_0 via a χ^2-test, the p-value is calculated. The p-value represents a measure of evidence of the plausibility of H_0. A high p-value indicates that

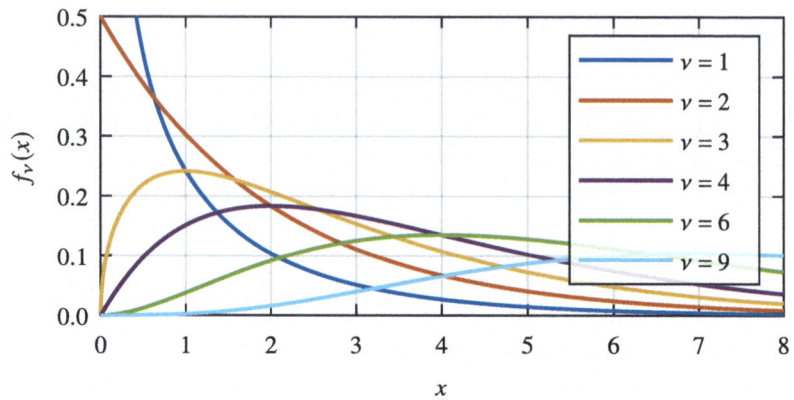

Fig. 8.4 Probability density function $f_\nu(x)$ of a χ^2-distributed random variable depending on the degrees of freedom ν

the estimations reinforce H_0. A low p-value indicates that the estimations do not reinforce H_0.

The calculation of the p-value is based on the calculation of the probability $P_h\big(\zeta(k), \nu\big) \in [0, 1]$ (index h: hypothesis), that any single estimation of ζ is in the interval $\big[0, \zeta(k)\big]$ assuming ζ is χ^2-distributed and H_0 is true. If $\zeta(k) \to \infty$, the probability equals one. The probability $P_h\big(\zeta(k), \nu\big)$ is calculated using the cumulative distribution function (CDF) representing the integral of $f_\nu(x)$, see Eq. (8.24).

$$P_h\big(\zeta(k), \nu\big) = \int_0^{\zeta(k)} f_\nu(x)\,\mathrm{d}x \qquad (8.24)$$

The p-value $p_h(k) \in [0, 1]$ (see [137]) is defined as the complementary value of $P_h\big(\zeta(k), \nu\big)$ to one, see Eq. (8.25a). Substituting Eq. (8.24) and Eq. (8.23) leads to Eq. (8.25b) and Eq. (8.25c).

$$p_h(k) = 1 - P_h\big(\zeta(k), \nu\big) \qquad (8.25a)$$

$$= 1 - \int_0^{\zeta(k)} f_\nu(x)\,\mathrm{d}x \qquad (8.25b)$$

$$= 1 - \int_0^{\zeta(k)} \frac{x^{\frac{1}{2}(\nu-2)} e^{-\frac{1}{2}x}}{2^{\frac{1}{2}\nu}\Gamma\big(\frac{1}{2}\nu\big)}\,\mathrm{d}x \qquad (8.25c)$$

8.3 Protection Criterion Based on Hypothesis Testing

In Fig. 8.5, the probability density function $f_3(x)$ of a χ^2-distributed random variable assuming $\nu = 3$ and the p-value $p_h(k)$ depending on $\zeta(k)$ is shown exemplary.

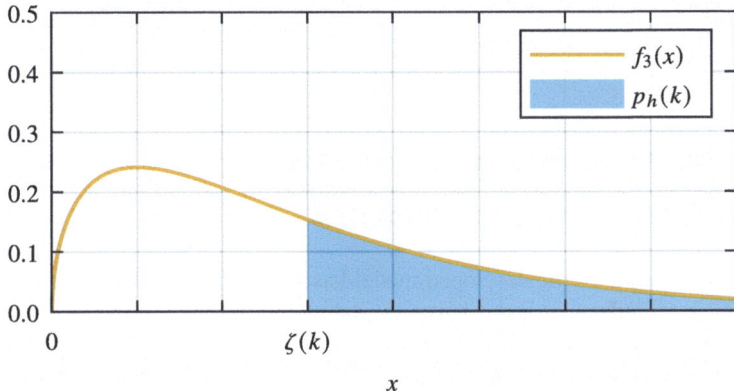

Fig. 8.5 Probability density function $f_3(x)$ of a χ^2-distributed random variable assuming $\nu = 3$ and the p-value $p_h(k)$ depending on $\zeta(k)$

If the value of $\zeta(k)$ is high, the calculated probability P_h is nearly one, the p-value is nearly zero and consequently H_0 is rejected. The model given by Eq. (8.7) and Eq. (8.8) is assumed not to reflect the real characteristics of the monitored grid area. If the value of $\zeta(k)$ is low, the calculated probability P_h is nearly zero, the p-value is nearly one and consequently H_0 is accepted. The model given by Eq. (8.7) and Eq. (8.8) is assumed to reflect the real characteristics of the monitored grid area.

The current p-value $p_h(k)$ (see Eq. (8.25c)) shows a non-linear dependence from current value of the random variable $\zeta(k)$ (see Eq. (8.21)). The p-value is independent of past values and thus reflects the current characteristics of the monitored grid area.

From this perspective, the p-value $p_h(k)$ of the null-hypothesis H_0 is selected as the protection criterion of the MBP, see also [124]. Comparing $p_h(k)$ with a predefined confidence level $\alpha_h \in [0, 1]$ allows to decide, whether a grid fault inside the monitored grid is present or not. Based on this decision, the tripping of circuit breakers is initiated or not. The matrices $\mathbf{A}_d$ and $\widetilde{\mathbf{B}}_d$ of the model given by Eq. (8.7) used inside H_0 describing healthy grid areas are given in Sect. 7.2. Two cases are distinguished:

- $p_h(k) \geq \alpha_h$: H_0 is accepted on the confidence level α_h. The model describing a healthy grid area is assumed to be correct and no grid fault is assumed to be present.
- $p_h(k) < \alpha_h$: H_0 is rejected on the confidence level α_h. The model describing a healthy grid area is assumed to be incorrect and a grid fault is assumed to be present.

8.4 Model-Based Protection Scheme

8.4.1 Basic Structure and Properties

The basic principle of the developed model-based protection (MBP) scheme is to compare measured values and calculated values based on models of corresponding output variables and to evaluate, if they match or not, see Sect. 7.1. The calculation is realized using the dynamic state estimation (DSE) principle, see Sect. 8.1. Kalman filters (KFs) are used, since they are suitable for the practical application of the state estimation, see Sect. 8.2.2 and [159].

The prerequisite to apply the MBP are state-space models of grid areas (see Sect. 7.2) and lines (see Sect. 7.4). The definition of the input variables and output variables of the models is crucial. The estimation error of the output variables is calculated. Based on this estimation error, the most suitable model is found and the fault analysis and the tripping decision are made.

The MBP consists of three layers: The fault identification (FI) layer (see Sect. 8.4.2), the fault characterization (FC) layer (see Sect. 8.4.3) and the fault localization (FL) layer (see Sect. 8.4.4). These three layers are hierarchically structured, see Fig. 8.6.

Future power grids will profit from advances in communication and digitalization. The MBP is based on one central data processing unit per monitored grid area. It is also conceivable, that multiple grid areas are agglomerated regarding the communication structure and a modular protection scheme is established. The Fig. 8.7 exemplary shows the central data processing and communication effort of the three layers of the MBP monitoring one grid area. For clarity reasons, the circuit breaker of the according protection relay R is not illustrated.

The FI is based on the measurements of voltages and currents at the border of the monitored grid area. In contrast, the FC and the FL additionally need measurements of voltages and currents inside the monitored grid area. Therefore, information about both sides of one line are necessary.

8.4 Model-Based Protection Scheme

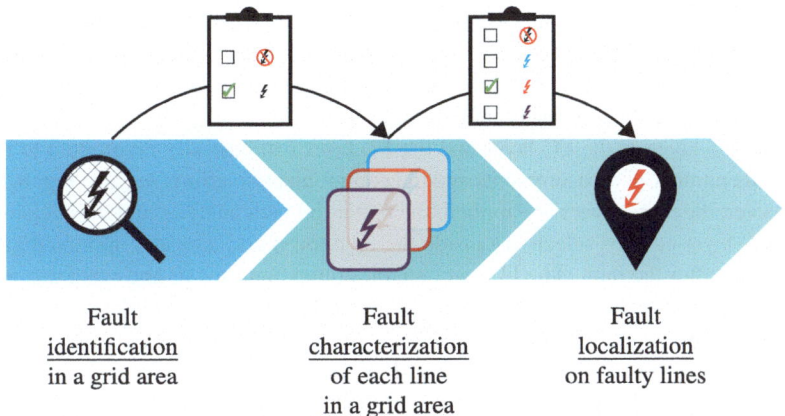

Fig. 8.6 Model-based protection (MBP) scheme as a three layer structure consisting of fault identification (FI), fault characterization (FC) and fault localization (FL)

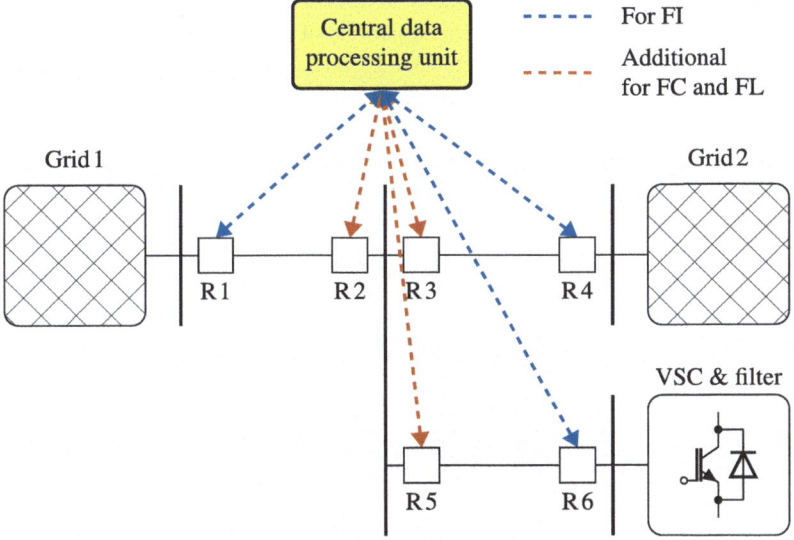

Fig. 8.7 Central data processing unit and communication effort of the fault identification (FI) layer, the fault characterization (FC) layer and the fault localization (FL) of the model-based protection (MBP) monitoring one grid area regarding six relays R 1 to R 6

The measurement equipment basically consists of voltage transformers (VTs) and current transformers (CTs). Time-synchronized measurements via the Global Positioning System (GPS) as well as high speed data transmission between the relay and the central data processing unit are necessary.

The FC layer or the FC layer and the FL layer can basically be omitted to reduce the number of measurements and the communication effort (see also Fig. 8.7). However, the more layers are applied, the better the fault analysis and the decision. The FC layer ensures selectivity inside the monitored grid area and improved grid operation after the fault. The FL layer enables quick fault repair and rapid restoration of the grid architecture. Therefore, all three layers are used in the subsequent investigations.

All three layers of the MBP are based on the application of KFs. KFs process sampled values of the measurement equipment and operate in the time domain. The models used in the KF equations are based on space vector components in the stationary reference frame, see Sect. 3.1.2.3, Sect. 7.2 and Sect. 7.4. Discretization of these state-space models is applied, see Sect. 3.2.3.

Space vector components in rotating reference frames (see Sect. 3.1.2.4 and Sect. 3.1.2.5) are not used to avoid the separation of positive and negative sequence components (see Sect. 4.3.3 and Sect. 4.3.4.1) and the enlargement of the state-space models. Furthermore, the calculation of transformations angles in the transient state during grid faults leads to inaccuracies.

The MBP accomplishes fundamental requirements on protection algorithms. These properties are summarized in Fig. 8.8 and discussed below.

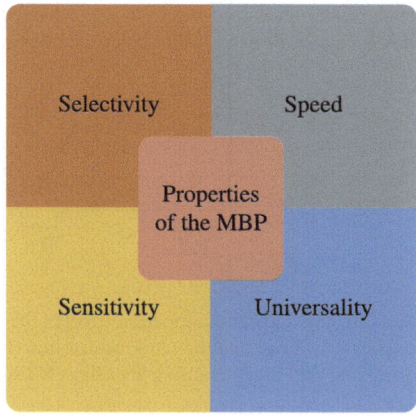

Fig. 8.8 Fundamental properties of the model-based protection (MBP) algorithm

8.4 Model-Based Protection Scheme

The selectivity regarding the faulty line is reached by monitoring each line of the faulty grid area inside the FC layer. Alike the differential protection, the MBP using the FC layer is inherently selective. This property is crucial for the operation of the grid after the fault. The FI layer is inherently selective regarding the monitored grid area.

Since instantaneous values are processed by KFs as points in time and no time interval is used as calculation basis (see Tab. 3.1), the MBP represents a high speed scheme. In this context, the tripping speed of the circuit breakers and further equipment is not regarded. The speed of the protection algorithm is important to avoid or limit damages to living beings and technical equipment as far as possible.

The usage of estimation errors in each layer of the MBP leads to good sensitivity, which is independent of absolute values of voltages or currents. The definition of the significance level as a measure of comparison of the calculated p-value (see Sect. 8.3) allows to adjust the sensitivity of the FI layer. Sensitivity, which is independent of absolute values of currents, is essential for the protection of 100 % inverter-based power systems (IBPS) because fault currents are limited by semiconductor power device, see also Sect. 2.1.

The MBP uses models that are based on physical laws (Kirchhoff's voltage law (KVL) and Kirchhoff's current law (KCL)) and V/I-characteristics. Consequently, the MBP does not restrict the flexibility options of VSCs, especially in the negative and the zero sequence during grid faults (see Sect. 5.2.3.2 and Sect. 6.2.3.1). Furthermore, multidirectional power flows (see Sect. 1.1.3) and greater load angles than today are handled.

8.4.2 Fault Identification

The objective of the fault identification (FI) layer is to decide, if normal operation conditions inside the monitored grid area are present or not. A grid area is investigated. Broad areas of a utility grid are divided into multiple smaller grid areas.

One model of a healthy grid area using $\alpha\beta$-components according to Sect. 7.2 is used for the FI. Zero sequence components are not considered because a healthy grid area with zero sequence currents, that equal zero, is assumed. The $\alpha\beta$-components of the external currents at the borderlines of the grid area are used as input variables $\tilde{\mathbf{u}}_{FI}$, see e.g. Eq. (7.4), Fig. 7.4 and Eq. (8.7). The $\alpha\beta$-components of the line currents and the busbar voltages are modelled as state variables $\mathbf{x}_{FI}$, see e.g. Eq. (7.4), Fig. 7.4 and Eq. (8.7). From these state variables, the busbar voltages are selected as measured output variables $\mathbf{y}_{FI}$ by defining the matrix $\mathbf{C}$, see Eq. (8.8).

To clarify the selection of variables, a topology consisting of two serial lines according to Sect. 7.2.2.2 is regarded. The vectors in accordance with Eq. (8.26), Eq. (8.27) and Eq. (8.28) reveal.

$$\widetilde{\mathbf{u}}_{FI} = (i_{A\alpha}, i_{A\beta}, i_{C\alpha}, i_{C\beta})^T \quad (8.26)$$

$$\mathbf{x}_{FI} = (i_{L1\alpha}, i_{L1\beta}, i_{L2\alpha}, i_{L2\beta}, v_{A\alpha}, v_{A\beta}, v_{B\alpha}, v_{B\beta}, v_{C\alpha}, v_{C\beta})^T \quad (8.27)$$

$$\mathbf{y}_{FI} = (v_{A\alpha}, v_{A\beta}, v_{C\alpha}, v_{C\beta})^T \quad (8.28)$$

The principle of the FI is explained based on Fig. 8.9. The input variables $\widetilde{\mathbf{u}}_{FI}$ and the output variables $\mathbf{y}_{FI}$ of the grid area are measured. The state variables $\mathbf{x}_{FI}$ are estimated using a Kalman filter (KF_{FI}) with $\widetilde{\mathbf{u}}_{FI}$ and $\mathbf{y}_{FI}$, see Sect. 8.2.2. The matrices $\mathbf{A}_d$ and $\widetilde{\mathbf{B}}_d$ are applied according to the model of a healthy grid area, see Sect. 7.2. The estimated output variables $\hat{\mathbf{y}}_{FI}$ are calculated from the estimation of $\mathbf{x}_{FI}$ via Eq. (8.4). The estimation error of the output variables $\epsilon_{\hat{y}}(k)$ is calculated from $\mathbf{y}_{FI}$ and $\hat{\mathbf{y}}_{FI}$, see Eq. (8.20).

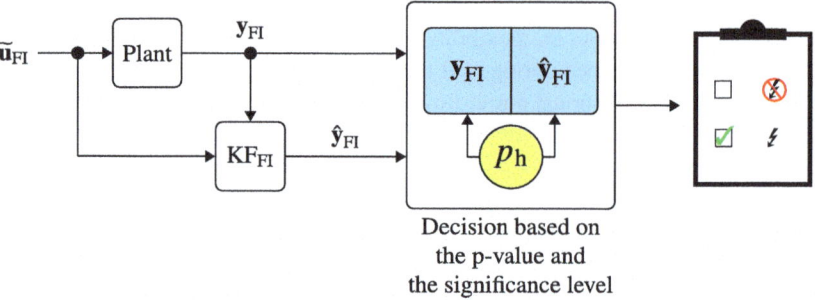

Fig. 8.9 Principle of the fault identification (FI) layer of the model-based protection (MBP) scheme investigating a grid area based on one Kalman filter KF_{FI}

A statistical hypothesis test (see Sect. 8.3) regarding the following null-hypothesis H_0 is executed: The model reflects the real characteristics of the monitored grid area. The p-value p_h is calculated according to Eq. (8.25c) using the weighted estimation error $\zeta(k)$, see Eq. (8.21). Comparing $p_h(k)$ with a predefined confidence level α_h allows to decide, if a grid fault inside the monitored grid area is present or not. If a grid fault is identified, the entire grid area is disconnected by circuit breakers or the FC layer (see Sect. 8.4.3) is activated.

8.4 Model-Based Protection Scheme

There are different ways to achieve the state-space model of a healthy grid area, that is used inside the FI. In this thesis, physical laws and V/I-characteristics are applied. In future research work, the state-space model can also be derived from artificial intelligence (AI) methods for parameter identification. The idea is to use the broad measurement data basis gained during normal operating conditions for the AI training.

8.4.3 Fault Characterization

The objective of the fault characterization (FC) layer is to decide, if monitored lines are healthy or not. Each line inside the monitored grid area of the FI layer (see Sect. 8.4.2) is investigated. All lines are investigated independent of each other. If a fault on one line is detected, this line is investigated in detail regarding the fault resistance configuration and the values of the appearing fault resistances, see Tab. 7.1 and Tab. 7.2. A few selected fault locations of one line are chosen.

In contrast to the FI, the FC uses one model of a healthy line (see Sect. 7.4.1) and multiple models of faulty lines (see Sect. 7.4.2). One model of a faulty line is characterized by the fault resistance configuration, the values of the appearing fault resistances and the fault location. These models are based on αβ-components and zero sequence components. The capacitances are neglected because faulty lines show dominant inductive characteristics during a fault [143].

The αβ0-components of the busbar voltages of the monitored line are used as input variables $\widetilde{\mathbf{u}}_{FC}$, see e.g. Sect. 7.4.3, Sect. 7.4.4 and Eq. (8.7). The αβ-components of the line currents on both ends of the line are modelled as state variables $\mathbf{x}_{FC}$ and are selected as measured output variables $\mathbf{y}_{FC}$ by defining the matrix $\mathbf{C}$, see Eq. (8.8).

The Eq. (8.29), Eq. (8.30) and Eq. (8.31) clarify the selection of variables, see also Fig. 7.11.

$$\widetilde{\mathbf{u}}_{FC} = (v_{A\alpha},\ v_{A\beta},\ v_{A0},\ v_{B\alpha},\ v_{B\beta},\ v_{B0})^T \tag{8.29}$$

$$\mathbf{x}_{FC} = (i_{a\alpha},\ i_{a\beta},\ i_{a0},\ i_{b\alpha},\ i_{b\beta},\ i_{b0})^T \tag{8.30}$$

$$\mathbf{y}_{FC} = \mathbf{x}_{FC} \tag{8.31}$$

The principle of the FC is explained based on Fig. 8.10. The input variables $\widetilde{\mathbf{u}}_{FC}$ and the output variables $\mathbf{y}_{FC}$ of the line are measured. The state variables $\mathbf{x}_{FC}$ are estimated using multiple Kalman filters (KF_{FCi}) with $\widetilde{\mathbf{u}}_{FC}$ and $\mathbf{y}_{FC}$, see Sect. 8.2.2. The matrices $\mathbf{A}_d$ and $\widetilde{\mathbf{B}}_d$ of each KF differ depending on the model, see Sect. 7.4. The estimated state variables equal the estimated output variables $\hat{\mathbf{y}}_{FCi}$.

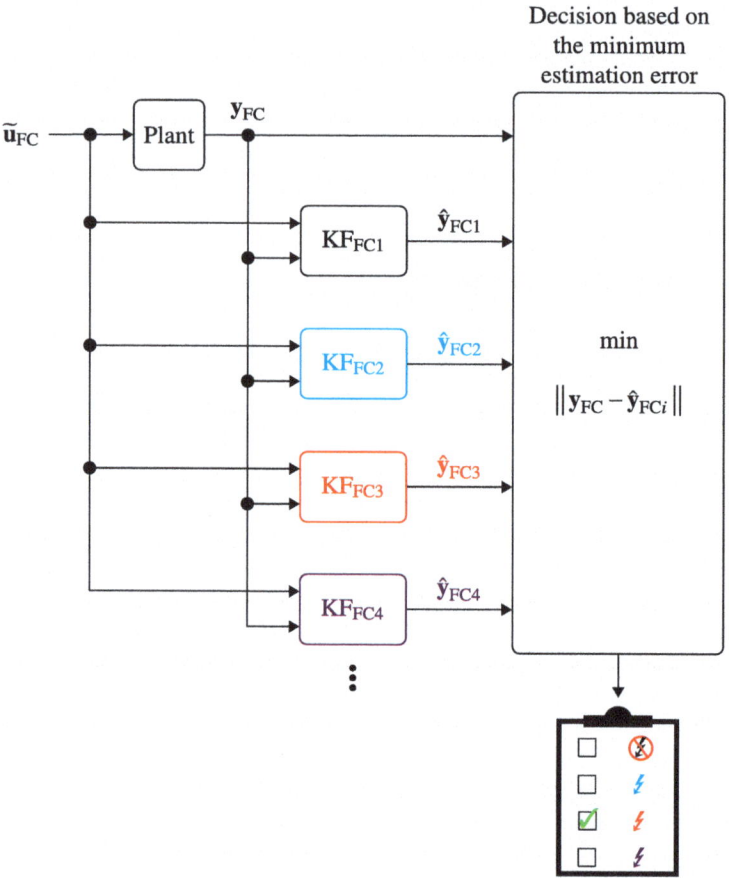

Fig. 8.10 Principle of the fault characterization (FC) layer of the model-based protection (MBP) scheme investigating one line based on multiple Kalman filters KF_{FCi} using different models. The models differ from each other regarding the fault resistance configuration, the values of the appearing fault resistances and the rough fault location

8.4 Model-Based Protection Scheme

The Euclidean norm of the estimation errors of the output variables $\| \mathbf{y}_{FC} - \hat{\mathbf{y}}_{FCi} \|$ is calculated. The model representing the minimum Euclidean norm is selected to decide, whether the line is healthy or not and optionally to characterize the fault. If one or multiple faulty lines inside the monitored grid area of the FI are identified, these lines are disconnected by circuit breakers. The information about the characterization of faulty lines are passed to the FL layer (see Sect. 8.4.4).

The FC is based on the selection of one model out of a pool of models. In contrast, the FI is based on the confidence level of one model of a healthy grid area.

In contrast to the FI, it is difficult to derive the models of the FC from AI methods. The reason is, that there are numerous parameter uncertainties and degrees of freedom. Moreover, the measurement data basis regarding grid faults is limited.

8.4.4 Fault Localization

The objective of the fault localization (FL) layer is to locate the fault on a faulty line based on one fault configuration and the values of the appearing fault resistances identified by the FC layer, see Sect. 8.4.3. Each line, on which a fault is identified in the FC layer, is investigated. In contrast to the FC, models based on many fault locations are chosen. The fault localization is not possible for longitudinal faults in principle, see Sect. 7.3.2. The fault localization on cables is often more crucial than the fault location on overhead lines. The FL is often not as time-sensitive as the FI and the FC.

The FL uses multiple models of faulty lines, which differ from each other regarding the fault location. The fault resistance configuration, and the values of the appearing fault resistances are equal in each model and predefined by the FC. The models of the FL are also based on $\alpha\beta$-components and zero sequence components. The capacitances are neglected because faulty lines show dominant inductive characteristics during a fault [143].

The input variables $\tilde{\mathbf{u}}_{FL}$, the state variables and the measured output variables $\mathbf{y}_{FL}$ are equal to the variables of the FC. The Eq. (8.32), Eq. (8.33) and Eq. (8.34) clarify the selection of variables, see also Fig. 7.11.

$$\tilde{\mathbf{u}}_{FL} = (v_{A\alpha},\ v_{A\beta},\ v_{A0},\ v_{B\alpha},\ v_{B\beta},\ v_{B0})^T \tag{8.32}$$

$$\mathbf{x}_{FL} = (i_{a\alpha},\ i_{a\beta},\ i_{a0},\ i_{b\alpha},\ i_{b\beta},\ i_{b0})^T \tag{8.33}$$

$$\mathbf{y}_{FL} = \mathbf{x}_{FL} \tag{8.34}$$

The principle of the FL is explained based on Fig. 8.11. It basically equals the principle of the FC, see Sect. 8.4.3. The input variables $\tilde{\mathbf{u}}_{FL}$ and the output variables

$\mathbf{y}_{FL}$ of the line are measured. Multiple Kalman filters (KF_{FLi}) are used to estimate the state variables $\mathbf{x}_{FL}$. The matrices $\mathbf{A}_d$ and $\widetilde{\mathbf{B}}_d$ of each KF differ depending on the model, see Sect. 7.4. The estimated state variables equal the estimated output variables $\hat{\mathbf{y}}_{FCi}$.

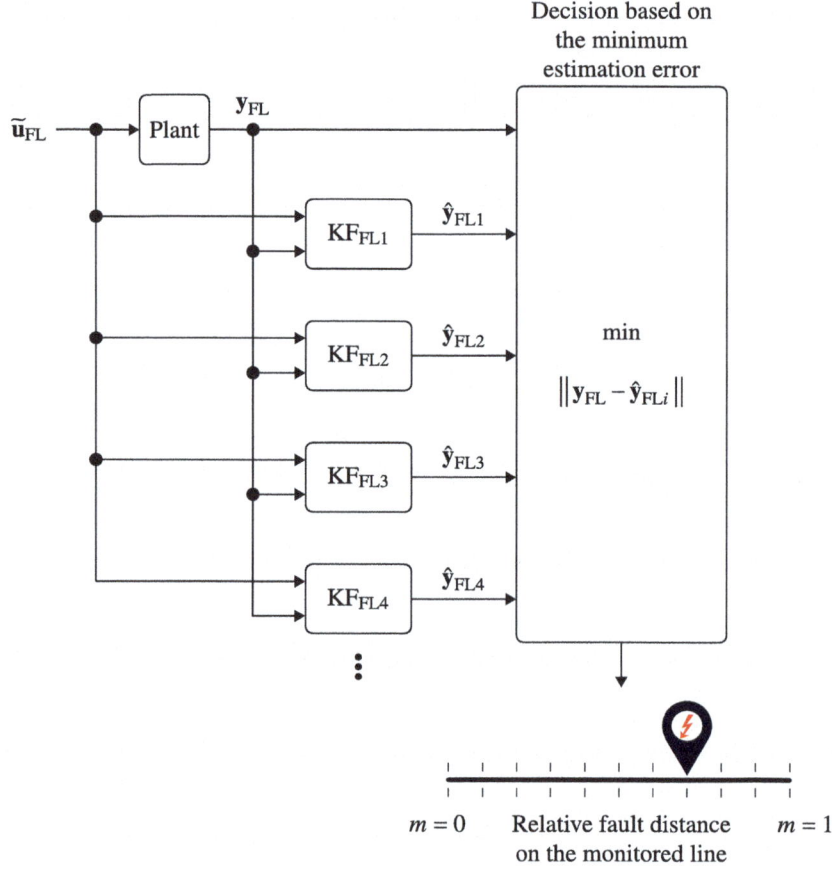

Fig. 8.11 Principle of the fault characterization (FL) layer of the model-based protection (MBP) scheme investigating one line based on multiple Kalman filters KF_{FLi} using different models. The models differ from each other regarding the precise fault location

8.4 Model-Based Protection Scheme

The Euclidean norm of the estimation errors of the output variables $\|\mathbf{y}_{FL} - \hat{\mathbf{y}}_{FLi}\|$ is calculated. The model representing the minimum Euclidean norm is selected to decide, where the fault is localised on the monitored line. If the fault on one line is localised, the repair work is initiated.

Similar to the FC, the derivation of models for the FL from AI methods is difficult, because of numerous parameter uncertainties, degrees of freedom and limited data basis.

8.4.5 Distinction From Alternative Protection Principles

8.4.5.1 State-of-the-art Protection Schemes

The differential protection scheme (see [123] and Fig. 7.2) is based on Kirchhoff's current law (KCL). It is often used to protect individual components, such as transformers. However, the differential protection scheme is also applied to high voltage overhead lines.

Several similarities between the differential protection scheme and the MBP are identified. Both protection schemes show inherent selectivity regarding the faulty line. Besides, both schemes use measurement information from both ends of one line. Therefore, communication effort between these ends is necessary. Moreover, both schemes do not provide backup protection for grid areas beyond the monitored grid area (FI) or line (FC).

In contrast to the MBP, the differential protection scheme solely evaluates KCL. Consequently, voltage transformers (VTs) are not needed. Since the differential protection scheme does not analyse characteristics of the line in detail, the characterization and the location of the fault is not possible. The fault characterization and the fault localization are explicit part of the MBP, see Sect. 8.4.3 and Sect. 8.4.4.

From the comparison above, the MBP is regarded as an extension and generalization of the differential protection scheme.

The distance protection scheme (see [41, 143] and Fig. 7.1) is based on the measurement of the electrical distance between the protection relay and the grid fault by calculating the impedance of the fault loop.

The distance protection scheme and the MBP show similarities. Both protection schemes use measurement data from VTs and current transformers (CTs). The characterization and the location of the fault are a fundamental part of both protection schemes. The distance protection scheme and the MBP rely on the coordination of different relays, although the principles of coordination differ fundamentally.

In contrast to the MBP, the distance protection scheme evaluates the V/I-characteristics of several fault loops and not a holistic electrical model of a grid area or a line. In this way, the protection area of the distance protection scheme is not limited to a certain grid area and backup protection is provided. The MBP does not provide backup protection capability. The selectivity regarding the faulty line is realized by the temporal stagger of different protection zones. However, this concept evokes drawbacks regarding the speed of the fault clearance. Furthermore, the calculations of the distance protection scheme are based on phasors (see also Sect. 3.1.1 and Sect. 3.1.2.1) and the communication between different measurement points is not strictly necessary.

Regarding the grid architectures and the grid operation of energy systems based on renewable energy sources (RES) (see also Sect. 1.1.3 and Sect. 1.1.4), different problems of the distance protection scheme are known. Due to the model-based fault analysis, the MBP can handle these problems. The intermediate infeed (see Sect. 7.2.2.3) and the feeding from both sides can lead to a malfunction of the distance protection relay [143]. Moreover, the value of the fault resistance can lead to calculation errors of the fault distance [143]. Due to the limited overcurrent capability of VSCs (see Sect. 2.1.1), the distinction between normal operating conditions and faulty operating conditions is often not clear [143]. The angle between voltages and currents induced by VSCs can exceed the predefined value inside the distance protection scheme [143].

Due to the complementary advantages described above, the combination of the MBP and the distance protection scheme appears to be an attractive approach for future research activities. The distance protection scheme can provide the backup protection of the MBP.

8.4.5.2 Alternative Model-Based Protection Schemes

The "Setting-Less Protection (SLP)" also known as the "Dynamic State Estimation (DSE) Based Protection" is presented in [124–127]. Further investigations are found in [138]. The analysis of KCL of the differential protection scheme is extended to the multi-criteria evaluation "All physical laws", see [126] and also Fig. 7.3. These publications represent the fundamental of the developed MBP in this thesis. The SLP and the MBP are both based on the dynamic state estimation. The χ^2-test is used in the SLP and the FI layer of the MBP. In contrast to [124–127], the focus of the MBP in this thesis are 100 % inverter-based power systems (IBPS).

In contrast to the SLP, the MBP is based on a three layer structure consisting of FI, FC and FL, see Fig. 8.6. The advantage of the three layer structure as a modular concept is that all layers are optimized individually from each other. For example, the model of the FI can be adjusted using artificial intelligence (AI) methods, see Sect. 8.4.2.

Furthermore, the models of the MBP are based on space vector components in the stationary reference frame. In this way, the models are clearer in their structure and the influence of the estimation error of zero sequence currents on the minimum estimation error of the FC layer (see Sect. 8.4.3) is separated. The advantage is, that depending on the zero sequence currents during ground faults, additional decision layers for the tripping of the circuit breakers can be added in the future. The objective is to make decisions for the tripping, that disconnect as few lines as possible and as many as necessary to maintain the grid operation, e.g. during high impedance ground faults.

Moreover, the models of faults lines of the SLP are derived from a model of a healthy line using fault matrices (see also [71]). In contrast to the SLP, the value of the fault resistance is estimated inside the FC layer of the MBP. The advantage is, that this information used inside the FL layer (see Sect. 8.4.4) can improve the precision of the fault location.

In contrast to the publications above, models of lines with a longitudinal fault (see Sect. 7.3.2) are derived. These models help to identify damaged circuit breakers or broken lines.

An "Observer-based Protection (OBP) System" is presented in [139, 140]. This concept significantly differs from the MBP and the SLP. The OBP is based on state observers (see also Sect. 8.2.1) using single phase models. State observers differ from state estimation techniques such as Kalman filters (KFs) in a way, that model uncertainties and measurement errors are not considered. Therefore, the quality of estimation in practical applications is worse and estimation errors are not reliably related to grid faults. Besides, no model of a faulty line is used in the OBP. In this way, the FC and the FL are not possible. The quality of decisions regarding tripping of circuit breakers and fault repair is decreased.

8.5 Summary

In this chapter, the model-based protection (MBP) algorithm is presented. The MBP represents an appropriate protection algorithm for 100 % inverter-based power systems (IBPS) without restricting the flexibility options of voltage source converters during grid faults. The limited overcurrent capability or the harmonics of voltages and currents of VSCs do not influence the MBP negatively, too.

Models of a healthy grid area and of healthy faulty lines are the fundament of the MBP. These models are derived in Chap. 7.

Measured values and estimated values of corresponding variables are compared inside the MBP algorithm. The estimation is based on models and thus the minimum estimation error allows to identify the most suitable model. There are different state estimation algorithms, e.g. the state observer and the Kalman filter (KF). State observers are often applied in the context of state-feedback controllers. The KF is based on statistical methods. It is more suitable for practical applications of the MBP than the state observer, since it considers model uncertainties and measurement errors. A chi-squared distribution test (χ^2-test) is used to test the goodness-of-fit of a model. The p-value is defined as the protection criterion.

The MBP algorithm uses multiple KFs and is based on a three layer structure: fault identification, fault characterization and fault localization. It accomplishes the fundamental requirements on protection algorithms selectivity, speed and sensitivity. The MBP algorithm works independent of characteristics of voltages and currents. Due to its universality, it meets the protection requirements of 100 % IBPS.

The objective of the fault identification layer is to decide, if normal operation conditions inside a monitored grid area are present or not. The objective of the fault characterization is the determination of the fault resistance configuration, the values of the appearing fault resistances and the rough fault location on one line. Each line inside the monitored grid area is investigated. Each faulty line is further investigated in detail by the fault localization layer to specify the fault location precisely.

The MBP algorithm distinguishes from alternative model-based protection algorithms regarding also its three layer structure and the nature of the applied state-space models, opening further advantages. Moreover, possibilities for hybrid protections schemes combining the MBP and the distance protection algorithm reveal.

The structure of the MBP algorithm is developed analytically in this chapter. It is validated by laboratory tests in Chap. 9 and simulations tests in Chap. 10. To follow the key aspects of the central theme of this thesis on a shortcut: The conclusion of the next chapter is provided in Sect. 9.6.

Open Access This chapter is licensed under the terms of the Creative Commons Attribution 4.0 International License (http://creativecommons.org/licenses/by/4.0/), which permits use, sharing, adaptation, distribution and reproduction in any medium or format, as long as you give appropriate credit to the original author(s) and the source, provide a link to the Creative Commons license and indicate if changes were made.

The images or other third party material in this chapter are included in the chapter's Creative Commons license, unless indicated otherwise in a credit line to the material. If material is not included in the chapter's Creative Commons license and your intended use is not permitted by statutory regulation or exceeds the permitted use, you will need to obtain permission directly from the copyright holder.

Validation of the MBP by Laboratory Tests 9

This chapter presents the validation of the model-based protection (MBP) algorithm by laboratory tests. The laboratory environment and the workflow is explained in Sect. 9.1. Four different laboratory single-phase-to-ground fault scenarios are presented in this thesis. The main focus of this chapter is the investigation of different grid topologies based on overhead lines. In Sect. 9.2, the Scenario 1: Double-sided infeed is evaluated. Sect. 9.3 contains the Scenario 2: Double-sided infeed with fault localization. Scenario 3: Parallel line topology is investigated in Sect. 9.4. In Sect. 9.5, the Scenario 4: Intermediate infeed topology is analysed.

9.1 Laboratory Environment and Workflow

The model-based protection (MBP) algorithm (see Chap. 8) is validated by laboratory tests using the analogue grid model at the Institute of Electrical Energy Systems in Erlangen. Detailed information about the analogue grid model is given in [141]. The laboratory environment is shown in Fig. 9.1.

A synchronous inner-pole generator with an excitation device and a nominal power of 12 kVA is the central element of the power plant emulation. The nominal voltage equals 220 V at 1500 min^{-1} and 50 Hz. The turbine is emulated by a speed-controlled externally excited shunt DC motor. The generator is connected via a unit transformer showing a voltage ratio of 1 : 1 to the analogue grid model.

The analogue grid model is connected via a delta-wye transformer to the 400 V low voltage utility grid. The star points of both transformers are solid grounded on their 220 V-sides during the laboratory tests. For clarity reasons, the transformers are not shown in the figures describing the laboratory setups.

Three Omicron DANEO 400 hybrid signal analysers for power utility automation systems with a sampling frequency of 40 kHz are used as multimeters and fault

© The Author(s) 2025
F. Mahr, *Control and Protection of 100% Inverter-based Power Systems*,
https://doi.org/10.1007/978-3-658-47217-7_9

Fig. 9.1 Laboratory environment at the Institute of Electrical Energy Systems in Erlangen

recorders. The technical specifications are given in [142]. The fault recorders directly measure the voltages. The currents are measured via current shunts.

The analogue grid model allows to emulate different 220 kV grid topologies based on overhead lines. The scales of power, voltage, current and impedance are given in Table 9.1. The emulated overhead line parameters are shown in Table 9.2.

The main focus of this chapter is the evaluation of the MBP algorithm in different grid topologies. Overall, 22 scenarios were successfully tested. The results of four selected scenarios with a strong significance for 100 % inverter-based power systems (IBPS) are presented in detail in this thesis. Since voltage source converters

Table 9.1 Scale of electrical quantities in the laboratory environment

Quantity	Scale
Power	$1 : 10^5$
Voltage	$1 : 10^3$
Current	$1 : 10^2$
Impedance	$1 : 10$

9.1 Laboratory Environment and Workflow

Table 9.2 Emulated overhead line parameters in the laboratory according to [141], see also Fig. 7.4

Variable	Value	Description
f_grid	50 Hz	nominal grid frequency
V_grid	220 kV	nominal grid voltage
R'	7 mΩ/km	phase resistance per km
R'_0	5 mΩ/km	neutral resistance per km
L'	95 μH/km	phase inductance per km
L'_0	73 μH/km	neutral inductance per km
C'	60 nF/km	phase-to-ground capacitance per km

(VSCs) connect renewable energy sources (RES) and innovative consumer technologies to the grid, the double-sided infeed of faults is an important issue. The fault localization is basically crucial for quick fault repair and rapid restoration of the grid architecture. The operational integration of location dependant RES into existing grid topologies requires additional energy transport capacity and intermediate connecting lines. Parallel lines topologies are an obvious and practical solution to increase the transport capacity. Besides, intermediate infeed topologies will appear. In this chapter, single-phase-to-ground faults are investigated due to their practical importance. The fault resistance in the analogue grid model is set to zero. The Fig. 9.2 exemplary shows the laboratory setup investigating a parallel line topology.

To test the MBP algorithm using the measurement data from the analogue grid model, the workflow according to Fig. 9.3 is applied. Analogue voltage and current signals are recorded by the Omicron DANEO 400 devices. These signals are saved as COMTRADE sampled values data files. These COMTRADE data files are converted using MathWorks MATLAB into MATLAB timeseries data files. These data files are used to create the appropriate input data for the MBP algorithm executed in MATLAB/Simulink.

Subsequently, the basic diagrams illustrating the achieved results of the four laboratory test scenarios and the shown variables are explained. The p-value $p_\text{h}(k) \in [0, 1]$ is calculated inside the fault identification (FI) layer (see Sect. 8.3 and Sect. 8.4.2) according to Eq. (8.25c). It characterizes the goodness-of-fit of a model of a healthy grid area. The fault decision of the FI layer regarding a grid area is made based on the comparison of the p-value p_h and a predefined confidence level $\alpha_\text{h} \in [0, 1]$. In this chapter, $\alpha_\text{h} = 0.8$ is chosen.

Fig. 9.2 Laboratory setup investigating a parallel line topology

The fault characterization (FC) layer (see Sect. 8.4.3) is described by the index of the model of one line evoking the minimum Euclidean norm of the estimation errors at each sample point. These diagrams show the stability and the speed of the fault decision regarding one line. The FC layer is additionally described by one spider plot per line. Here, the fault decision is visualized regarding selected models characterized by the fault resistance configuration, the values of the appearing fault resistances and the rough fault location. To quantify and to visualize the decision at different points in time before and after the fault, the inverse relative estimation error $IRE_{\text{FC}i} \in (0, 1]$ for the model i is introduced, see Eq. (9.1).

$$IRE_{\text{FC}i}(k) = \frac{\min_i \left\| \mathbf{y}_{\text{FC}}(k) - \hat{\mathbf{y}}_{\text{FC}i}(k) \right\|}{\left\| \mathbf{y}_{\text{FC}}(k) - \hat{\mathbf{y}}_{\text{FC}i}(k) \right\|} \tag{9.1}$$

The most suitable model of one line evokes the minimum Euclidean norm of the estimation errors at one sample point. Consequently, the IRE_{FC}-value of this model equals one. The IRE_{FC}-values of all other investigated models are less than one. The smaller the IRE_{FC}-value of one model, the more unsuitable the regarded model appears to be.

9.2 Scenario 1: Double-sided Infeed

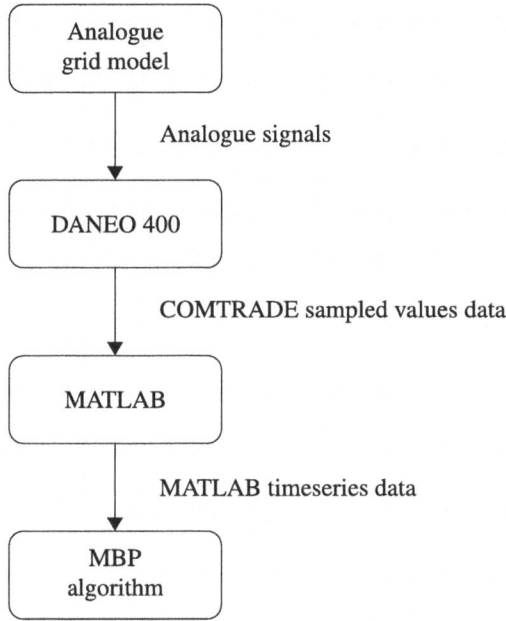

Fig. 9.3 Flow chart of the laboratory tests validating the model-based protection (MBP) algorithm

The result of the fault localization (FL) layer is illustrated by a slider. In contrast to the FI and the FC layer, the evaluation of the FL layer is not time-sensitive. Here, different models are derived from the most suitable model of the FC layer regarding the fault resistance configuration and the values of the appearing fault resistances. The models of the FL layer differ from each other by the fault location.

9.2 Scenario 1: Double-sided Infeed

In this laboratory scenario, the MBP algorithm is validated by the investigation of a double-sided infeed topology. Table 9.3 gives an overview of the relevant figures of this section, providing also the possibility for quick navigation in the digital document.

Table 9.3 Overview of the relevant figures of Sect. 9.2 containing the validation of the model-based protection (MBP) algorithm in a laboratory test investigating a single-phase-to-ground fault in phase 1 with a relative fault distance of $m = 0.5$ at $t = 0.1$ s

Content	Figure
Scenario under investigation	Fig. 9.4
Model of the grid topology for the fault identification	Fig. 7.5
Model of the healthy line for the fault characterization	Fig. 7.11
Models of the faulty line for the fault characterization and localization	Fig. 7.12 & Fig. 7.13
Phase-to-ground voltages at busbar A	Fig. 9.5
Phase-to-ground voltages at busbar B	Fig. 9.6
Phase currents at busbar A	Fig. 9.7
Phase currents at busbar B	Fig. 9.8
Result of the FI: P-value p_h	Fig. 9.9
Result of the FC: Index of the identified model	Fig. 9.10
Result of the FC: Inverse relative estimation errors IRE_{FCi}	Fig. 9.11

The generator G is connected via an overhead line between the busbars A and B with an emulated electrical length of 200 km to the grid. A single-phase-to-ground fault in phase 1 with a relative fault distance m of 0.5 is applied at $t = 0.1$ s. The fault resistance R_F equals $0\,\Omega$. The relays of this section are named R 1 and R 2. Fig. 9.4 shows the laboratory scenario.

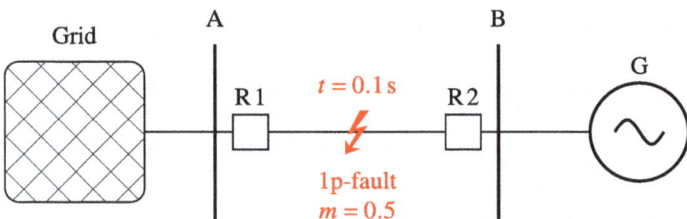

Fig. 9.4 Laboratory scenario for validating the MBP algorithm during a single-phase-to-ground fault in phase 1 with a relative fault distance of $m = 0.5$ at $t = 0.1$ s

The voltages at the busbars A and B are given in Fig. 9.5 and Fig. 9.6. It is observed, that all voltages contain notable harmonic distortions. Nonlinearities in

9.2 Scenario 1: Double-sided Infeed

the analogue grid model are the reason for this effect. The distorted signals are used as an opportunity to test the MBP algorithm under these conditions. VSCs also lead to distorted voltages which show other characteristics, see Fig. 2.2.

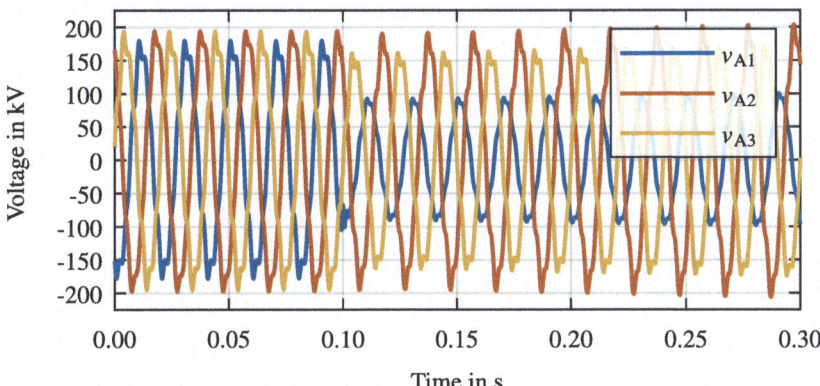

Fig. 9.5 Phase-to-ground voltages at busbar A before and after a single-phase-to-ground fault in phase 1 with a relative fault distance of $m = 0.5$ at $t = 0.1$ s

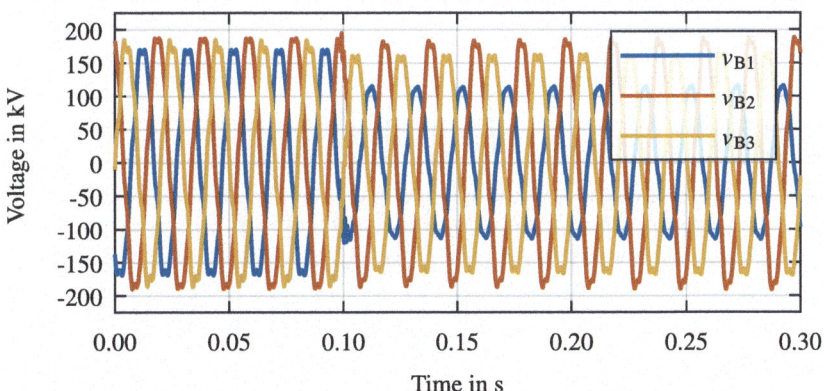

Fig. 9.6 Phase-to-ground voltages at busbar B before and after a single-phase-to-ground fault in phase 1 with a relative fault distance of $m = 0.5$ at $t = 0.1$ s

The phase currents at the busbars A and B before and after the fault are shown in Fig. 9.7 and Fig. 9.8. The amplitude of the current i_{b1} of the generator at busbar B in the faulty phase increases to multiples of the nominal value immediately after the fault, see Fig. 9.8. The limited overcurrent capability of VSCs would not allow comparable high values. Furthermore, the decreasing DC-component (see also [143]) and the phase step become obvious.

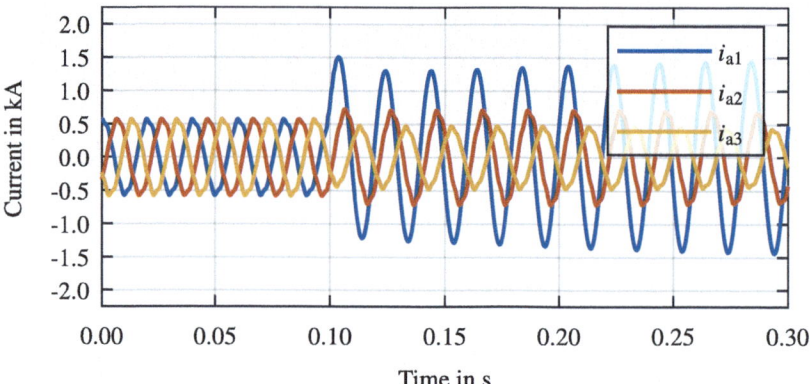

Fig. 9.7 Phase currents at busbar A before and after a single-phase-to-ground fault in phase 1 with a relative fault distance of $m = 0.5$ at $t = 0.1$ s

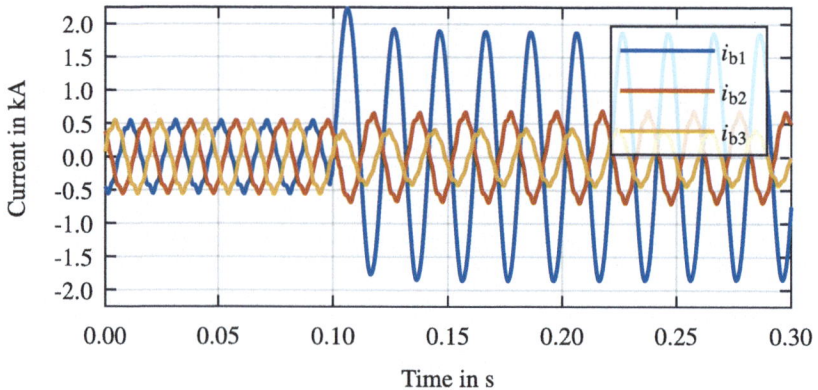

Fig. 9.8 Phase currents at busbar B before and after a single-phase-to-ground fault in phase 1 with a relative fault distance of $m = 0.5$ at $t = 0.1$ s

9.2 Scenario 1: Double-sided Infeed

The p-value p_h calculated in the FI layer is shown in Fig. 9.9. Before the fault at $t < 0.1$ s, the p-value is greater than the predefined confidence level α_h. Consequently, no fault is identified. After the fault at $t > 0.1$ s, $p_h < \alpha_h$ is valid and the fault is correctly identified 2 ms after the fault occurred. In this time interval, periodic peaks showing a frequency of 100 Hz appear. They are caused by the minima of the weighted sum $\zeta(k)$ of the squared elements of the vector of the estimation errors, see Eq. (8.21). This effect does not influence the functionality of the FI layer negatively.

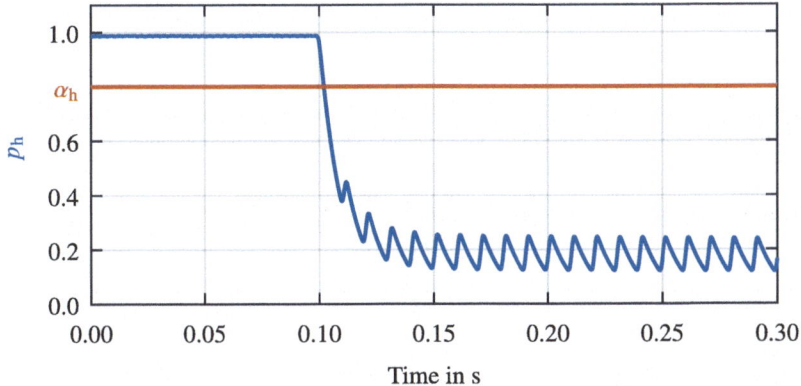

Fig. 9.9 P-value p_h and confidence level α_h of the FI layer of the MBP algorithm before and after a single-phase-to-ground fault in phase 1 with a relative fault distance of $m = 0.5$ at $t = 0.1$ s

Figure 9.10 depicts the index of the model evoking the minimum Euclidean norm of the estimation errors in the FC layer. The healthy model is described by the index one. The fault is identified at high speed 10 ms after the fault occurred. Since the index stays constant before and after the fault, the stability of decision is high.

The fault decision of the FC layer is additionally visualized regarding selected models by the spider plot in Fig. 9.11 and quantified by the IRE_{FC}-value (see Eq. (9.1)). Two models of the single-phase-to-ground fault at $m = 0.5$ and $m = 1$ are chosen. Different models with different fault resistance configurations with a relative fault distance of $m = 0.5$ are additionally selected for comparison in this figure. At $t = 0.05$ s no fault is identified. At $t = 0.15$ s, the single-phase-to-ground fault at $m = 0.5$ with a fault resistance of $0\,\Omega$ is correctly identified.

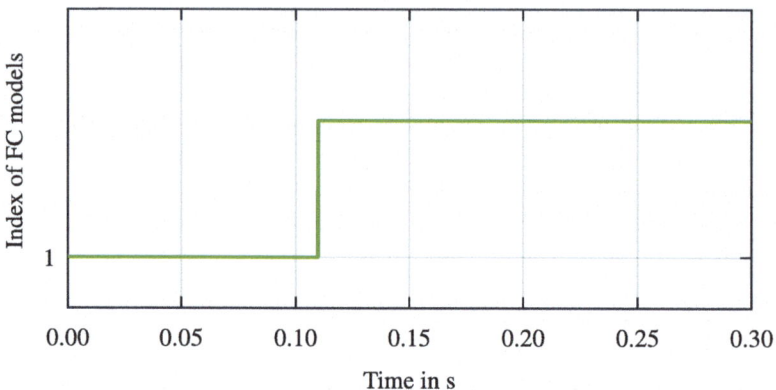

Fig. 9.10 Index of the models evoking the minimum Euclidean norm of the estimation errors of the FC layer of the MBP algorithm before and after a single-phase-to-ground fault in phase 1 with a relative fault distance of $m = 0.5$ at $t = 0.1$ s

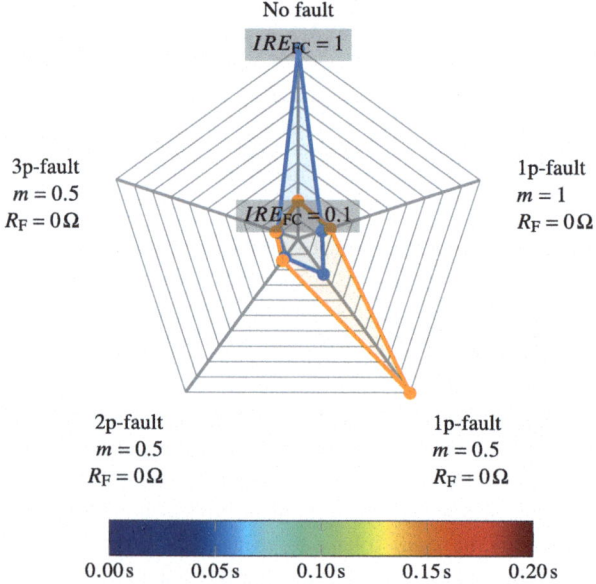

Fig. 9.11 Inverse relative estimation errors IRE_{FCi} at $t = 0.05$ s and $t = 0.15$ s of selected models of the FC layer of the MBP algorithm before and after a single-phase-to-ground fault in phase 1 with a relative fault distance of $m = 0.5$ at $t = 0.1$ s

The scenario in this section demonstrates, that a single-phase-to-ground fault in the laboratory is identified fast and reliable by the MBP. The fault resistance configuration and the value of the fault resistance is detected correctly.

9.3 Scenario 2: Double-sided Infeed with Fault Localization

In this laboratory scenario, the MBP algorithm is validated by the investigation of a double-sided infeed topology with a focus on fault localization. An overview of the relevant figures of this section, providing also the possibility for quick navigation in the digital document, is given in Table 9.4.

Table 9.4 Overview of the relevant figures of Sect. 9.3 containing the validation of the model-based protection (MBP) algorithm in a laboratory test investigating a single-phase-to-ground fault in phase 1 with a relative fault distance of $m = 0.75$ at $t = 0.1$ s

Content	Figure
Scenario under investigation	Fig. 9.12
Model of the grid topology for the fault identification	Fig. 7.5
Model of the healthy line for the fault characterization	Fig. 7.11
Models of the faulty line for the fault characterization and localization	Fig. 7.12 & Fig. 7.13
Result of the FI: P-value p_h	Fig. 9.13
Result of the FC: Index of the identified model	Fig. 9.14
Result of the FC: Inverse relative estimation errors IRE_{FCi}	Fig. 9.15
Result of the FL in a slider diagram	Fig. 9.16

The scenario in this section basically equals the scenario in Sect. 9.2. The difference is, that the single-phase-to-ground fault in phase 1 is applied at a relative fault distance of $m = 0.75$ seen from busbar A. Due to the different fault location compared to Sect. 9.2, the fault currents at the busbars A and B change. The laboratory scenario is presented in Fig. 9.12.

Figure 9.13 depicts the p-value p_h calculated in the FI layer. The characteristics are very similar to the characteristics in Fig. 9.9 of Sect. 9.2. The p-value drops below the confidence level α_h 9 ms after the fault occurred. The FI layer of the MBP identifies the fault correctly.

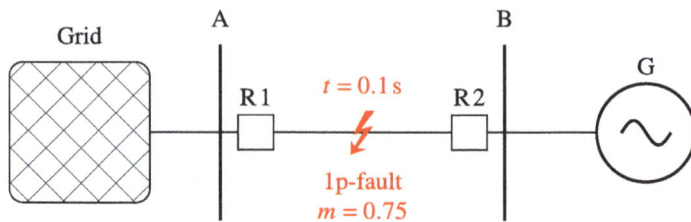

Fig. 9.12 Laboratory scenario for validating the MBP algorithm during a single-phase-to-ground fault in phase 1 with a relative fault distance of $m = 0.75$ at $t = 0.1$ s

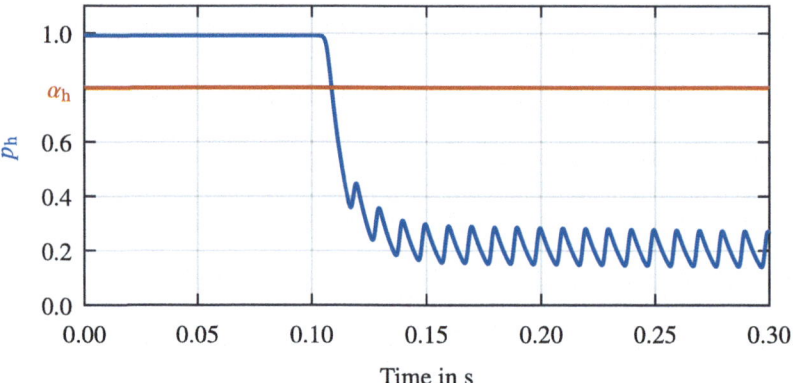

Fig. 9.13 P-value p_h and confidence level α_h of the FI layer of the MBP algorithm before and after a single-phase-to-ground fault in phase 1 with a relative fault distance of $m = 0.75$ at $t = 0.1$ s

The evolution of the index of the model evoking the minimum Euclidean norm of the estimation errors in the FC layer is illustrated in Fig. 9.14. The index changes from one to another value immediately after the fault occurred and the fault is identified. At $t = 0.18$ s, the index changes. In this chase, the most suitable model changes from a single-phase-to-ground fault in phase 1 with $R_F = 1\,\Omega$ to $R_F = 0\,\Omega$. This effect does not influence the fault decision significantly.

Figure 9.15 contains the spider plot of the FC layer. Different models with different fault resistance configurations with a relative fault distance of $m = 0.5$ are chosen for comparison in this figure. Two models of the single-phase-to-ground fault at $m = 0.5$ and $m = 1$ are chosen. At $t = 0.05$ s, no fault is identified. At $t = 0.15$ s, the single-phase-to-ground fault is identified. According to the calcu-

9.3 Scenario 2: Double-sided Infeed with Fault Localization

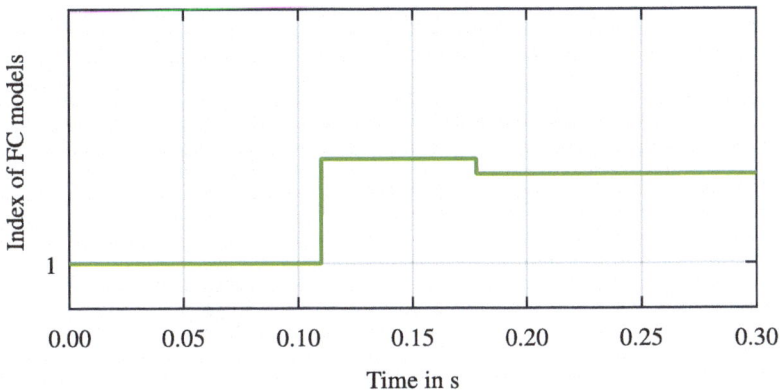

Fig. 9.14 Index of the models evoking the minimum Euclidean norm of the estimation errors of the FC layer of the MBP algorithm before and after a single-phase-to-ground fault in phase 1 with a relative fault distance of $m = 0.75$ at $t = 0.1$ s

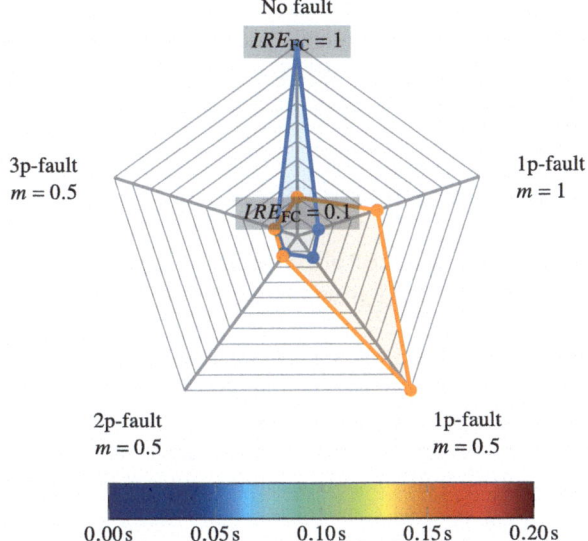

Fig. 9.15 Inverse relative estimation errors IRE_{FCi} at $t = 0.05$ s and $t = 0.15$ s of selected models of the FC layer of the MBP algorithm before and after a single-phase-to-ground fault in phase 1 with a relative fault distance of $m = 0.75$ at $t = 0.1$ s

lated IRE_{FC}-values, the distinction between $m = 0.5$ and $m = 1$ is not definitely clear. A subsequent fault localization is necessary.

In the FL layer of the MBP, models described by a single-phase-to-ground fault with $R_F = 0\,\Omega$ are investigated at different relative fault distances. The variable m is varied from 0 to 1 in 0.1-steps. The model described by a single-phase-to-ground fault with $R_F = 0\,\Omega$ at $m = 0.8$ is identified as the best suitable model. The result of the FL layer is visualized in Fig. 9.16.

Fig. 9.16 Result of the fault localization of the MBP algorithm after a single-phase-to-ground fault in phase 1 with a relative fault distance of $m = 0.75$ at $t = 0.1$ s

This scenario demonstrates, that a single-phase-to-ground fault in the laboratory is identified fast and reliable by the MBP. The fault characterization is sufficiently good. The subsequent FL layer localizes the fault correctly.

9.4 Scenario 3: Parallel Line Topology

In this laboratory scenario, the MBP algorithm is validated by the investigation of a parallel line topology. Table 9.5 gives an overview of the relevant figures of this section.

Two grids are connected via two parallel overhead lines between the busbars A and B with an emulated electrical length of 200 km. A single-phase-to-ground fault on line 1 in phase 1 with $m = 0.5$ is applied at $t = 0.1$ s. The fault resistance R_F equals $0\,\Omega$. After the fault, the power flow direction calculated at the position of the relay R 1 changes. This effect can cause problems for distance protection algorithms. The laboratory scenario is illustrated in Fig. 9.17.

The p-value p_h calculated in the FI layer is shown in Fig. 9.18. The p-value drops below the confidence level α_h 7 ms after the fault occurred and the fault is correctly identified. Compared to Fig. 9.9 and Fig. 9.13, the p-value is approximately increased by factor two. However, the fault identification based on $\alpha_h = 0.8$ is clear.

The evolution of the indices of the models of line 1 and line 2 of the FC layer is illustrated in Fig. 9.19. The change of the index from one to another value indicates

9.4 Scenario 3: Parallel Line Topology

Table 9.5 Overview of the relevant figures of Sect. 9.4 containing the validation of the model-based protection (MBP) algorithm in a laboratory test investigating a single-phase-to-ground fault on line 1 in phase 1 with a relative fault distance of $m = 0.5$ at $t = 0.1$ s based on a parallel line topology

Content	Figure
Scenario under investigation	Fig. 9.17
Model of the grid topology for the fault identification	Fig. 7.6
Model of the healthy line for the fault characterization	Fig. 7.11
Models of the faulty line for the fault characterization and localization	Fig. 7.12 & Fig. 7.13
Result of the FI: P-value p_h	Fig. 9.18
Result of the FC: Index of the identified model	Fig. 9.19
Result of the FC: Inverse relative estimation errors IRE_{FCi} of line 1	Fig. 9.20
Result of the FC: Inverse relative estimation errors IRE_{FCi} of line 2	Fig. 9.21

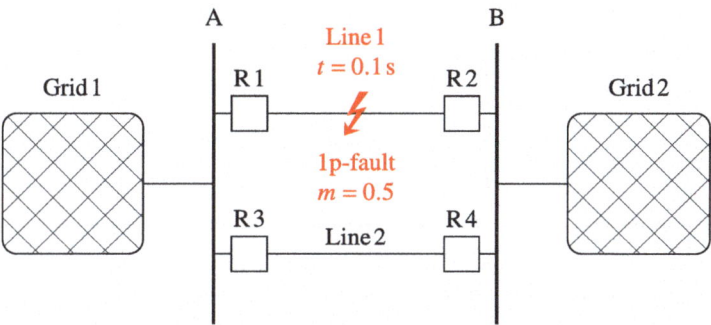

Fig. 9.17 Laboratory scenario for validating the MBP algorithm during a single-phase-to-ground fault on line 1 in phase 1 with a relative fault distance of $m = 0.5$ at $t = 0.1$ s based on a parallel line topology

the fault on line 1. The line 2 is correctly characterized as healthy, since the model index does not change.

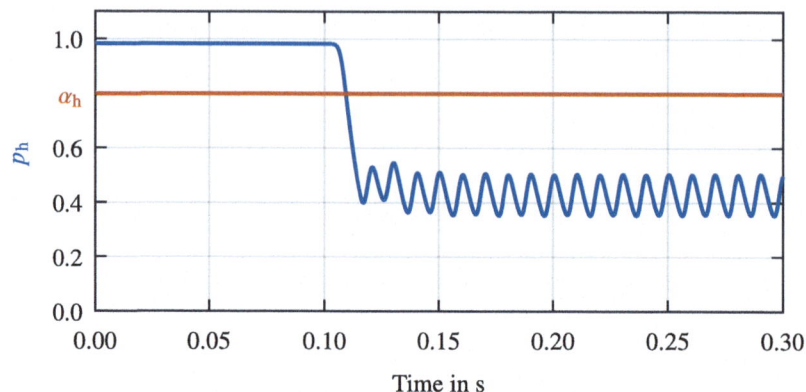

Fig. 9.18 P-value p_h and confidence level α_h of the FI layer of the MBP algorithm before and after a single-phase-to-ground fault on line 1 in phase 1 with a relative fault distance of $m = 0.5$ at $t = 0.1$ s based on a parallel line topology

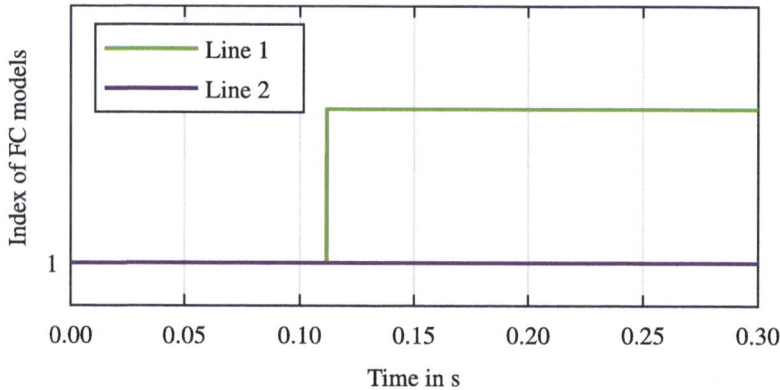

Fig. 9.19 Index of the models of line 1 and line 2 evoking the minimum Euclidean norm of the estimation errors of the FC layer of the MBP algorithm before and after a single-phase-to-ground fault on line 1 in phase 1 with a relative fault distance of $m = 0.5$ at $t = 0.1$ s based on a parallel line topology

9.4 Scenario 3: Parallel Line Topology

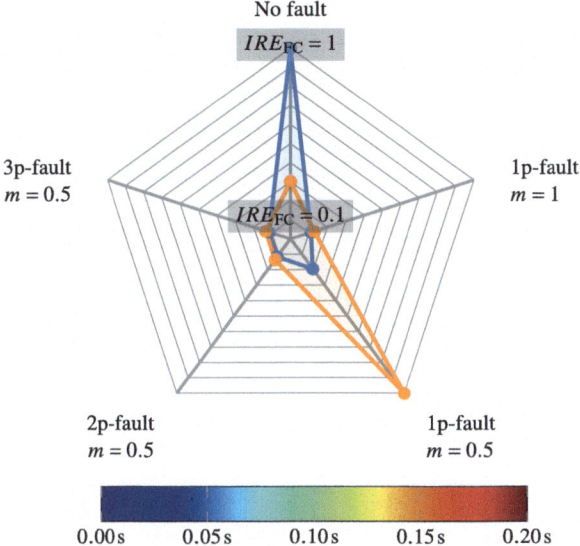

Fig. 9.20 Inverse relative estimation errors IRE_{FCi} at $t = 0.05$ s and $t = 0.15$ s of selected models of line 1 of the FC layer of the MBP algorithm before and after a single-phase-to-ground fault on line 1 in phase 1 with a relative fault distance of $m = 0.5$ at $t = 0.1$ s based on a parallel line topology

The spider plot of the FC layer of line 1 is depicted in Fig. 9.20. At $t = 0.05$ s no fault is identified. At $t = 0.15$ s, the single-phase-to-ground fault at $m = 0.5$ is correctly identified. Figure 9.21 contains the spider plot of line 2. Here, no fault is identified.

This scenario demonstrates, that a single-phase-to-ground fault in a parallel line topology in the laboratory is identified fast and reliable by the MBP. The FC layer enables the selective disconnection of the faulty line 1. After this disconnection, the power on line 2 has possibly to be reduced to avoid, that the transport capacity is exceeded.

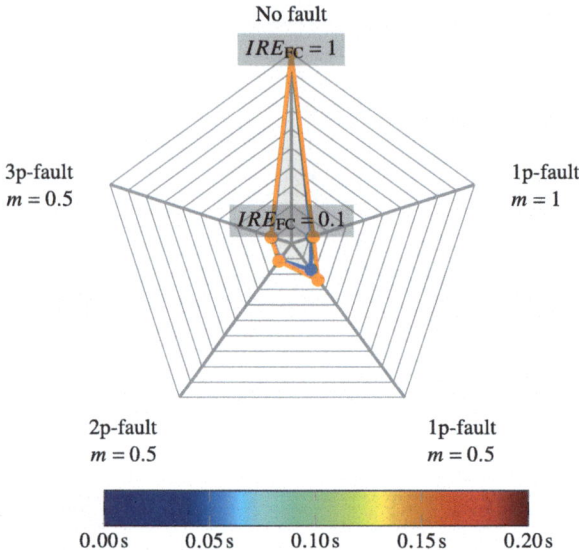

Fig. 9.21 Inverse relative estimation errors IRE_{FCi} at $t = 0.05$ s and $t = 0.15$ s of selected models of line 2 of the FC layer of the MBP algorithm before and after a single-phase-to-ground fault on line 1 in phase 1 with a relative fault distance of $m = 0.5$ at $t = 0.1$ s based on a parallel line topology

9.5 Scenario 4: Intermediate Infeed Topology

In this laboratory scenario, the MBP algorithm is validated by the investigation of an intermediate infeed topology. Table 9.6 gives an overview of the relevant figures of this section.

The investigated grid area consists of three lines with an emulated electrical length of 100 km per line. A single-phase-to-ground fault on line 2 in phase 1 with $m = 0.5$ is applied at $t = 0.1$ s. The fault resistance R_F equals $0\,\Omega$. The intermediate infeed can cause problems for distance protection algorithms. The laboratory scenario is illustrated in Fig. 9.22.

9.5 Scenario 4: Intermediate Infeed Topology

Table 9.6 Overview of the relevant figures of Sect. 9.5 containing the validation of the model-based protection (MBP) algorithm in a laboratory test investigating a single-phase-to-ground fault on line 2 in phase 1 with a relative fault distance of $m = 0.5$ at $t = 0.1$ s based on an intermediate infeed topology

Content	Figure
Scenario under investigation	Fig. 9.22
Model of the grid topology for the fault identification	Fig. 7.8
Model of the healthy line for the fault characterization	Fig. 7.11
Models of the faulty line for the fault characterization and localization	Fig. 7.12 & Fig. 7.13
Result of the FI: P-value p_h	Fig. 9.23
Result of the FC: Index of the identified model	Fig. 9.24
Result of the FC: Inverse relative estimation errors IRE_{FCi} of line 1	Fig. 9.25
Result of the FC: Inverse relative estimation errors IRE_{FCi} of line 2	Fig. 9.26
Heatmap visualizing IRE_{FCi} at $t = 0.15$ s of line 2	Fig. 9.28
Result of the FC: Inverse relative estimation errors IRE_{FCi} of line 3	Fig. 9.27

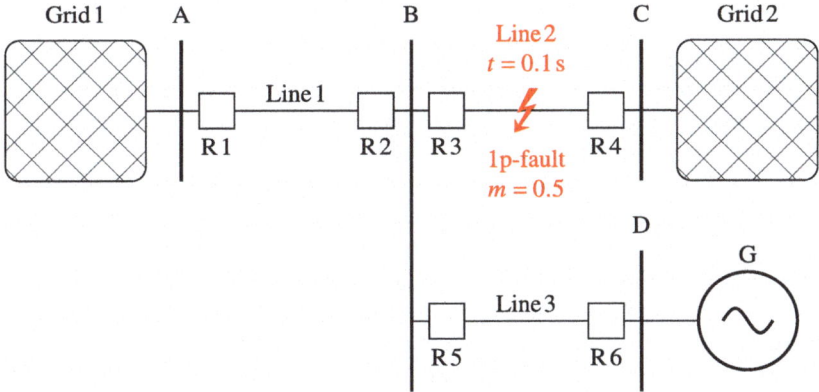

Fig. 9.22 Laboratory scenario for validating the MBP algorithm during a single-phase-to-ground fault on line 2 in phase 1 with a relative fault distance of $m = 0.5$ at $t = 0.1$ s based on an intermediate infeed topology

Figure 9.23 presents the p-value p_h calculated in the FI layer. It drops below the confidence level α_h 5 ms after the fault occurred. The FI layer of the MBP identifies the fault correctly.

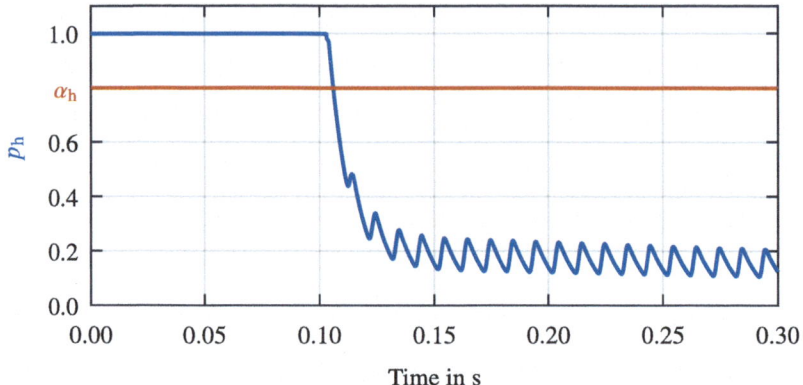

Fig. 9.23 P-value p_h and confidence level α_h of the FI layer of the MBP algorithm before and after a single-phase-to-ground fault on line 2 in phase 1 with a relative fault distance of $m = 0.5$ at $t = 0.1$ s based on an intermediate infeed topology

The evolution of the indices of the models of line 1, line 2 and line 3 of the FC layer is depicted in Fig. 9.24. The line 1 and line 3 are correctly characterized as healthy, since their model indices do not change. The fault on line 2 is indicated by the change from index one to another value.

Figure 9.25 shows the spider plot of the FC layer of line 1 and Fig. 9.27 illustrates the spider plot of the FC layer of line 3. On both lines, no fault is identified. The spider plot of line 2 is visualized in Fig. 9.26. Here, no fault is identified at $t = 0.05$ s. At $t = 0.15$ s, the single-phase-to-ground fault at $m = 0.5$ is correctly identified.

A different visualisation to Fig. 9.26 of the inverse relative estimation error $IRE_{FCi} \in (0, 1]$ for the model i (see Eq. (9.1)) of line 2 at $t = 0.15$ s is given in Fig. 9.28. Different fault locations on line 2 characterized by the relative fault distance m seen from busbar B are considered. Furthermore, different fault types (see also Table 7.1) in different phases (ph) are regarded. The results applying models of single-phase-to-ground (1p) faults, phase-to-phase (2p) faults and three-phase (3p) faults are presented. In contrast to Fig. 9.26, the focus of the heatmap in Fig. 9.28 is not on exact quantification. In fact, the dependence of IRE_{FCi} from the fault location and the fault type is visualized. It is shown, that the IRE_{FCi} regarding

9.5 Scenario 4: Intermediate Infeed Topology

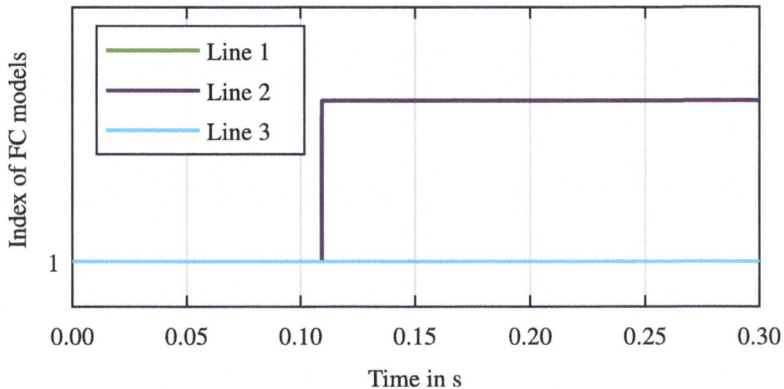

Fig. 9.24 Index of the models of line 1, line 2 and line 3 evoking the minimum Euclidean norm of the estimation errors of the FC layer of the MBP algorithm before and after a single-phase-to-ground fault on line 2 in phase 1 with a relative fault distance of $m = 0.5$ at $t = 0.1$ s based on an intermediate infeed topology

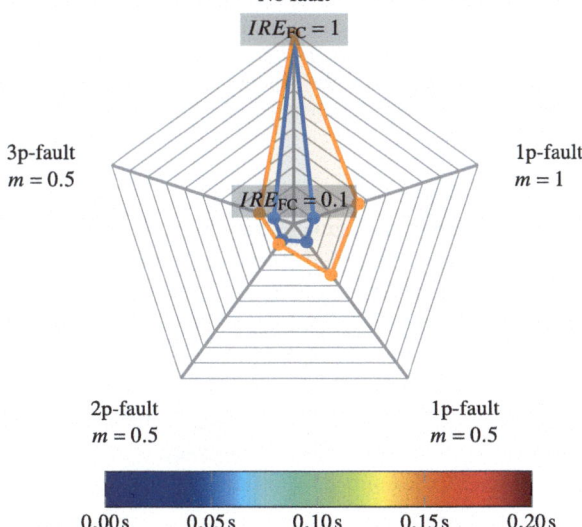

Fig. 9.25 Inverse relative estimation errors IRE_{FCi} at $t = 0.05$ s and $t = 0.15$ s of selected models of line 1 of the FC layer of the MBP algorithm before and after a single-phase-to-ground fault on line 2 in phase 1 with a relative fault distance of $m = 0.5$ at $t = 0.1$ s based on an intermediate infeed topology

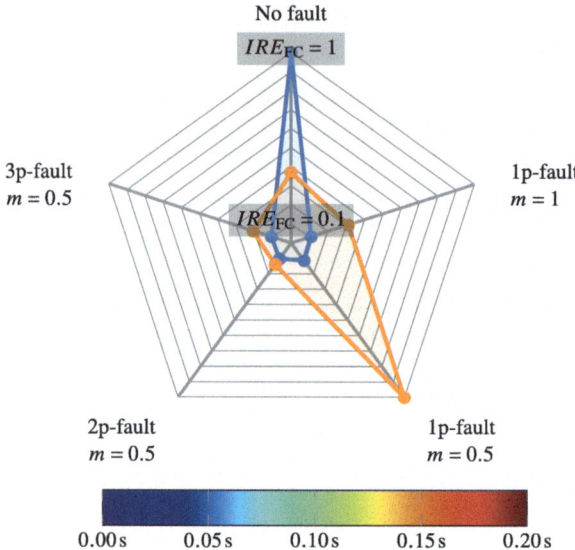

Fig. 9.26 Inverse relative estimation errors IRE_{FCi} at $t = 0.05$ s and $t = 0.15$ s of selected models of line 2 of the FC layer of the MBP algorithm before and after a single-phase-to-ground fault on line 2 in phase 1 with a relative fault distance of $m = 0.5$ at $t = 0.1$ s based on an intermediate infeed topology

a model of a 1p-fault in phase 1 is greater the closer the fault location assumed in the model approaches the actual fault location. Moreover, a 2p-fault between phase 1 and phase 3 appears to be the second most likely fault type after the 1p-fault in phase 1. The reason is, that this 2p-fault also includes the characteristics of the faulty phase 1.

This scenario demonstrates, that a single-phase-to-ground fault in an intermediate infeed topology in the laboratory is identified fast and reliable by the MBP. The FC layer enables the selective disconnection of the faulty line 2. After this disconnection, new operating points have to be defined by the tertiary grid control (see Sect. 2.2.3).

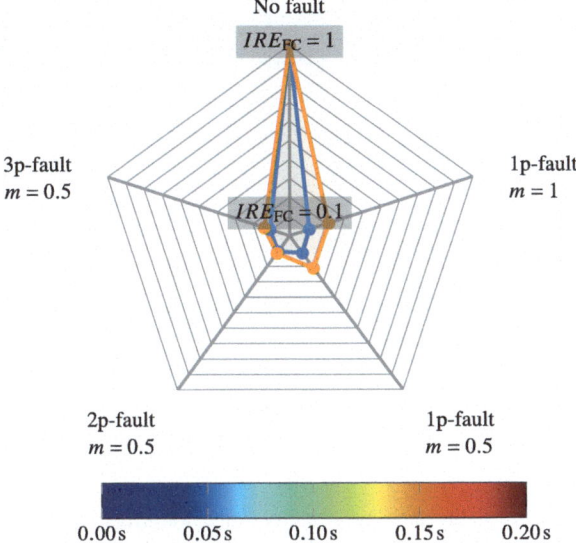

Fig. 9.27 Inverse relative estimation errors IRE_{FCi} at $t = 0.05$ s and $t = 0.15$ s of selected models of line 3 of the FC layer of the MBP algorithm before and after a single-phase-to-ground fault on line 2 in phase 1 with a relative fault distance of $m = 0.5$ at $t = 0.1$ s based on an intermediate infeed topology

9.6 Summary

In this chapter, the validation of the model-based protection (MBP) algorithm by laboratory tests is presented. Single-phase-to-ground faults are investigated in different grid topologies, which are especially relevant for 100 % inverter-based power systems. The validation of the MBP algorithm based on real hardware equipment represents a suitability test and reinforces the theoretical design.

The analytical design from Chap. 8 is the basic prerequisite of the laboratory tests in this chapter.

The laboratory environment of the Institute of Electrical Energy Systems in Erlangen including an analogous grid model, a synchronous generator and the connection to the low voltage utility grid enables the validation of the MBP algorithm. The use of a synchronous generator instead of a voltage source converter underlines the universality of application of the MBP algorithm. In a multistage workflow, three Omicron DANEO 400 hybrid signal analysers are used to record the mea-

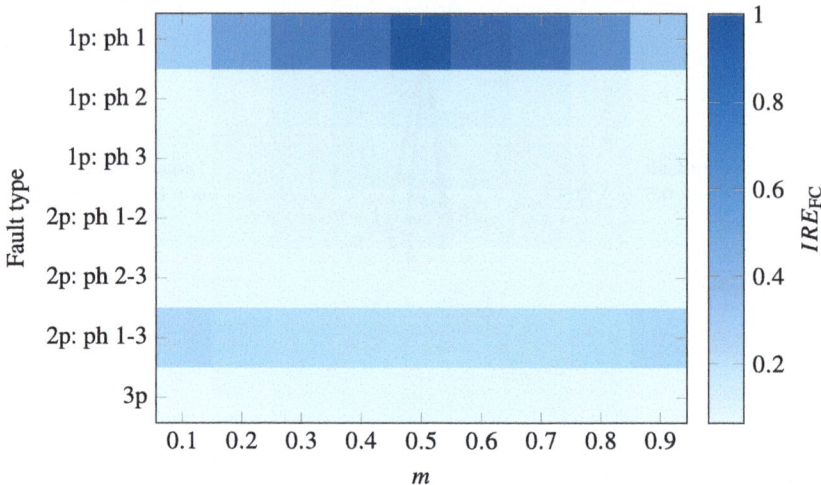

Fig. 9.28 Heatmap visualizing the inverse relative estimation errors IRE_{FCi} at $t = 0.15$ s of selected models of line 2 of the MBP algorithm after a single-phase-to-ground fault on line 2 in phase 1 with a relative fault distance of $m = 0.5$ at $t = 0.1$ s based on an intermediate infeed topology

surement data. The MBP algorithm is tested based on the converted data files in MATLAB/Simulink.

Several presentation methods are used to visualize the features of the MBP algorithm. The p-value is defined as the protection criterion of the fault identification layer. The inverse relative estimation error is introduced as a new key figure to quantify the result of the fault characterization layer.

The MBP algorithm is successfully validated by the investigation of a double-sided infeed topology. The applied single-phase-to-ground fault is identified fast and reliable and characterized correctly by the MBP algorithm.

Furthermore, the MBP algorithm is successfully validated by the investigation of a double-sided infeed topology with fault localization. The single-phase-to-ground fault is identified fast and reliable and localised correctly by the MBP algorithm.

Moreover, the MBP algorithm is successfully validated by the investigation of a parallel line topology and an intermediate infeed topology. Again, the single-phase-to-ground faults are identified fast and reliable. In both scenarios, the selective disconnection of the faulty line is enabled.

9.6 Summary

In Chap. 10, the MBP algorithm is additionally validated based on the simulations including enhanced grid fault characteristics of voltage source converters (VSCs) from Chap. 5 and Chap. 6. To follow the key aspects of the central theme of this thesis on a shortcut: The conclusion of the next chapter is provided in Sect. 10.6.

Open Access This chapter is licensed under the terms of the Creative Commons Attribution 4.0 International License (http://creativecommons.org/licenses/by/4.0/), which permits use, sharing, adaptation, distribution and reproduction in any medium or format, as long as you give appropriate credit to the original author(s) and the source, provide a link to the Creative Commons license and indicate if changes were made.

The images or other third party material in this chapter are included in the chapter's Creative Commons license, unless indicated otherwise in a credit line to the material. If material is not included in the chapter's Creative Commons license and your intended use is not permitted by statutory regulation or exceeds the permitted use, you will need to obtain permission directly from the copyright holder.

Model-based Protection Scenarios in 100% IBPS

10

This chapter presents the investigation of the model-based protection (MBP) algorithm in 100% inverter-based power systems (IBPS). The main focus of this chapter is the evaluation of different grid fault types and enhanced grid fault characteristics of voltage source converters (VSCs). In Sect. 10.1, the VSC-specific grid fault characteristics and the four selected simulation scenarios are described. The scenarios are numbered, continuing the numbering from Chap. 9. Sect. 10.2 contains the Scenario 5: 3p-fault and voltage reduction. In Sect. 10.3, the Scenario 6: 2p-fault and reactive current injection is analysed. Scenario 7: 1p-fault and RGS-mode is investigated in Sect. 10.4. In Sect. 10.5, the Scenario 8: 1p-fault and transient current limiting is evaluated.

10.1 VSC-specific Fault Characteristics and Simulation Scenarios

The model-based protection (MBP) algorithm (see Chap. 8) is additionally validated by simulation tests of 100% inverter-based power systems (IBPS) in MATLAB/Simulink. The control of voltage source converters (VSCs) and the MBP algorithm are merged. The main focus of this chapter is the evaluation of the MBP algorithm applying different grid fault types and enhanced grid fault characteristics of VSCs. The VSC grid fault characteristics are presented in Chap. 5 and Chap. 6. Moreover, the specific characteristics of VSCs are considered. Especially, the limited overcurrent capability of VSCs is considered.

Four selected scenarios are investigated. Since this chapter—as well as Chap. 9—validates the developed MBP algorithm, the scenarios are numbered, continuing the numbering from Chap. 9. In Scenario 5, a three-phase fault (see Tab. 7.1) inside a serial lines topology is investigated. The superordinate objective of the VSC

© The Author(s) 2025
F. Mahr, *Control and Protection of 100% Inverter-based Power Systems*,
https://doi.org/10.1007/978-3-658-47217-7_10

controller is to avoid steady state overcurrents and to stay connected to the grid after the fault. Consequently, the amplitude of the positive sequence filter capacitor voltages of the VSC is reduced according to Sect. 6.3.2.3. The subordinate objective is to continue the grid operation to supply consumers parallel to the VSC—possibly with lower power and lower voltage amplitude compared to normal grid operation conditions.

In Scenario 6, a phase-to-phase fault (see Tab. 7.1) is investigated. The primary objective of the VSC is to continue the grid operation under symmetrical voltages. Therefore, the amplitude of the negative sequence voltages after the fault has to be reduced. Consequently, the VSC injects reactive current in the negative sequence according to Sect. 5.2.3.2.

Scenario 7 contains the investigation of a single-phase-to-ground fault (see Tab. 7.1). The superordinate objective of the VSC is to ensure a symmetrical grid operation, the reduction of the fault current and the arc suppression. Furthermore, the VSC itself is protected by low fault currents. Consequently, the RGS (resonant grounding system) mode of the VSC according to Sect. 6.2.3.1 is activated.

In Scenario 8, a single-phase-to-ground fault is investigated, too. Extending Scenario 7, the transient current limiting according to Sect. 6.3.1 is additionally regarded.

The four simulation scenarios are based on 10 kV grid topologies. Each overhead line shows a length of 10 km. The simulated line parameters are shown in Tab. 10.1.

Table 10.1 Overhead line parameters in the simulation (see also Fig. 7.4)

Variable	Value	Description
f_{grid}	50 Hz	nominal grid frequency
V_{grid}	10 kV	nominal grid voltage
R'	150 mΩ/km	phase resistance per km
L'	1 mH/km	phase inductance per km
C'	10 nF/km	phase-to-ground capacitance per km

The basic diagrams illustrating the achieved results of the four simulation test scenarios basically correspond to the diagrams of Chap. 9. The p-value $p_h(k) \in [0, 1]$ characterizes the goodness-of-fit of a model of a healthy grid area. It is calculated inside the fault identification (FI) layer (see Sect. 8.3 and Sect. 8.4.2) according to Eq. (8.25c). The fault decision of the FI layer regarding a grid area is made based on the comparison of the p-value p_h and a predefined confidence level $\alpha_h \in [0, 1]$. In this chapter, $\alpha_h = 0.8$ is chosen.

The index of the model of one line evoking the minimum Euclidean norm of the estimation errors at each sample point describes the fault characterization (FC) layer (see Sect. 8.4.3). These diagrams show the stability and the speed of the fault decision regarding one line. The FC layer is additionally described by one spider plot per line. Here, the fault decision is visualized regarding selected models characterized by the fault resistance configuration, the values of the appearing fault resistances and the rough fault location. The inverse relative estimation error $IRE_{FCi} \in (0, 1]$ for the model i quantifies and visualizes the decision at different points in time before and after the fault according to Eq. (9.1). The IRE_{FC}-value of the most suitable model equals one. The IRE_{FC}-values of all other investigated models are less than one. The smaller the IRE_{FC}-value of one model, the more unsuitable the regarded model appears to be.

A slider illustrates the result of the fault localization (FL) layer. The evaluation of the FL layer is not time-sensitive—in contrast to the FI and the FC layer. In the FL layer, different models are derived from the most suitable model of the FC layer regarding the fault resistance configuration and the values of the appearing fault resistances. The models of the FL layer differ from each other by the fault location.

10.2 Scenario 5: 3p-fault and Voltage Reduction

This simulation scenario contains the investigation of a three-phase fault inside a serial lines topology. The amplitude of the positive sequence output voltages of the VSC is reduced after the fault according to Sect. 6.3.2.3 to ensure steady state current limiting. Tab. 10.2 gives an overview of the relevant figures of this section, providing also the possibility for quick navigation in the digital document.

The VSC at busbar C is connected via two serial lines to the grid at busbar A. Line 1 connects busbar A and busbar B. Busbar B and busbar C are connected via line 2. Both lines show a length of 10 km. The nominal active power of the VSC equals 7.5 MW. A three-phase fault on line 1 at the relative fault location $m = 0.3$ seen from busbar A is applied at $t = 0.1$ s. The fault resistances equal zero. The amplitude of the positive sequence output voltages of the VSC is reduced from $t = 0.2$ s to $t = 0.3$ s. Fig. 10.1 shows the simulation scenario.

The voltages at busbar C are given in Fig. 10.2. Immediately after the fault, the amplitude of the voltages is reduced according to Kirchhoff's voltage law (KVL). From $t = 0.2$ s, the voltage amplitude is actively reduced by the VSC by reducing the amplitude of the reference voltages of the filter capacitor voltages. In this way,

Table 10.2 Overview of the relevant figures of Sect. 10.2 containing the validation of the model-based protection (MBP) algorithm in a simulation test investigating a three-phase fault on line 1 with a relative fault distance of $m = 0.3$ at $t = 0.1$ s based on a two serial lines topology

Content	Figure
Scenario under investigation	Fig. 10.1
Model of the grid topology for the fault identification	Fig. 7.7
Model of the healthy line for the fault characterization	Fig. 7.11
Models of the faulty line for the fault characterization and localization	Fig. 7.12 & Fig. 7.13
Phase-to-ground voltages at busbar C	Fig. 10.2
Phase currents at busbar C	Fig. 10.3
Result of the FI: P-value p_h	Fig. 10.4
Result of the FC: Index of the identified model	Fig. 10.5
Result of the FC: Inverse relative estimation errors IRE_{FCi} of line 1	Fig. 10.6
Result of the FC: Inverse relative estimation errors IRE_{FCi} of line 2	Fig. 10.7
Result of the FL of line 1 in a slider diagram	Fig. 10.8

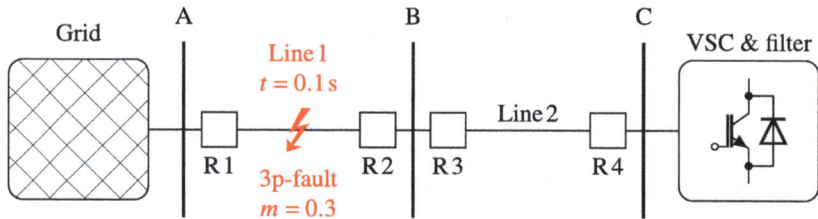

Fig. 10.1 Simulation scenario for validating the MBP algorithm during a three-phase fault on line 1 with a relative fault distance of $m = 0.3$ at $t = 0.1$ s based on a two serial lines topology

the steady state current limiting is achieved. The time interval between the fault occurrence and the activation of the steady state current limiting of 100 ms is chosen to demonstrate the reaction of the MBP with and without current limiting.

The phase currents at busbar C before and after the fault are shown in Fig. 10.3. The amplitude of the currents increases immediately after the fault $t = 0.1$ s due to KVL and the V/I-characteristics of passive grid elements. For practical applications

10.2 Scenario 5: 3p-fault and Voltage Reduction

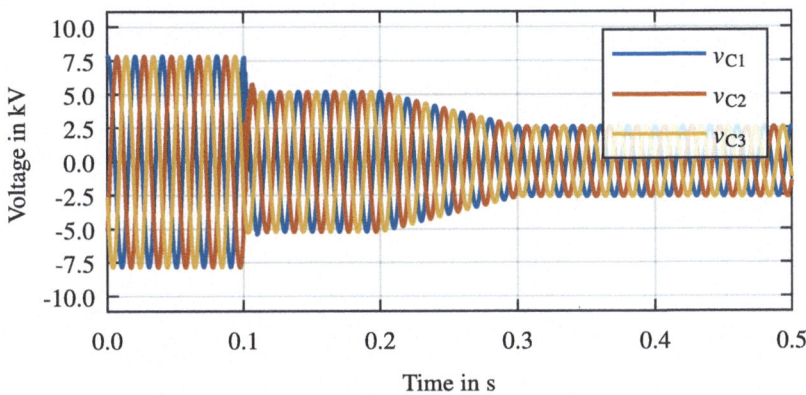

Fig. 10.2 Phase-to-ground voltages at busbar C before and after a three-phase fault on line 1 with a relative fault distance of $m = 0.3$ at $t = 0.1$ s based on a two serial lines topology

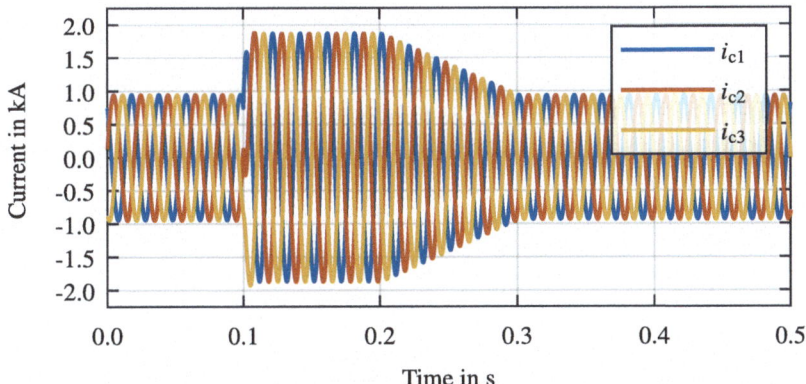

Fig. 10.3 Phase currents at busbar C before and after a three-phase fault on line 1 with a relative fault distance of $m = 0.3$ at $t = 0.1$ s based on a two serial lines topology

it has to be checked, if the resulting factor of 2 is allowed. From $t = 0.2$ s, the amplitude of the currents is decreased approximately to its nominal value.

The p-value p_h calculated in the FI layer is shown in Fig. 10.4. The p-value is greater than the predefined confidence level α_h before the fault at $t < 0.1$ s. Consequently, no fault is identified. After the fault at $t > 0.1$ s, $p_h < \alpha_h$ is valid. The fault is correctly identified 1.5 ms after the fault occurred.

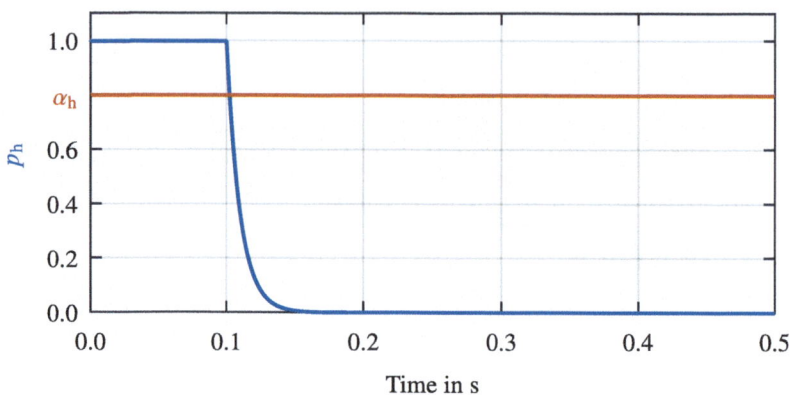

Fig. 10.4 P-value p_h and confidence level α_h of the FI layer of the MBP algorithm before and after a three-phase fault on line 1 with a relative fault distance of $m = 0.3$ at $t = 0.1$ s based on a two serial lines topology

The evolution of the indices of the models of line 1 and line 2 of the FC layer is visualized in Fig. 10.5. The change from index one to another value indicates the fault on line 1. Since the model index does not change, the line 2 is correctly characterized as healthy.

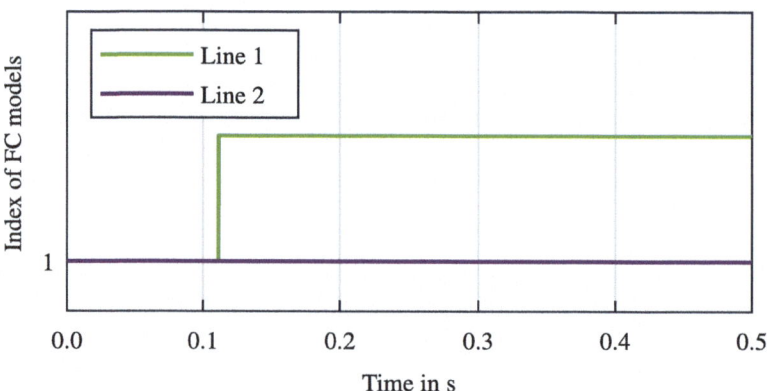

Fig. 10.5 Index of the models of line 1 and line 2 evoking the minimum Euclidean norm of the estimation errors of the FC layer of the MBP algorithm before and after a three-phase fault on line 1 with a relative fault distance of $m = 0.3$ at $t = 0.1$ s based on a two serial lines topology

10.2 Scenario 5: 3p-fault and Voltage Reduction

The spider plot of the FC layer of line 1 is depicted in Fig. 10.6. Two models of three-phase faults at $m = 0$ and $m = 0.5$ with fault resistances of $0\,\Omega$ are chosen for comparison in this figure. Different models with different fault resistance configurations with a relative fault distance of $m = 0.5$ are additionally chosen for comparison. Here, the fault resistances are also assumed to be zero. At $t = 0.05\,\text{s}$ no fault is identified. At $t = 0.15\,\text{s}$, the three-phase fault at $m = 0.5$ is identified. According to the calculated IRE_{FC}-values, the distinction between $m = 0$ and $m = 0.5$ is not definitely clear. A subsequent fault localization is necessary. Fig. 10.7 contains the spider plot of line 2. Here, no fault is identified.

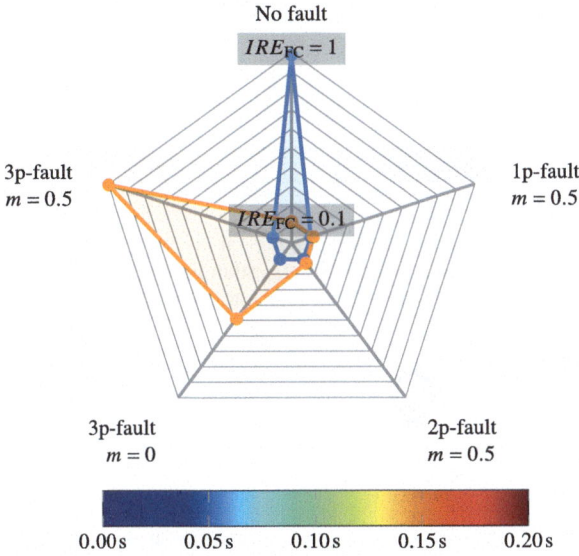

Fig. 10.6 Inverse relative estimation errors IRE_{FCi} at $t = 0.05\,\text{s}$ and $t = 0.15\,\text{s}$ of selected models of line 1 of the FC layer of the MBP algorithm before and after a three-phase fault on line 1 with a relative fault distance of $m = 0.3$ at $t = 0.1\,\text{s}$ based on a two serial lines topology

In the FL layer of the MBP, models described by a three-phase fault are investigated at different relative fault distances with $R_F = 0\,\Omega$. The variable m is varied from 0 to 1 in 0.1-steps. The model described by a three-phase fault with $R_F = 0\,\Omega$ at $m = 0.3$ is identified as the most suitable model. The result of the FL layer is visualized in Fig. 10.8.

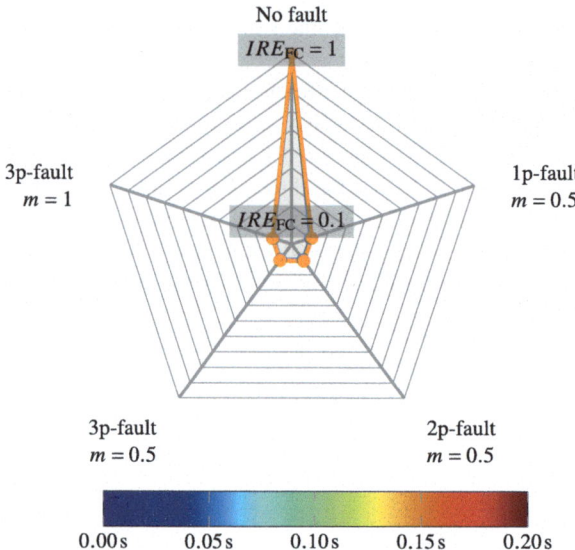

Fig. 10.7 Inverse relative estimation errors IRE_{FCi} at $t = 0.05$ s and $t = 0.15$ s of selected models of line 2 of the FC layer of the MBP algorithm before and after a three-phase fault on line 1 with a relative fault distance of $m = 0.3$ at $t = 0.1$ s based on a two serial lines topology

Fig. 10.8 Result of the fault localization on line 1 of the MBP algorithm after a three-phase fault on line 1 with a relative fault distance of $m = 0.3$ at $t = 0.1$ s based on a two serial lines topology

The scenario in this section shows that a three-phase fault followed by voltage reduction in the positive sequence by the VSC is identified fast and reliable by the MBP in the simulation. The FC layer identifies the faulty line 1. The fault resistance configuration and the value of the fault resistance are detected correctly. The FL layer localizes the fault precisely.

10.3 Scenario 6: 2p-fault and Reactive Current Injection

In this simulation scenario, a phase-to-phase fault is investigated. Reactive current is injected in the negative sequence by the VSC after the fault according to Sect. 5.2.3.2 to ensure symmetrical grid voltages. The MBP algorithm is validated based on the resulting voltage and current characteristics of Sect. 5.2.3.3. An overview of the relevant figures of this section, providing also the possibility for quick navigation in the digital document, is given in Tab. 10.3.

Table 10.3 Overview of the relevant figures of Sect. 10.3 containing the validation of the model-based protection (MBP) algorithm in a simulation test investigating a phase-to-phase fault between phase 2 and phase 3 with a relative fault distance of $m = 0.8$ at $t = 0.1$ s

Content	Figure
Scenario under investigation	Fig. 10.9
Model of the grid topology for the fault identification	Fig. 7.5
Model of the healthy line for the fault characterization	Fig. 7.11
Models of the faulty line for the fault characterization and localization	Fig. 7.12 & Fig. 7.13
Phase-to-ground voltages at busbar B	Fig. 5.14
Phase currents at busbar B	Fig. 5.15
Result of the FI: P-value p_h	Fig. 10.10
Result of the FC: Index of the identified model	Fig. 10.11
Result of the FC: Inverse relative estimation errors IRE_{FCi}	Fig. 10.12
Result of the FL in a slider diagram	Fig. 10.13

The VSC at busbar B is connected to the grid at busbar A. The overhead line has a length of 10 km. The nominal active power of the VSC equals 8.5 MW. A phase-to-phase fault between phase 2 and phase 3 at the relative fault location $m = 0.8$ seen from busbar A is applied at $t = 0.1$ s. The fault resistance R_F equals $2\,\Omega$. The reactive current injection in the negative sequence is activated at $t = 0.3$ s. The time interval between the fault occurrence and the activation of the reactive current injection of 200 ms is chosen to demonstrate the reaction of the MBP with and without reactive current injection. The simulation scenario is shown in Fig. 10.9.

The voltages at busbar B are shown in Fig. 5.14 and the phase currents at busbar B are presented in Fig. 5.15. Detailed descriptions and explanations of these figures are given in Sect. 5.2.3.3.

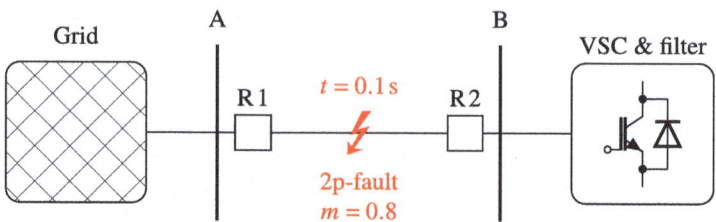

Fig. 10.9 Simulation scenario for validating the MBP algorithm during a phase-to-phase fault between phase 2 and phase 3 with a relative fault distance of $m = 0.8$ at $t = 0.1\,\text{s}$

The p-value p_h calculated in the FI layer is shown in Fig. 10.10. Before the fault at $t < 0.1\,\text{s}$, the p-value is greater than the predefined confidence level α_h and no fault is identified. After the fault at $t > 0.1\,\text{s}$, $p_\text{h} < \alpha_\text{h}$ is valid and the fault is correctly identified 5 ms after the fault occurred. Periodic peaks showing a frequency of 100 Hz appear in this time interval. They are caused by the minima of the weighted sum $\zeta(k)$ of the squared elements of the vector of the estimation errors, see Eq. (8.21). This effect does not influence the functionality of the FI layer negatively. Compared to Fig. 10.4, the p-value is increased after the fault. However, the fault identification based on $\alpha_\text{h} = 0.8$ is clear.

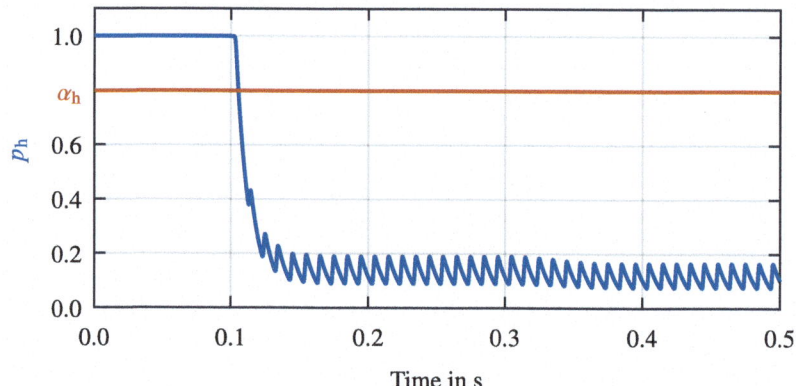

Fig. 10.10 P-value p_h and confidence level α_h of the FI layer of the MBP algorithm before and after a phase-to-phase fault between phase 2 and phase 3 with a relative fault distance of $m = 0.8$ at $t = 0.1\,\text{s}$

10.3 Scenario 6: 2p-fault and Reactive Current Injection

Fig. 10.11 depicts the index of the model evoking the minimum Euclidean norm of the estimation errors in the FC layer. The healthy model is described by the index one. The fault is identified at high speed 7 ms after the fault occurred. The stability of decision is high because the index stays constant before and after the fault.

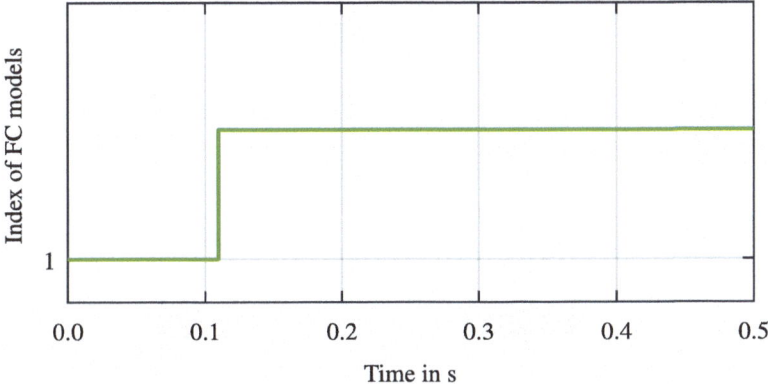

Fig. 10.11 Index of the models evoking the minimum Euclidean norm of the estimation errors of the FC layer of the MBP algorithm before and after a phase-to-phase fault between phase 2 and phase 3 with a relative fault distance of $m = 0.8$ at $t = 0.1$ s

The Fig. 10.12 shows the spider plot of the FC layer. Two models of phase-to-phase faults between phase 2 and phase 3 at $m = 0$ and $m = 0.75$ with fault resistances of $2\,\Omega$ are chosen for comparison. Different models with different fault resistance configurations with a relative fault distance of $m = 0.5$ and fault resistances of $2\,\Omega$ are additionally chosen for comparison in this figure. At $t = 0.05$ s no fault is identified. At $t = 0.15$ s, the phase-to-phase fault between phase 2 and phase 3 at $m = 0.75$ is identified. According to the calculated IRE_{FC}-values, the distinction between $m = 0.5$ and $m = 0.75$ is not definitely clear. A subsequent fault localization is necessary.

In the FL layer of the MBP, models described by a phase-to-phase faults between phase 2 and phase 3 with $R_F = 2\,\Omega$ are investigated at different relative fault distances. The variable m is varied from 0 to 1 in 0.1-steps. The model described by a phase-to-phase faults between phase 2 and phase 3 with $R_F = 2\,\Omega$ at $m = 0.8$ is identified as the most suitable model. The result of the FL layer is visualized in Fig. 10.13.

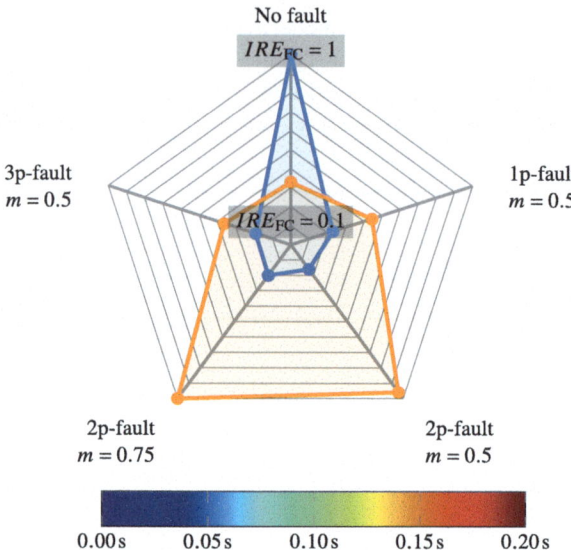

Fig. 10.12 Inverse relative estimation errors IRE_{FCi} at $t = 0.05$ s and $t = 0.15$ s of selected models of the FC layer of the MBP algorithm before and after a phase-to-phase fault between phase 2 and phase 3 with a relative fault distance of $m = 0.8$ at $t = 0.1$ s

Fig. 10.13 Result of the fault localization of the MBP algorithm after a phase-to-phase fault between phase 2 and phase 3 with a relative fault distance of $m = 0.8$ at $t = 0.1$ s

This scenario demonstrates, that a phase-to-phase faults between phase 2 and phase 3 followed by reactive current injection in the negative sequence by the VSC is identified fast and reliable by the MBP. The fault characterization is sufficiently good. The subsequent FL layer localizes the fault correctly.

10.4 Scenario 7: 1p-fault and RGS-mode

This simulation scenario contains the investigation of a single-phase-to-ground fault. The RGS-mode of the VSC according to Sect. 6.2.3.1 is applied to ensure a symmetrical grid operation, the reduction of the fault current and the arc suppression. The MBP algorithm is validated based on the resulting voltage and current characteristics of Sect. 6.2.3.2. Tab. 10.4 gives an overview of the relevant figures of this section.

Table 10.4 Overview of the relevant figures of Sect. 10.4 containing the validation of the model-based protection (MBP) algorithm in a simulation test investigating a single-phase-to-ground fault in phase 1 with a relative fault distance of $m = 0.8$ at $t = 0.1$ s

Content	Figure
Scenario under investigation	Fig. 10.14
Model of the grid topology for the fault identification	Fig. 7.5
Model of the healthy line for the fault characterization	Fig. 7.11
Models of the faulty line for the fault characterization and localization	Fig. 7.12 & Fig. 7.13
Phase-to-ground voltages at busbar B	Fig. 6.15
Phase currents at busbar B	Fig. 6.14
Result of the FI: P-value p_h	Fig. 10.15
Space vector components of the phase-to-ground voltages at busbar B	Fig. 10.16
Space vector components of the phase currents at busbar B	Fig. 10.17
Result of the FC: Index of the identified model	Fig. 10.18
Result of the FC: Inverse relative estimation errors IRE_{FCi}	Fig. 10.19
Result of the FL in a slider diagram	Fig. 10.20

The scenario is similar to the scenario of Sect. 10.3. The VSC at busbar B is connected to the grid at busbar A via an overhead line of 10 km. The nominal active power of the VSC equals 8.5 MW. A single-phase-to-ground fault in phase 1 with the fault resistance $R_F = 0\,\Omega$ is applied at $t = 0.1$ s. The relative fault distance m equals 0.8 seen from busbar A. The reference value of the virtual inductance in the zero sequence is reached approximately at $t = 0.25$ s, see Fig. 6.17. From $t = 0.25$ s, steady state characteristics are observed. The simulation scenario is shown in Fig. 10.14.

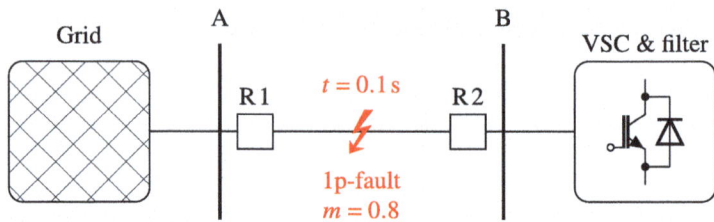

Fig. 10.14 Simulation scenario for validating the MBP algorithm during a single-phase-to-ground fault in phase 1 with a relative fault distance of $m = 0.8$ at $t = 0.1$ s

Fig. 6.15 visualizes the voltages at busbar B and Fig. 6.14 presents the phase currents at busbar B. Detailed descriptions and explanations of these figures are given in Sect. 6.2.3.2.

Fig. 10.15 presents the p-value p_h calculated in the FI layer. The p-value is greater than the predefined confidence level α_h before the fault at $t < 0.1$ s. Consequently, no fault is identified. Immediately after the fault at $t > 0.1$ s, $p_h < \alpha_h$ is valid and the fault is identified.

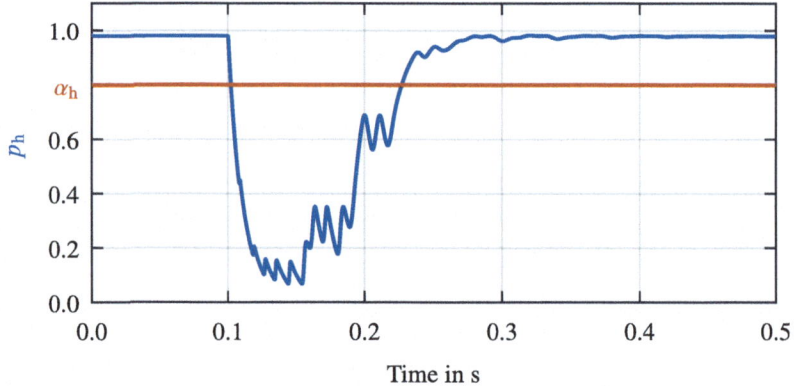

Fig. 10.15 P-value p_h and confidence level α_h of the FI layer of the MBP algorithm before and after a single-phase-to-ground fault in phase 1 with a relative fault distance of $m = 0.8$ at $t = 0.1$ s

Approximately at $t = 0.225$ s, p_h is greater than α_h and the fault decision of the FI layer is reversed. The reference value of the virtual inductance in the zero sequence is

10.4 Scenario 7: 1p-fault and RGS-mode

nearly reached and the absolute value of the zero sequence impedance becomes very high compared to the time interval immediately after the fault. The positive sequence, the negative sequence and the zero sequence become decoupled again from the VSC perspective (see also Fig. 6.13). The space vector components of voltages and currents before the fault do not distinguish significantly from their characteristics in the steady state after the fault. Exemplary, the space vector components of voltages and currents at busbar B are visualized in Fig. 10.16 and Fig. 10.17. These variables represent subsets of the input variables of the Kalman filter and the measurement variables of the FI layer. Since zero sequence components are not considered in the FI layer (see Sect. 7.2), the fault decision of the FI layer is reversed after the reference value of the virtual inductance in the zero sequence is reached. The low-impedance time interval depends on the dynamic performance and the settings of the V/I-controller. It can for example be increased, see Sect. 6.2.3.4.

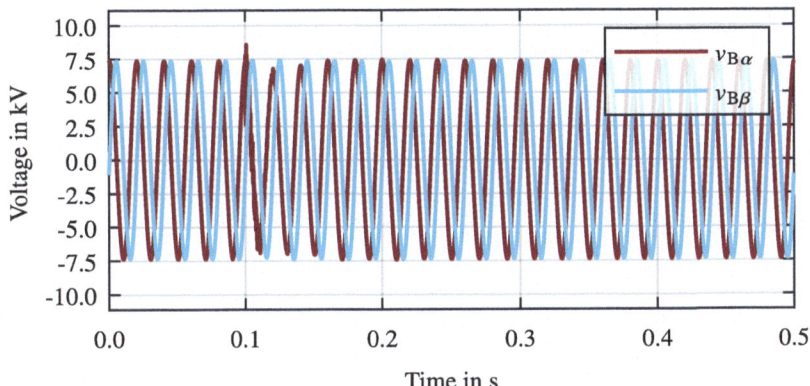

Fig. 10.16 Space vector components in the stationary reference frame of the phase-to-ground voltages at busbar B before and after a single-phase-to-ground fault in phase 1 with a relative fault distance of $m = 0.8$ at $t = 0.1$ s

The evolution of the indices of the models of the FC layer is shown in Fig. 10.18. The index changes from one to another value immediately after the fault occurred and the fault is identified. At $t = 0.18$ s, the index changes. In this chase, the most suitable model changes from a single-phase-to-ground fault with $R_F = 1\,\Omega$ to $R_F = 0\,\Omega$. This effect does not influence the fault decision significantly. The changing value of the realized value of the virtual inductance in the zero sequence does not influence the FC layer. The reason is that the zero sequence is modelled in

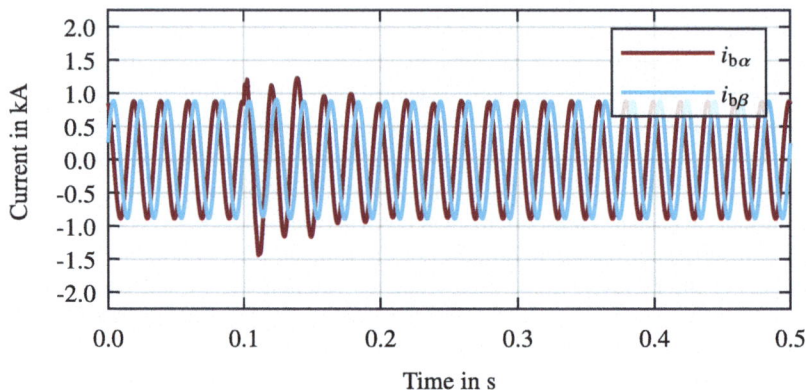

Fig. 10.17 Space vector components in the stationary reference frame of the phase currents at busbar B before and after a single-phase-to-ground fault in phase 1 with a relative fault distance of $m = 0.8$ at $t = 0.1$ s

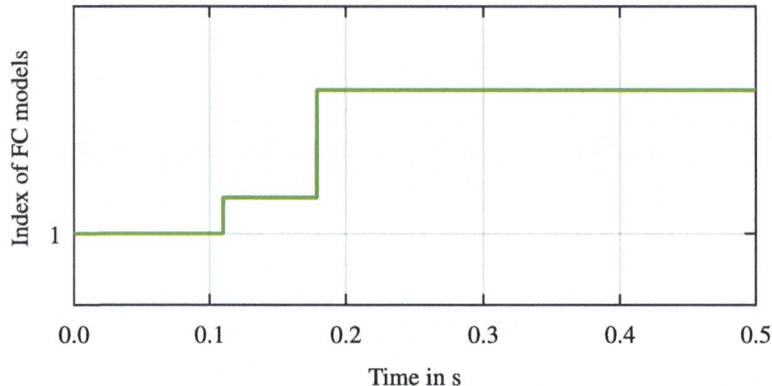

Fig. 10.18 Index of the models evoking the minimum Euclidean norm of the estimation errors of the FC layer of the MBP algorithm before and after a single-phase-to-ground fault in phase 1 with a relative fault distance of $m = 0.8$ at $t = 0.1$ s

the FC layer, see Sect. 7.4. Since the FI layer reverses its decision, the FC layer as the second layer of the MBP is crucial.

The spider plot of the FC layer is shown in Fig. 10.19. Two models of single-phase-to-ground faults in phase 1 at $m = 0.5$ and $m = 0.75$ with fault resistances of $0\,\Omega$ are chosen for comparison in this figure. Different models with different

10.4 Scenario 7: 1p-fault and RGS-mode

fault resistance configurations with a relative fault distance of $m = 0.5$ with zero fault resistances are additionally chosen for comparison. At $t = 0.05$ s, no fault is identified. At $t = 0.15$ s, the single-phase-to ground fault in phase 1 at $m = 0.75$ is identified. According to the calculated IRE_{FC}-values, the distinction between $m = 0.5$ and $m = 0.75$ is not definitely clear. A subsequent fault localization is necessary.

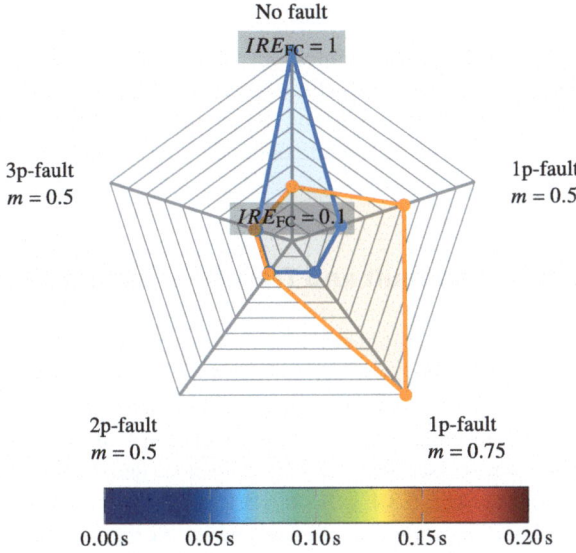

Fig. 10.19 Inverse relative estimation errors IRE_{FCi} at $t = 0.05$ s and $t = 0.15$ s of selected models of the FC layer of the MBP algorithm before and after a single-phase-to-ground fault in phase 1 with a relative fault distance of $m = 0.8$ at $t = 0.1$ s

Models described by a single-phase-to ground fault in phase 1 with $R_F = 0\,\Omega$ are investigated at different relative fault distances in the FL layer of the MBP. The variable m is varied from 0 to 1 in 0.1-steps. The model described by a three-phase fault with $R_F = 0\,\Omega$ at $m = 0.8$ is identified as the best suitable model. The result of the FL layer is presented in Fig. 10.20.

This scenario demonstrates that the FI layer identifies a single-phase-to ground fault in phase 1 fed by a VSC applying the RGS-mode immediately after the fault. The FC layer ensures the fault characterization even after the reference value of the virtual inductance in the zero sequence is reached. The multilayer principle

Fig. 10.20 Result of the fault localization of the MBP algorithm after a single-phase-to-ground fault in phase 1 with a relative fault distance of $m = 0.8$ at $t = 0.1$ s

of the MBP allows the fast and reliable fault identification, characterization and localization. Opposite decisions of the FI layer and the FC layer in the steady state after the fault (e.g. at $t = 0.3$ s, see Fig. 10.15 and Fig. 10.18) are used as an indicator, that the grid operation can be continued after a single-phase-to ground fault.

10.5 Scenario 8: 1p-fault and Transient Current Limiting

In this simulation scenario, a single-phase-to-ground fault with transient current limiting by the VSC is investigated. As in Sect. 10.4, the RGS-mode of the VSC according to Sect. 6.2.3.1 is applied to ensure a symmetrical grid operation, the reduction of the fault current and the arc suppression. Furthermore, non-zero mean values of the phase currents (see Sect. 6.2.1.2) are avoided and the constant switching frequency according to Sect. 6.3.1.1, Sect. 6.3.1.2, Sect. 6.3.1.3 and Sect. 6.3.1.4 is activated. The MBP algorithm is validated based on the resulting voltage and current characteristics of Sect. 6.3.1.5. An overview of the relevant figures of this section, providing also the possibility for quick navigation in the digital document, is given in Tab. 10.5.

The VSC at busbar B is connected to the grid at busbar A via an overhead line of 10 km. The nominal active power of the VSC equals 4 MW and the nominal reactive power of the VSC equals −3.5 Mvar. Voltage stability will be a growing challenge in future distribution grids due to multidirectional power flows. VSCs offer the possibility to adjust reactive power independently of the active power. This scenario shows a VSC drawing inductive reactive power for reducing the grid voltage with simultaneous active power injection. The current limit of the phase currents is set to ±600 A. At $t = 0.1$ s, single-phase-to-ground fault in phase 1 with the fault resistance $R_\mathrm{F} = 0\,\Omega$ is applied. The relative fault distance m equals 0.8 seen from busbar A. The simulation scenario is shown in Fig. 10.21.

10.5 Scenario 8: 1p-fault and Transient Current Limiting

Table 10.5 Overview of the relevant figures of Sect. 10.5 containing the validation of the model-based protection (MBP) algorithm in a simulation test investigating a single-phase-to-ground fault in phase 1 with a relative fault distance of $m = 0.8$ at $t = 0.1$ s

Content	Figure
Scenario under investigation	Fig. 10.21
Model of the grid topology for the fault identification	Fig. 7.5
Model of the healthy line for the fault characterization	Fig. 7.11
Models of the faulty line for the fault characterization and localization	Fig. 7.12 & Fig. 7.13
Phase-to-ground voltages at busbar B	Fig. 6.31
Phase currents at busbar B	Fig. 6.30
Result of the FI: P-value p_h	Fig. 10.22
Result of the FC: Index of the identified model	Fig. 10.23
Result of the FC: Inverse relative estimation errors IRE_{FCi}	Fig. 10.24
Result of the FL in a slider diagram	Fig. 10.25

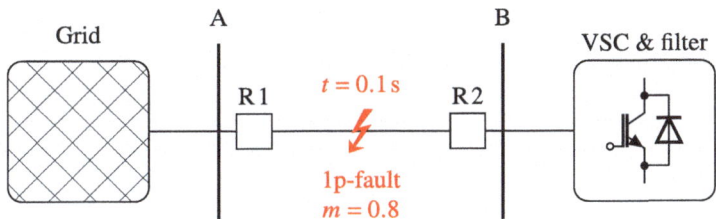

Fig. 10.21 Simulation scenario for validating the MBP algorithm during a single-phase-to-ground fault in phase 1 with a relative fault distance of $m = 0.8$ at $t = 0.1$ s

The voltages at busbar B are shown in Fig. 6.31 and the phase currents at busbar B are presented in Fig. 6.30. Detailed descriptions and explanations of these figures are given in Sect. 6.3.1.5.

The p-value p_h calculated in the FI layer is shown in Fig. 10.22. The p-value is greater than the predefined confidence level α_h before the fault at $t < 0.1$ s. Consequently, no fault is identified. $p_h < \alpha_h$ is valid after the fault at $t > 0.1$ s. The fault is correctly identified 3 ms after the fault occurred. As in Sect. 10.3, periodic peaks showing a frequency of 100 Hz appear in this time interval. They are caused by the minima of the weighted sum $\zeta(k)$ of the squared elements of the vector of the estimation errors, see Eq. (8.21). This effect does not influence the functionality of the FI layer negatively. Compared to the previous three sections, the p-value is increased after the fault showing values from 0.5 to 0.6. However, the fault identification based on $\alpha_h = 0.8$ is clear.

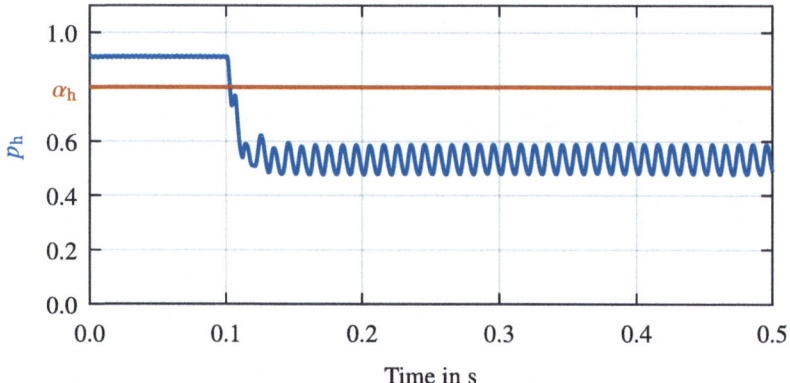

Fig. 10.22 P-value p_h and confidence level α_h of the FI layer of the MBP algorithm before and after a single-phase-to-ground fault in phase 1 with a relative fault distance of $m = 0.8$ at $t = 0.1$ s

The evolution of the index of the FC layer is demonstrated in Fig. 10.23. The grid fault is indicated by the change of the value of the index from one to another value. The fault is identified 28 ms after the fault occurred. The stability of decision is high because the index stays constant before and after the fault.

The Fig. 10.24 shows the spider plot of the FC layer. Two models of single-phase-to-ground faults in phase 1 at $m = 0.5$ and $m = 0.75$ with a fault resistance of $0\,\Omega$ are chosen for comparison in this figure. Different models with different fault resistance configurations with a relative fault distance of $m = 0.5$ and fault resistances of $0\,\Omega$ are additionally chosen for comparison. At $t = 0.05$ s no fault is identified. At $t = 0.15$ s, the single-phase-to-ground fault in phase 1 at $m = 0.75$ is identified. According to the calculated IRE_{FC}-values, the distinction between

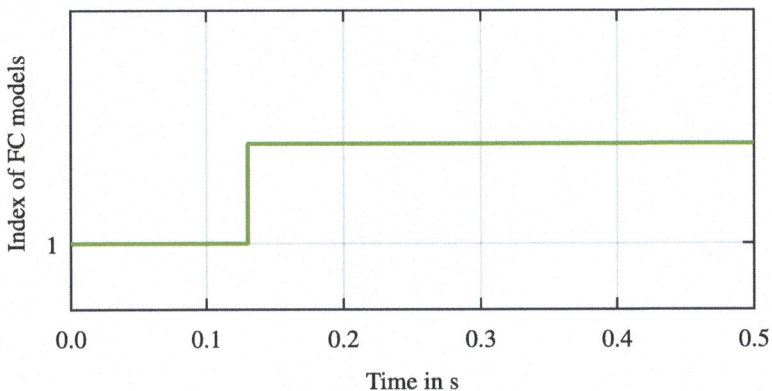

Fig. 10.23 Index of the models evoking the minimum Euclidean norm of the estimation errors of the FC layer of the MBP algorithm before and after a single-phase-to-ground fault in phase 1 with a relative fault distance of $m = 0.8$ at $t = 0.1$ s

$m = 0.5$ and $m = 0.75$ is not definitely clear. A subsequent fault localization is necessary.

In the FL layer of the MBP, models described by a single-phase-to-ground fault in phase 1 with $R_F = 0\,\Omega$ are investigated at different relative fault distances. The variable m is varied from 0 to 1 in 0.1-steps. The model described by a single-phase-to-ground fault in phase 1 with $R_F = 0\,\Omega$ at $m = 0.8$ is identified as the best suitable model. Fig. 10.25 shows the result of the FL layer.

This scenario demonstrates, that a single-phase-to ground fault in phase 1 fed by a VSC applying the RGS-mode and transient current limiting is identified fast and reliable by the MBP. The fault characterization is sufficiently good. The subsequent FL layer localizes the fault correctly.

10.6 Summary

In this chapter, the investigation of the model-based protection (MBP) algorithm in 100% inverter-based power systems (IBPS) applying different grid fault types is presented. The control algorithm of voltage source converters (VSCs) and the MBP algorithm are merged and a perspective of the resilient operation of 100% IBPS is pointed out.

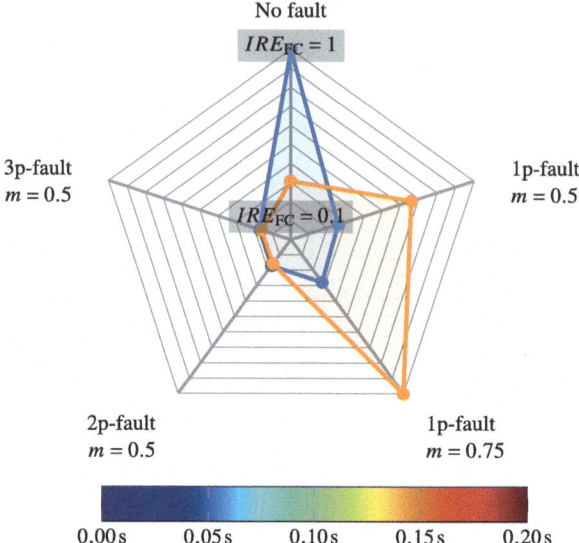

Fig. 10.24 Inverse relative estimation errors IRE_{FCi} at $t = 0.05$ s and $t = 0.15$ s of selected models of the FC layer of the MBP algorithm before and after a single-phase-to-ground fault in phase 1 with a relative fault distance of $m = 0.8$ at $t = 0.1$ s

Fig. 10.25 Result of the fault localization of the MBP algorithm after a single-phase-to-ground fault in phase 1 with a relative fault distance of $m = 0.8$ at $t = 0.1$ s

The MBP algorithm from Chap. 8 is tested, realizing the state-of-the-art and enhanced grid fault characteristics from Chap. 5 and Chap. 6 in simulations. In this way, the MBP algorithm is validated by simulation tests in addition to the validation by laboratory tests in Chap. 9.

The objective of the grid-forming VSC controller as a reaction to the grid fault depends on the grid fault type. Based on the control algorithms presented in Chap. 5 and Chap. 6, different favourable voltage/current (V/I) characteristics appear. The MBP algorithm is tested based on these V/I-characteristics applying a three-phase fault, a phase-to-phase fault and a single-phase-to-ground fault.

10.6 Summary

The MBP algorithm is successfully validated by the investigation of a three-phase fault inside a serial lines topology. The amplitude of the positive sequence voltages is reduced by the VSC after the fault to ensure steady state current limiting. The fault is identified fast and reliable, characterized and localised correctly by the MBP algorithm. The faulty line is detected properly.

Moreover, the MBP algorithm is successfully validated by the investigation of a phase-to-phase fault. Reactive current is injected in the negative sequence by the VSC after the fault to ensure symmetrical grid voltages. The phase-to-phase fault is identified fast and reliable and localised correctly.

Furthermore, the MBP algorithm is successfully validated by the investigation of a single-phase-to-ground fault. The resonant grounding system (RGS) mode of the VSC is activated to ensure a symmetrical grid operation, the reduction of the fault current and the arc suppression. The multilayer principle of the MBP algorithm allows a fast and reliable detection of the fault—even after the reference value of the virtual inductance in the zero sequence is reached.

Finally, the MBP algorithm is successfully validated by the investigation of a single-phase-to-ground fault with transient current limiting by the VSC. In addition, non-zero mean values of the phase currents are avoided and the constant switching frequency mode is activated. The fault is identified fast and reliable, characterized and localised correctly by the MBP algorithm.

Open Access This chapter is licensed under the terms of the Creative Commons Attribution 4.0 International License (http://creativecommons.org/licenses/by/4.0/), which permits use, sharing, adaptation, distribution and reproduction in any medium or format, as long as you give appropriate credit to the original author(s) and the source, provide a link to the Creative Commons license and indicate if changes were made.

The images or other third party material in this chapter are included in the chapter's Creative Commons license, unless indicated otherwise in a credit line to the material. If material is not included in the chapter's Creative Commons license and your intended use is not permitted by statutory regulation or exceeds the permitted use, you will need to obtain permission directly from the copyright holder.

Conclusion and Research Perspectives 11

11.1 Conclusion

In this thesis, a holistic control and protection concept for 100% inverter-based power systems (IBPS) based on voltage source converters (VSCs) is presented. A perspective of sustainable energy systems based on renewable energy sources (RES) is pointed out.

Energy systems are in transition. VSCs are key technologies of future electrical power grids. New grid architectures and grid operation strategies with multidirectional power flows based on storage-based RES arise. The grid synchronization, the grid operation and the grid fault characteristics have to be re-evaluated regarding the increasing penetration of the grid by VSCs. The characteristics of VSCs regarding grid synchronization, grid operation and grid fault characteristics depend on their control algorithm. New flexibility options to increase the reliability of energy supply after grid faults reveal. These flexibility options lead to characteristics of voltages and currents, that differ from conventional characteristics caused by synchronous machines. The fault characteristics are challenging for state-of-the-art grid protection algorithms. Sophisticated solutions are missing up to now. New concepts for the control and protection concept of 100% IBPS have to be developed.

The key points of the own contribution refer to the **research objectives** in Sec. 1.2.3. The basic principles and impacts of these contributions are presented subsequently.

Results regarding research objective 1

In this thesis, advances regarding the post-fault characteristics and power synchronization applying the grid-forming mode in the positive sequence regarding **research objective 1** (see Sec. 1.2.3) are achieved. A design guideline of the configurable natural droop (CND) controller for the active power synchronization is developed. Reflecting the transfer function of the CND-controller of a grid-connected VSC to the transfer function of a second order lag element allows defining the parameters of the CND-controller. In this way, the desired frequency inertia, the power damping coefficient and the frequency droop constant are realized independently of each other via a VSC. These properties define the frequency and active power characteristics after a grid fault. They are essential for the post-fault operation of 100% IBPS.

Moreover, a state-space model of a grid-connected VSC using a CND-controller and a reactive power droop controller is derived. Consequently, the small-signal power synchronization stability can be evaluated. It also depends on the parameters of the CND-controller and the reactive power droop constant. In this way, the grid integration of VSCs into 100% IBPS is formalized.

A hybrid synchronization method is presented. A CND-controller and a multi-sequence harmonic decoupling cell (MSHDC) phase-locked loop (PLL) are combined to calculate the reference frequency of the filter capacitor voltages to adjust the active power. Consequently, properties of the voltage-based synchronization and the power-based synchronization are merged. In this way, the provision of instantaneous power is maintained while the essential control of the DC-link voltage works independent of the grid frequency. These characteristics are for example used for wind power plants and photovoltaic power plants inside 100% IBPS. The active power is controlled by the second VSC on the grid-averted side, e.g. based on maximum power point tracking.

11.1 Conclusion

Results regarding research objective 2

In this thesis, the connection of a VSC to the grid as a three-phase four-wire system to open new flexibility options during grid faults regarding **research objective 2** (see Sec. 1.2.3) is presented. Different characteristics in the zero sequence during grid faults are realized. Therefore, the hardware configuration and the software design of the grid-connected VSC are extended.

A conventional three-phase three-wire VSCs is extended by a fourth half-bridge. The fourth half-bridge operates as a DC-link voltage balancer. The extended VSC is connected to the grid as a three-phase four-wire system and enables zero sequence currents.

By controlling the zero sequence current or voltage of the VSC, the grid-side zero sequence characteristics are determined. A state-feedback controller based on the linear-quadratic regulator (LQR) algorithm is used. Realizing a virtual inductance, a parallel resonant circuit in the zero sequence appears. Adjusting the resonance frequency of the parallel resonant circuit to the operating frequency of the grid, the absolute value of the impedance of the zero sequence increases significantly. The fault current induced by a single-phase-to-ground fault is consequently reduced and symmetrical grid operation can often be continued. Besides, arc suppression is achieved.

Moreover, the transient state current limiting is realized by including aspects of the model predictive control (MPC) and designing a suitable cost function. Steady state current limiting is achieved by a pre-review of the targeted operating point.

Danger to living beings and damage on hardware equipment due to high fault currents is reduced. Besides, the resiliency of 100% IBPS against grid fault and the reliability of energy supply is increased. Advanced current limiting techniques raise the resilience of VSCs against grid faults.

Results regarding research objective 3

In this thesis, a novel model-based protection (MBP) algorithm regarding **research objective 3** (see Sec. 1.2.3) is developed. The basic principle of the MBP algorithm is to compare, if measured values and estimated values based on state-space models of corresponding variables match or not. The estimation error indicates the goodness-of-fit of the models and conclusions of the real grid state are drawn. Decisions based on this analysis for the tripping of circuit breakers to protect living beings and hardware equipment are taken.

Like the design of the state-feedback LQR-controllers of VSCs, the design of the developed MBP algorithm is also based on models emulating real physical characteristics. Existing protection algorithms like the distance protection and the differential protection are also based on models. These models are of different nature and are less detailed compared to the state-space models of the MBP algorithm.

The Kalman filter (KF) is used as the state estimation algorithm of the MBP. KFs take stochastic process and measurement noise into account and are thus suitable for practical applications.

The MBP algorithm consists of three layers: The fault identification (FI) layer, the fault characterization (FC) layer and the fault localization (FL) layer. The objective of the FI layer is to decide, if normal operation conditions inside the monitored grid area are present or not. The objective of the fault characterization is the determination of the fault resistance configuration, the values of the appearing fault resistances and the rough fault location on one line. Each line inside the monitored grid area is investigated. Each faulty line is further investigated in detail by the fault localization layer to specify the fault location precisely.

Inherent selectivity regarding the faulty line, speed and sensitivity as fundamental properties of a protection algorithm are ensured. Since the MBP algorithm operates independently of signal characteristics and the direction of power flows, it is additionally characterized by its universality.

The developed MBP algorithm represents an appropriate protection algorithm for 100% IBPS without restricting the flexibility options of VSCs during grid faults. The limited overcurrent capability or the harmonics of voltages and currents of VSCs do not influence the MBP negatively, too. For safety reasons, the MBP algorithm is an indispensable part of 100% IBPS.

All control and protection algorithms in this thesis are developed analytically, with a clear focus on mathematical precision and evidence. Furthermore, they are thoroughly validated in simulations using MathWorks MATLAB/Simulink software suite. In addition, the MBP algorithm is successfully validated by real hardware laboratory tests.

The sustainable and reliable energy supply is an essential cornerstone of human societies. This thesis pointing out a holistic approach for the control and protection of 100% IBPS contributes along this way.

11.2 Research Perspectives

Based on the key contributions in Sec. 11.1, different starting points of future research work reveal. The most significant aspects—from the author's perspective—are described subsequently.

Post-fault characteristics of VSCs
The state-space model of a grid-connected VSC using a CND-controller and a reactive power droop controller to evaluate the small-signal power synchronization stability can be enhanced by the dynamics of the closed-loop V/I-state-space controller. In this way, the impact of the parameters of the state-space controller is investigated and further design guidelines can be defined. The state-space model of the VSC can additionally be extended to microgrids or larger grid areas. Investigations of the hybrid synchronization method can include the effects on the DC-side and optionally the characteristics of wind turbines and photovoltaic modules to include additional effect appearing in practical applications. Moreover, future work can include investigations regarding the transient stability of the hybrid synchronization method.

Neutral point treatment and resonant grounding via VSCs
Today, state-of-the-art VSCs are still often connected via a transformer to the grid. To open the application range of three-phase four-wire VSCs for neutral point treatment to conventional hardware setups, VSCs connected via a wye-wye transformer to the grid can be evaluated. To further reduce the residual current after a single-phase-to-ground fault, the controller of the VSC can be extended to adjust the characteristics of multiple frequencies independently of each other at the same time. A polyfrequency three-phase four-wire system appears. Since the requirements on the maximum values of voltages and currents of a three-phase four-wire VSC differ from the ones on a conventional VSC, new standards have to be defined. Besides, a decentralized

resonant grounding system (RGS) can be investigated. The necessary value of the virtual inductance is realized in a cooperation of multiple VSCs. In this way, the additional current is shared upon multiple VSCs.

Model-based protection algorithm

The MBP algorithm can be investigated in combination with the distance protection algorithm as a backup. Complementary advantages can lead to a high efficacy of the hybrid protection scheme. Furthermore, the state-space model of the FI layer can be derived from artificial intelligence (AI) methods for parameter identification. A broad measurement data basis gained during normal operating conditions for the AI training can be used. Models of further hardware equipment can be derived to be used inside the FC layer. Non-linear models (e.g. of a transformer) can increase the necessity for the use of the extended Kalman filter (EKF) or the unscented Kalman filter (UKF) instead of the standard KF. Models of superconducting components can include further physical laws, e.g. to characterize the dynamics of the temperature. Finally, the MBP algorithm can be adapted for DC-grids. The models can be formulated based on common mode (CM) and differential mode (DM) components, that replace the space-vector components in the stationary reference frame.

Open Access This chapter is licensed under the terms of the Creative Commons Attribution 4.0 International License (http://creativecommons.org/licenses/by/4.0/), which permits use, sharing, adaptation, distribution and reproduction in any medium or format, as long as you give appropriate credit to the original author(s) and the source, provide a link to the Creative Commons license and indicate if changes were made.

The images or other third party material in this chapter are included in the chapter's Creative Commons license, unless indicated otherwise in a credit line to the material. If material is not included in the chapter's Creative Commons license and your intended use is not permitted by statutory regulation or exceeds the permitted use, you will need to obtain permission directly from the copyright holder.

References

1. Martin Kaltschmitt, Wolfgang Streicher, and Andreas Wiese. *Erneuerbare Energien*. Berlin, Heidelberg: Springer Berlin Heidelberg, 2020. ISBN: 978-3-662-61189-0. https://doi.org/10.1007/978-3-662-61190-6.
2. Michael Sterner and Ingo Stadler. *Handbook of Energy Storage*. Berlin, Heidelberg: Springer Berlin Heidelberg, 2019. ISBN: 978-3-662-55503-3. https://doi.org/10.1007/978-3-662-55504-0.
3. Achim Kampker and Heiner Hans Heimes. *Elektromobilität*. Berlin, Heidelberg: Springer Berlin Heidelberg, 2024. ISBN: 978-3-662-65811-6. https://doi.org/10.1007/978-3-662-65812-3.
4. Nicolas Glaesmann. *Wärmepumpenheizungen*. Wiesbaden: Springer Fachmedien Wiesbaden, 2022. ISBN: 978-3-658-39030-3. https://doi.org/10.1007/978-3-658-39031-0.
5. Dierk Schröder and Rainer Marquardt. *Leistungselektronische Schaltungen*. Berlin, Heidelberg: Springer Berlin Heidelberg, 2019. ISBN: 978-3-662-55324-4. https://doi.org/10.1007/978-3-662-55325-1.
6. Sudipta Chakraborty, Marcelo G. Simões, and William E. Kramer. *Power Electronics for Renewable and Distributed Energy Systems*. London: Springer London, 2013. ISBN: 978-1-4471-5103-6. https://doi.org/10.1007/978-1-4471-5104-3.
7. S. B. Kjaer, J. K. Pedersen, and F. Blaabjerg. "A Review of Single-Phase Grid-Connected Inverters for Photovoltaic Modules". In: *IEEE Transactions on Industry Applications* 41.5 (2005), pp. 1292–1306. ISSN: 0093-9994. https://doi.org/10.1109/TIA.2005.853371.
8. Bidyadhar Subudhi and Raseswari Pradhan. "A Comparative Study on Maximum Power Point Tracking Techniques for Photovoltaic Power Systems". In: *IEEE Transactions on Sustainable Energy* 4.1 (2013), pp. 89–98. ISSN: 1949-3029. https://doi.org/10.1109/TSTE.2012.2202294.
9. Frede Blaabjerg and Ke Ma. "Future on Power Electronics for Wind Turbine Systems". In: *IEEE Journal of Emerging and Selected Topics in Power Electronics* 1.3 (2013), pp. 139–152. ISSN: 2168–6777. https://doi.org/10.1109/JESTPE.2013.2275978.
10. Thomas Holzer and Annette Muetze. "A comparative survey of power converter topologies for full-size converter operation of medium-voltage hydropower generators". In: *e & i Elektrotechnik und Informationstechnik* 136.6 (2019), pp. 263–270. ISSN: 0932-383X. https://doi.org/10.1007/s00502-019-00731-6.

11. Thomas Holzer et al. "Generator Design Possibilities for Full-Size Converter Operation of Large Pumped Storage Power Plants". In: *IEEE Transactions on Industry Applications* (2020), p. 1. ISSN: 0093-9994. https://doi.org/10.1109/TIA.2020.2989074.
12. Alexandre Christe, Alexander Faulstich, Michail Vasiladiotis, and Peter Steinmann. "World's First Fully Rated Direct ac/ac MMC for Variable-Speed Pumped-Storage Hydropower Plants". In: *IEEE Transactions on Industrial Electronics* 70.7 (2023), pp. 6898–6907. ISSN: 0278-0046. https://doi.org/10.1109/TIE.2022.3204858.
13. Deepak Ravi, Shimi Sudha Letha, Paulson Samuel, and Bandi Mallikarjuna Reddy. "An Overview of Various DC-DC Converter Techniques used for Fuel Cell based Applications". In: *2018 International Conference on Power Energy, Environment and Intelligent Control (PEEIC)*. IEEE, 2018, pp. 16–21. ISBN: 978-1-5386-2341-1. https://doi.org/10.1109/PEEIC.2018.8665465.
14. Mengxing Chen, Shih-Feng Chou, Frede Blaabjerg, and Pooya Davari. "Overview of Power Electronic Converter Topologies Enabling Large-Scale Hydrogen Production via Water Electrolysis". In: *Applied Sciences* 12.4 (2022), p. 1906. https://doi.org/10.3390/app12041906.
15. Hao Tu, Hao Feng, Srdjan Srdic, and Srdjan Lukic. "Extreme Fast Charging of Electric Vehicles: A Technology Overview". In: *IEEE Transactions on Transportation Electrification* 5.4 (2019), pp. 861–878. https://doi.org/10.1109/TTE.2019.2958709.
16. Xu She, Alex Q. Huang, Oscar Lucia, and Burak Ozpineci. "Review of Silicon Carbide Power Devices and Their Applications". In: *IEEE Transactions on Industrial Electronics* 64.10 (2017), pp. 8193–8205. ISSN: 0278-0046. https://doi.org/10.1109/TIE.2017.2652401.
17. Xuan Li et al. "Understanding switching losses in SiC MOSFET: Toward lossless switching". In: *2015 IEEE 3rd Workshop on Wide Bandgap Power Devices and Applications (WiPDA)*. IEEE, 2015, pp. 257–262. ISBN: 978-1-4673-7885-7. https://doi.org/10.1109/WiPDA.2015.7369295.
18. Jih-Sheng Lai and Fang Zheng Peng. "Multilevel converters-a new breed of power converters". In: *IEEE Transactions on Industry Applications* 32.3 (1996), pp. 509–517. ISSN: 0093-9994. https://doi.org/10.1109/28.502161.
19. J. Rodriguez, Jih-Sheng Lai, and Fang Zheng Peng. "Multilevel inverters: a survey of topologies, controls, and applications". In: *IEEE Transactions on Industrial Electronics* 49.4 (2002), pp. 724–738. ISSN: 0278-0046. https://doi.org/10.1109/TIE.2002.801052.
20. A. Lesnicar and R. Marquardt. "An innovative modular multilevel converter topology suitable for a wide power range". In: *2003 IEEE Bologna Power Tech Conference Proceedings*. IEEE, 2003, pp. 272–277. ISBN: 0-7803-7967-5. https://doi.org/10.1109/PTC.2003.1304403.
21. Hirofumi Akagi. "Classification, Terminology, and Application of the Modular Multilevel Cascade Converter (MMCC)". In: *IEEE Transactions on Power Electronics* 26.11 (2011), pp. 3119–3130. ISSN: 0885-8993. https://doi.org/10.1109/TPEL.2011.2143431.
22. Markus Schröder. "Multidirektionale Energieflusssteuerung eines Modularen Multilevel-Umrichters mit integrierten Batteriespeichern in elektrischen Netzen". PhD thesis. FAU University Press, 2019. https://doi.org/10.25593/978-3-96147-199-7.
23. Markus Schroeder and Johann Jaeger. "Advanced Energy Flow Control Concept of an MMC for Unrestricted Operation as a Multiport Device". In: *IEEE Transactions on*

Power Electronics 34.11 (2019), pp. 11496–11512. ISSN: 0885-8993. https://doi.org/10.1109/TPEL.2019.2902098.
24. Felix Kammerer, Johannes Kolb, and Michael Braun. "Fully decoupled current control and energy balancing of the Modular Multilevel Matrix Converter". In: *2012 15th International Power Electronics and Motion Control Conference (EPE/PEMC)*. IEEE, 2012, LS2a.3–1–LS2a.3–8. ISBN: 978-1-4673-1972-0. https://doi.org/10.1109/EPEPEMC.2012.6397408.
25. Fei Zhang, Geza Joos, and Wei Li. "A multiport modular multilevel DC-DC converter". In: *2016 IEEE 7th International Symposium on Power Electronics for Distributed Generation Systems (PEDG)*. IEEE, 2016, pp. 1–7. ISBN: 978-1-4673-8617-3. https://doi.org/10.1109/PEDG.2016.7527061.
26. J. S. Siva Prasad and G. Narayanan. "Minimization of Grid Current Distortion in Parallel-Connected Converters Through Carrier Interleaving". In: *IEEE Transactions on Industrial Electronics* 61.1 (2014), pp. 76–91. ISSN: 0278-0046. https://doi.org/10.1109/TIE.2013.2245620.
27. Levy Ferreira Costa, Giovanni de Carne, Giampaolo Buticchi, and Marco Liserre. "The Smart Transformer: A solid-state transformer tailored to provide ancillary services to the distribution grid". In: *IEEE Power Electronics Magazine* 4.2 (2017), pp. 56–67. ISSN: 2329-9207. https://doi.org/10.1109/MPEL.2017.2692381.
28. R. W. de Doncker, D. M. Divan, and M. H. Kheraluwala. "A three-phase soft-switched high power density DC/DC converter for high power applications". In: *Conference Record of the 1988 IEEE Industry Applications Society Annual Meeting*. IEEE, 1988, pp. 796–805. https://doi.org/10.1109/IAS.1988.25153.
29. Biao Zhao, Qiang Song, Wenhua Liu, and Yandong Sun. "Overview of Dual-Active-Bridge Isolated Bidirectional DC-DC Converter for High-Frequency-Link Power-Conversion System". In: *IEEE Transactions on Power Electronics* 29.8 (2014), pp. 4091–4106. ISSN: 0885-8993. https://doi.org/10.1109/TPEL.2013.2289913.
30. Julia Matevosyan et al. "A Future With Inverter-Based Resources: Finding Strength From Traditional Weakness". In: *IEEE Power and Energy Magazine* 19.6 (2021), pp. 18–28. ISSN: 1540-7977. https://doi.org/10.1109/MPE.2021.3104075.
31. Deepak Ramasubramanian, Evangelos Farantatos, Saleh Ziaeinejad, and Ali Mehrizi-Sani. "Operation paradigm of an all converter interfaced generation bulk power system". In: *IET Generation, Transmission & Distribution* 12.19 (2018), pp. 4240–4248. ISSN: 1751-8695. https://doi.org/10.1049/iet-gtd.2018.5179.
32. Johann Jäger, Christian Romeis, and Edmond Petrossian. *Duale Netzplanung*. Wiesbaden: Springer Fachmedien Wiesbaden, 2016. ISBN: 978-3-658-12729-9. https://doi.org/10.1007/978-3-658-12730-5.
33. F. Bignucolo et al. "The Voltage Control on MV Distribution Networks with Aggregated DG Units (VPP)". In: *Proceedings of the 41st International Universities Power Engineering Conference*. IEEE, 2006, pp. 187–192. ISBN: 978-186135-342-9. https://doi.org/10.1109/UPEC.2006.367741.
34. N. Ruiz, I. Cobelo, and J. Oyarzabal. "A Direct Load Control Model for Virtual Power Plant Management". In: *IEEE Transactions on Power Systems* 24.2 (2009), pp. 959–966. ISSN: 0885-8950. https://doi.org/10.1109/TPWRS.2009.2016607.
35. VDE Verband der Elektrotechnik, Elektronik und Informationstechnik e.V. "Der zellulare Ansatz: Grundlage einer erfolgreichen, regionenübergreifenden Energiewende". In: (2015).

36. VDE Verband der Elektrotechnik, Elektronik und Informationstechnik e.V. "Zellulares Energiesystem: Ein Beitrag zur Konkretisierung des zellularen Ansatzes mit Handlungsempfehlungen". In: (2019).
37. Björn Uhlemeyer et al. "Cellular approach as a principle in integrated energy system planning and operation". In: *CIRED—Open Access Proceedings Journal* 2020.1 (2020), pp. 58–61. https://doi.org/10.1049/oap-cired.2021.0021.
38. Dirk van Hertem, Mehrdad Ghandhari, and Marko Delimar. "Technical limitations towards a SuperGrid—A European prospective". In: *2010 IEEE International Energy Conference*. IEEE, 2010, pp. 302–309. ISBN: 978-1-4244-9378-4. https://doi.org/10.1109/ENERGYCON.2010.5771696.
39. Dushan Boroyevich et al. "Future electronic power distribution systems a contemplative view". In: *2010 12th International Conference on Optimization of Electrical and Electronic Equipment*. IEEE, 2010, pp. 1369–1380. ISBN: 978-1-4244-7019-8. https://doi.org/10.1109/OPTIM.2010.5510477.
40. Wencong Su. *The Energy Internet: An Open Energy Platform to Transform Legacy Power Systems into Open Innovation and Global Economic Engines*. Woodhead Publishing Series in Energy Ser. San Diego: Elsevier Science & Technology, 2019. ISBN: 978-0-08-102215-3. URL: https://ebookcentral.proquest.com/lib/kxp/detail.action?docID=5568986.
41. Gerhard Ziegler. *Digitaler Distanzschutz: Grundlagen und Anwendung*. 2. Aufl. Erlangen: Publicis Corp. Publ, 2008. ISBN: 978-3-89578-320-3.
42. Jan Machowski, Janusz W. Bialek, and James R. Bumby. *Power system dynamics: Stability and control*. 2. ed. Chichester: Wiley, 2008. ISBN: 978-0-470-72558-0.
43. F. Blaabjerg, R. Teodorescu, M. Liserre, and A. V. Timbus. "Overview of Control and Grid Synchronization for Distributed Power Generation Systems". In: *IEEE Transactions on Industrial Electronics* 53.5 (2006), pp. 1398–1409. ISSN: 0278-0046. https://doi.org/10.1109/TIE.2006.881997.
44. A. Timbus et al. "Evaluation of Current Controllers for Distributed Power Generation Systems". In: *IEEE Transactions on Power Electronics* 24.3 (2009), pp. 654–664. ISSN: 0885-8993. https://doi.org/10.1109/TPEL.2009.2012527.
45. Julia Matevosyan et al. "Grid-Forming Inverters: Are They the Key for High Renewable Penetration?" In: *IEEE Power and Energy Magazine* 17.6 (2019), pp. 89–98. ISSN: 1540-7977. https://doi.org/10.1109/MPE.2019.2933072.
46. Joan Rocabert, Alvaro Luna, Frede Blaabjerg, and Pedro Rodríguez. "Control of Power Converters in AC Microgrids". In: *IEEE Transactions on Power Electronics* 27.11 (2012), pp. 4734–4749. ISSN: 0885-8993. https://doi.org/10.1109/TPEL.2012.2199334.
47. VDE Verband der Elektrotechnik, Elektronik und Informationstechnik e.V. "VDE-AR-N 4110 (VDE-AR-N 4110) Anwendungsregel: 2018-11: Technische Regeln für den Anschluss von Kundenanlagen an das Mittelspannungsnetz und deren Betrieb (TAR Mittelspannung)". In: (2018).
48. Qing-Chang Zhong and George Weiss. "Synchronverters: Inverters That Mimic Synchronous Generators". In: *IEEE Transactions on Industrial Electronics* 58.4 (2011), pp. 1259–1267. ISSN: 0278-0046. https://doi.org/10.1109/TIE.2010.2048839.
49. Salvatore D'Arco and Jon Are Suul. "Virtual synchronous machines—Classification of implementations and analysis of equivalence to droop controllers for microgrids". In: *2013 IEEE Grenoble Conference*. IEEE, 2013, pp. 1–7. ISBN: 978-1-4673-5669-5. https://doi.org/10.1109/PTC.2013.6652456.

50. Dayan B. Rathnayake et al. "Grid Forming Inverter Modeling, Control, and Applications". In: *IEEE Access* 9 (2021), pp. 114781–114807. https://doi.org/10.1109/ACCESS.2021.3104617.
51. Roberto Rosso et al. "Grid-Forming Converters: Control Approaches, Grid-Synchronization, and Future Trends—A Review". In: *IEEE Open Journal of Industry Applications* 2 (2021), pp. 93–109. https://doi.org/10.1109/OJIA.2021.3074028.
52. Felix Rojas et al. "An Overview of Four-Leg Converters: Topologies, Modulations, Control and Applications". In: *IEEE Access* 10 (2022), pp. 61277–61325. https://doi.org/10.1109/ACCESS.2022.3180746.
53. Jundi Jia, Guangya Yang, Arne Hejde Nielsen, and Peter Roenne-Hansen. "Hardware-in-the–loop tests on distance protection considering VSC fault–ride–through control strategies". In: *The Journal of Engineering* 2018.15 (2018), pp. 824–829. ISSN: 2051-3305. https://doi.org/10.1049/joe.2018.0248.
54. Jundi Jia, Guangya Yang, Arne Hejde Nielsen, and Peter Ronne-Hansen. "Impact of VSC Control Strategies and Incorporation of Synchronous Condensers on Distance Protection Under Unbalanced Faults". In: *IEEE Transactions on Industrial Electronics* 66.2 (2019), pp. 1108–1118. ISSN: 0278-0046. https://doi.org/10.1109/TIE.2018.2835389.
55. Subhadeep Paladhi and Ashok Kumar Pradhan. "Adaptive Distance Protection for Lines Connecting Converter-Interfaced Renewable Plants". In: *IEEE Journal of Emerging and Selected Topics in Power Electronics* 9.6 (2021), pp. 7088–7098. ISSN: 2168-6777. https://doi.org/10.1109/JESTPE.2020.3000276.
56. Mohammad Meraj Alam, Helder Leite, Jun Liang, and Adriano Da Silva Carvalho. "Effects of VSC based HVDC system on distance protection of transmission lines". In: *International Journal of Electrical Power&Energy Systems* 92 (2017), pp. 245–260. ISSN: 01420615. https://doi.org/10.1016/j.ijepes.2017.04.012.
57. Ali Hooshyar, Maher A. Azzouz, and Ehab F. El-Saadany. "Distance Protection of Lines Emanating From Full-Scale Converter-Interfaced Renewable Energy Power Plants—Part I: Problem Statement". In: *IEEE Transactions on Power Delivery* 30.4 (2015), pp. 1770–1780. ISSN: 0885-8977. https://doi.org/10.1109/TPWRD.2014.2369479.
58. Michael Kleemann and Nathan Baeckeland. "Converter Meets Distance Protection: A Good Match?" In: *2021 23rd European Conference on Power Electronics and Applications (EPE'21 ECCE Europe)*. IEEE, 2021, pp. 1–10. ISBN: 978-9-0758-1537-5. https://doi.org/10.23919/EPE21ECCEEurope50061.2021.9570424.
59. Lina He, Chen-Ching Liu, Andrea Pitto, and Diego Cirio. "Distance Protection of AC Grid With HVDC-Connected Offshore Wind Generators". In: *IEEE Transactions on Power Delivery* 29.2 (2014), pp. 493–501. ISSN: 0885-8977. https://doi.org/10.1109/TPWRD.2013.2271761.
60. M. Mohan and K. Panduranga Vittal. "Performance Evaluation of Distance Relay in the Presence of Voltage Source Converters-Based HVDC Systems". In: *Journal of Electrical Engineering & Technology* 14.1 (2019), pp. 69–83. ISSN: 1975-0102. https://doi.org/10.1007/s42835-018-00026-4.
61. Amin Banaiemoqadam, Ali Hooshyar, and Maher A. Azzouz. "A Control-Based Solution for Distance Protection of Lines Connected to Converter-Interfaced Sources During Asymmetrical Faults". In: *IEEE Transactions on Power Delivery* 35.3 (2020), pp. 1455–1466. ISSN: 0885-8977. https://doi.org/10.1109/TPWRD.2019.2946757.
62. Neethu George, O. D. Naidu, and Ashok Kumar Pradhan. "Distance Protection for Lines Connecting Converter Interfaced Renewable Power Plants: Adaptive to Grid-

end Structural Changes". In: *IEEE Transactions on Power Delivery* 38.3 (2023), pp. 2011–2021. ISSN: 0885-8977. https://doi.org/10.1109/TPWRD.2022.3231403.
63. Subhadeep Paladhi and Ashok Kumar Pradhan. "Adaptive Zone-1 Setting Following Structural and Operational Changes in Power System". In: *IEEE Transactions on Power Delivery* 33.2 (2018), pp. 560–569. ISSN: 0885-8977. https://doi.org/10.1109/TPWRD.2017.2728682.
64. Akira Nabae, Isao Takahashi, and Hirofumi Akagi. "A New Neutral-Point-Clamped PWM Inverter". In: *IEEE Transactions on Industry Applications* IA-17.5 (1981), pp. 518–523. ISSN: 0093-9994. https://doi.org/10.1109/TIA.1981.4503992.
65. Jose I. Leon et al. "The Essential Role and the Continuous Evolution of Modulation Techniques for Voltage-Source Inverters in the Past, Present, and Future Power Electronics". In: *IEEE Transactions on Industrial Electronics* 63.5 (2016), pp. 2688–2701. ISSN: 0278-0046. https://doi.org/10.1109/TIE.2016.2519321.
66. Thomas Morstyn, Branislav Hredzak, and Vassilios G. Agelidis. "Control Strategies for Microgrids With Distributed Energy Storage Systems: An Overview". In: *IEEE Transactions on Smart Grid* 9.4 (2018), pp. 3652–3666. ISSN: 1949-3053. https://doi.org/10.1109/TSG.2016.2637958.
67. Michael Jaworski. "Schutztechnische Verfahren zur Erhöhung der Systemsicherheit beim Netzwiederaufbau". PhD thesis. FAU University Press, 2023. https://doi.org/10.25593/978-3-96147-636-7.
68. Federico Milano et al. "Foundations and Challenges of Low-Inertia Systems (Invited Paper)". In: *2018 Power Systems Computation Conference (PSCC)*. IEEE, 2018, pp. 1–25. ISBN: 978-1-910963-10-4. https://doi.org/10.23919/PSCC.2018.8450880.
69. Josep M. Guerrero et al. "Hierarchical Control of Droop-Controlled AC and DC Microgrids—A General Approach Toward Standardization". In: *IEEE Transactions on Industrial Electronics* 58.1 (2011), pp. 158–172. ISSN: 0278-0046. https://doi.org/10.1109/TIE.2010.2066534.
70. Josep M. Guerrero, Mukul Chandorkar, Tzung-Lin Lee, and Poh Chiang Loh. "Advanced Control Architectures for Intelligent Microgrids—Part I: Decentralized and Hierachical Control". In: *IEEE Transactions on Industrial Electronics* 60.4 (2013), pp. 1254–1262. ISSN: 0278-0046. https://doi.org/10.1109/TIE.2012.2194969.
71. Bernd R. Oswald. *Berechnung von Drehstromnetzen*. Wiesbaden: Springer FachmedienWiesbaden, 2017. ISBN: 978-3-658-14404-3. https://doi.org/10.1007/978-3-658-14405-0.
72. Manfred Albach. *Periodische und nichtperiodische Signalformen*. Vol. 2. Elektrotechnik Grundlagen. München: Pearson Studium, 2005. ISBN: 978-3-8273-7108-9.
73. Lothar Papula. *Mathematik für Ingenieure und Naturwissenschaftler Band 2*. Wiesbaden: Springer Fachmedien Wiesbaden, 2015. ISBN: 978-3-658-07789-1. https://doi.org/10.1007/978-3-658-07790-7.
74. Manfred Albach. *Erfahrungssätze, Bauelemente, Gleichstromschaltungen*. 2., aktualisierte Aufl. Vol. 1. Ing—Elektrotechnik. München: Pearson Studium, 2008. ISBN: 978-3-8273-7341-0.
75. Ilja N. Bronstejn, Konstantin A. Semendjaev, Gerhard Musiol, and Heiner Mühlig. *Taschenbuch der Mathematik*. 7., vollst. überarb. und erg. Aufl. Frankfurt am Main: Deutsch, 2008. ISBN: 978-3-8171-2007-9.
76. Gerhard Herold. *Drehstromsysteme, Leistungen, Wirtschaftlichkeit*. 2., überarb. und erw. Aufl. Vol. 1. Elektrische Energieversorgung / von Gerhard Herold. Wilburgstetten: Schlembach, 2005. ISBN: 3-935340-46-X.

References

77. Remus Teodorescu. *Grid Converters for Photovoltaic and Wind Power Systems*. 2nd ed. Vol. v.29. Wiley—IEEE. Hoboken: John Wiley & Sons, 2010. ISBN: 978-0-470-05751-3. URL: http://gbv.eblib.com/patron/FullRecord.aspx?p=698560.
78. Prof. Jan Lunze. *Regelungstechnik 1*. Berlin, Heidelberg: Springer Berlin Heidelberg, 2020. ISBN: 978-3-662-60745-9. https://doi.org/10.1007/978-3-662-60746-6.
79. Jan Lunze. *Regelungstechnik 2*. Berlin, Heidelberg: Springer Berlin Heidelberg, 2020. ISBN: 978-3-662-60759-6. https://doi.org/10.1007/978-3-662-60760-2.
80. Venkata Yaramasu. *Model Predictive Control of Wind Energy Conversion Systems*. Place of publication not identified and s.l.: John Wiley and Sons Inc, 2016. ISBN: 978-1-118-98858-9. URL: https://search.ebscohost.com/login.aspx?direct=true&scope=site&db=nlebk&db=nlabk&AN=1427460.
81. Otto Föllinger. *Regelungstechnik: Einführung in die Methoden und ihre Anwendung ; [aktualisierter Lehrbuch-Klassiker*. 11., völlig neu bearb. Aufl. Berlin: VDE-Verl., 2013. ISBN: 978-3800732319.
82. Wei Dong, Huanhai Xin, Di Wu, and Linbin Huang. "Small Signal Stability Analysis of Multi-Infeed Power Electronic Systems Based on Grid Strength Assessment". In: *IEEE Transactions on Power Systems* 34.2 (2019), pp. 1393–1403. ISSN: 0885-8950. https://doi.org/10.1109/TPWRS.2018.2875305.
83. Fuyilong Ma, Huanhai Xin, Di Wu, and Yun Liu. *Grid Strength Assessment for 100% Inverter-Based Power Systems via Generalized Short-Circuit Ratio*. 2022. https://doi.org/10.36227/techrxiv.21120517.v2.
84. Jiabing Hu et al. "Small Signal Instability of PLL-Synchronized Type-4 Wind Turbines Connected to High-Impedance AC Grid During LVRT". In: *IEEE Transactions on Energy Conversion* 31.4 (2016), pp. 1676–1687. ISSN: 0885-8969. https://doi.org/10.1109/TEC.2016.2577606.
85. Karel de Brabandere et al. "A Voltage and Frequency Droop Control Method for Parallel Inverters". In: *IEEE Transactions on Power Electronics* 22.4 (2007), pp. 1107–1115. ISSN: 0885-8993. https://doi.org/10.1109/TPEL.2007.900456.
86. Hirofumi Akagi, Edson Hirokazu Watanabe, and Mauricio Aredes. *Instantaneous power theory and applications to power conditioning*. Vol. 10. IEEE Press series on power engineering. Hoboken, NJ: Wiley-Interscience, 2007. ISBN: 978-0-470-10761-4.
87. Jia Liu, Yushi Miura, Hassan Bevrani, and Toshifumi Ise. "A Unified Modeling Method of Virtual Synchronous Generator for Multi-Operation-Mode Analyses". In: *IEEE Journal of Emerging and Selected Topics in Power Electronics* 9.2 (2021), pp. 2394–2409. ISSN: 2168-6777. https://doi.org/10.1109/JESTPE.2020.2970025.
88. Jia Liu, Hassan Bevrani, and Toshifumi Ise. "A Design-Oriented Q—V Response Modeling Approach for Grid-Forming Distributed Generators Considering Different Operation Modes". In: *IEEE Journal of Emerging and Selected Topics in Power Electronics* 10.1 (2022), pp. 387–401. ISSN: 2168-6777. https://doi.org/10.1109/JESTPE.2021.3057517.
89. Nikos Hatziargyriou et al. "Definition and Classification of Power System Stability—Revisited & Extended". In: *IEEE Transactions on Power Systems* 36.4 (2021), pp. 3271–3281. ISSN: 0885-8950. https://doi.org/10.1109/TPWRS.2020.3041774.
90. Zunaib Ali et al. "Three-phase phase-locked loop synchronization algorithms for grid-connected renewable energy systems: A review". In: *Renewable and Sustainable Energy Reviews* 90 (2018), pp. 434–452. ISSN: 13640321. https://doi.org/10.1016/j.rser.2018.03.086.

91. Eliabe Duarte Queiroz, Joao Inacio Yutaka Ota, and Jose Antenor Pomilio. "State-Space Representation Model of Phase-Lock Loop Systems for Stability Analysis of Grid-connected Converters". In: *2021 14th IEEE International Conference on Industry Applications (INDUSCON)*. IEEE, 2021, pp. 387–394. ISBN: 978-1-6654-4118-6. https://doi.org/10.1109/INDUSCON51756.2021.9529609.
92. Steffen Bernet. *Selbstgeführte Stromrichter am Gleichspannungszwischenkreis*. Berlin, Heidelberg: Springer Berlin Heidelberg, 2012. ISBN: 978-3-540-23656-6. https://doi.org/10.1007/978-3-540-68861-7.
93. W. Zhang et al. "Comparison of different power loop controllers for synchronous power controlled grid-interactive converters". In: *2015 IEEE Energy Conversion Congress and Exposition (ECCE)*. IEEE, 2015, pp. 3780–3787. ISBN: 978-1-4673-7151-3. https://doi.org/10.1109/ECCE.2015.7310194.
94. Ritwik Majumder, Arindam Ghosh, Gerard Ledwich, and Firuz Zare. "Angle droop versus frequency droop in a voltage source converter based autonomous microgrid". In: *2009 IEEE Power & Energy Society General Meeting*. IEEE, 2009, pp. 1–8. ISBN: 978-1-4244-4241-6. https://doi.org/10.1109/PES.2009.5275987.
95. Yao Sun et al. "New Perspectives on Droop Control in AC Microgrid". In: *IEEE Transactions on Industrial Electronics* 64.7 (2017), pp. 5741–5745. ISSN: 0278-0046. https://doi.org/10.1109/TIE.2017.2677328.
96. Helmut Ulrich and Hubert Weber. *Laplace-, Fourier- und z-Transformation*. Wiesbaden: Springer Fachmedien Wiesbaden, 2017. ISBN: 978-3-658-03449-8. https://doi.org/10.1007/978-3-658-03450-4.
97. Ebrahim Rokrok et al. "Effect of Using PLL-Based Grid-Forming Control on Active Power Dynamics Under Various SCR". In: *IECON 2019—45th Annual Conference of the IEEE Industrial Electronics Society*. IEEE, 2019, pp. 4799–4804. ISBN: 978-1-7281-4878-6. https://doi.org/10.1109/IECON.2019.8927648.
98. Lennart Harnefors et al. "Generic PLL-Based Grid-Forming Control". In: *IEEE Transactions on Power Electronics* (2021), p. 1. ISSN: 0885-8993. https://doi.org/10.1109/TPEL.2021.3106045.
99. Lennart Harnefors et al. "A Universal Controller for Grid-Connected Voltage-Source Converters". In: *IEEE Journal of Emerging and Selected Topics in Power Electronics* 9.5 (2021), pp. 5761–5770. ISSN: 2168-6777. https://doi.org/10.1109/JESTPE.2020.3039407.
100. Dietrich Oeding and Bernd R. Oswald. *Elektrische Kraftwerke und Netze*. Berlin, Heidelberg: Springer Berlin Heidelberg, 2016. ISBN: 978-3-662-52702-3. https://doi.org/10.1007/978-3-662-52703-0.
101. Jinwei He et al. "An Islanding Microgrid Power Sharing Approach Using Enhanced Virtual Impedance Control Scheme". In: *IEEE Transactions on Power Electronics* 28.11 (2013), pp. 5272–5282. ISSN: 0885-8993. https://doi.org/10.1109/TPEL.2013.2243757.
102. Andrew D. Paquette and Deepak M. Divan. "Virtual Impedance Current Limiting for Inverters in Microgrids With Synchronous Generators". In: *IEEE Transactions on Industry Applications* 51.2 (2015), pp. 1630–1638. ISSN: 0093-9994. https://doi.org/10.1109/TIA.2014.2345877.
103. Markos Papageorgiou, Marion Leibold, and Martin Buss. *Optimierung*. Berlin, Heidelberg: Springer Berlin Heidelberg, 2015. ISBN: 978-3-662-46935-4. https://doi.org/10.1007/978-3-662-46936-1.

104. Liuping Wang. *PID and predictive control of electrical drives and power supplies using MATLAB/Simulink*. Solaris South Tower, Singapore:Wiley-IEEE Press, 2015. ISBN: 978-1-118-33947-3. URL https://search.ebscohost.com/login.aspx?direct=true&scope=site&db=nlebk&db=nlabk&AN=931662.
105. Stanley H. Horowitz and Arun G. Phadke. *Power system relaying*. 3. ed., reprinted. Chichester: Wiley, 2008. ISBN: 978-0-470-05712-4.
106. Qing-Chang Zhong and Tomas Hornik. *Control of Power Inverters in Renewable Energy and Smart Grid Integration*. 1. Aufl. Vol. v.97. IEEE Press Ser. s.l.: Wiley-IEEE Press, 2012. ISBN: 978-0-470-66709-5. URL https://ebookcentral.proquest.com/lib/kxp/detail.action?docID=1118505.
107. H. Kakigano, Y. Miura, T. Ise, and R. Uchida. "DC Voltage Control of the DC Micro-grid for Super High Quality Distribution". In: *2007 Power Conversion Conference—Nagoya*. IEEE, 2007, pp. 518–525. ISBN: 1-4244-0843-1. https://doi.org/10.1109/PCCON.2007.373016.
108. Jenni Rekola and Heikki Tuusa. "Comparison of line and load converter topologies in a bipolar LVDC distribution". In: *Proceedings of the 2011 14th European Conference on Power Electronics and Applications* (2011).
109. Surin Khomfoi and Leon M. Tolbert. "Multilevel Power Converters". In: *Power Electronics Handbook*. Elsevier, 2011, pp. 455–486. ISBN: 9780123820365. https://doi.org/10.1016/B978-0-12-382036-5.00017-3.
110. Christina M. DiMarino et al. "10-kV SiC MOSFET Power Module With Reduced Common-Mode Noise and Electric Field". In: *IEEE Transactions on Power Electronics* 35.6 (2020), pp. 6050–6060. ISSN: 0885-8993. https://doi.org/10.1109/TPEL.2019.2952633.
111. L. M. Tolbert, Fang Zheng Peng, and T. G. Habetler. "Multilevel converters for large electric drives". In: *IEEE Transactions on Industry Applications* 35.1 (1999), pp. 36–44. ISSN: 0093-9994. https://doi.org/10.1109/28.740843.
112. Dirk Kranzer et al. "Development of a 10 kV three-phase transformerless inverter with 15 kV Silicon Carbide MOSFETs for grid stabilization and active filtering of harmonics". In: *2017 19th European Conference on Power Electronics and Applications (EPE'17 ECCE Europe)*. IEEE, 2017, P.1–P.8. ISBN: 978-90-75815-27-6. https://doi.org/10.23919/EPE17ECCEEurope.2017.8099222.
113. Jürgen Thoma et al. "Design and Commissioning of a 10 kV Three Phase Transformerless Inverter with 15 kV Silicon Carbide MOSFETs". In: *2018 20th European Conference on Power Electronics and Applications (EPE'18 ECCE Europe)* (2018).
114. Sachin Madhusoodhanan, Subhashish Bhattacharya, and Kamalesh Hatua. "A unified control scheme for harmonic elimination in the front end converter of a 13.8 kV, 100 kVA transformerless intelligent power substation grid tied with LCL filter". In: *2014 IEEE Applied Power Electronics Conference and Exposition—APEC 2014*. IEEE, 2014, pp. 964–971. ISBN: 978-1-4799-2325-0. https://doi.org/10.1109/APEC.2014.6803424.
115. U. A. Miranda, L.G.B. Rolim, and M. Aredes. "A DQ Synchronous Reference Frame Current Control for Single-Phase Converters". In: *IEEE 36th Conference on Power Electronics Specialists, 2005*. IEEE, 2005, pp. 1377–1381. ISBN: 0-7803-9033-4. https://doi.org/10.1109/PESC.2005.1581809.
116. Klaus Heuck, Klaus-Dieter Dettmann, and Detlef Schulz. *Elektrische Energieversorgung*. Wiesbaden: Springer Fachmedien Wiesbaden, 2013. ISBN: 978-3-8348-1699-3. https://doi.org/10.1007/978-3-8348-2174-4.

117. Chihchiang Hua, Jongrong Lin, and Chihming Shen. "Implementation of a DSP-controlled photovoltaic system with peak power tracking". In: *IEEE Transactions on Industrial Electronics* 45.1 (1998), pp. 99–107. ISSN: 0278-0046. https://doi.org/10.1109/41.661310.
118. Marcelo G. Villalva and Ernesto Ruppert F. "Analysis and simulation of the P&O MPPT algorithm using a linearized PV array model". In: *2009 35th Annual Conference of IEEE Industrial Electronics*. IEEE, 2009, pp. 231–236. ISBN: 978-1-4244-4648-3. https://doi.org/10.1109/IECON.2009.5414780.
119. Felipe Donoso et al. "Finite-Set Model-Predictive Control Strategies for a 3L-NPC Inverter Operating With Fixed Switching Frequency". In: *IEEE Transactions on Industrial Electronics* 65.5 (2018), pp. 3954–3965. ISSN: 0278-0046. https://doi.org/10.1109/TIE.2017.2760840.
120. Jordan P. Zucuni et al. "Cost Function Design for Stability Assessment of Modulated Model Predictive Control". In: *2020 22nd European Conference on Power Electronics and Applications (EPE'20 ECCE Europe)*. IEEE, 2020, P.1–P.9. ISBN: 978-9-0758-1536-8. https://doi.org/10.23919/EPE20ECCEEurope43536.2020.9215797.
121. Tobias Geyer. *Model predictive control of high power converters and industrial drives*. First edition. Chichester, West Sussex, United Kingdom: JohnWiley & Sons Incorporated, 2017. ISBN: 9781119010869. URL: https://ebookcentral.proquest.com/lib/kxp/detail.action?docID=4688951.
122. Bo Fan et al. "A Review of Current-Limiting Control of Grid-Forming Inverters Under Symmetrical Disturbances". In: *IEEE Open Journal of Power Electronics* 3 (2022), pp. 955–969. https://doi.org/10.1109/OJPEL.2022.3227507.
123. Gerhard Ziegler. *Digitaler Differentialschutz: Grundlagen und Anwendung*. 2., überarb. und erw. Aufl. Erlangen: Publicis Publishing, 2013. ISBN: 978-3895784163.
124. A. SakisP. Meliopoulos et al. "Setting-Less Protection: Feasibility Study". In: *2013 46th Hawaii International Conference on System Sciences*. IEEE, 2013, pp. 2345–2353. ISBN: 978-1-4673-5933-7. https://doi.org/10.1109/HICSS.2013.628.
125. A. P. Sakis Meliopoulos et al. "Setting-less Protection: Laboratory Testing: Final Project Report". In: *PSERC Publication 14-03* (2014).
126. A. P. Sakis Meliopoulos et al. "Dynamic State Estimation-Based Protection: Status and Promise". In: *IEEE Transactions on Power Delivery* 32.1 (2017), pp. 320–330. ISSN: 0885-8977. https://doi.org/10.1109/TPWRD.2016.2613411.
127. Sakis Meliopoulos et al. "Dynamic Estimation-Based Protection and Hidden Failure Detection and Identification: Inverter-Dominated Power Systems". In: *IEEE Power and Energy Magazine* 21.1 (2023), pp. 59–72. ISSN: 1540-7977. https://doi.org/10.1109/MPE.2022.3219180.
128. F. Schweppe and R. Masiello. "A Tracking Static State Estimator". In: *IEEE Transactions on Power Apparatus and Systems* PAS-90.3 (1971), pp. 1025–1033. ISSN: 0018-9510. https://doi.org/10.1109/TPAS.1971.292844.
129. David G. Luenberger. "Observing the State of a Linear System". In: *IEEE Transactions on Military Electronics* 8.2 (1964), pp. 74–80. ISSN: 0536-1559. https://doi.org/10.1109/TME.1964.4323124.
130. D. Luenberger. "Observers for multivariable systems". In: *IEEE Transactions on Automatic Control* 11.2 (1966), pp. 190–197. ISSN: 0018-9286. https://doi.org/10.1109/TAC.1966.1098323.

131. D. Luenberger. "An introduction to observers". In: *IEEE Transactions on Automatic Control* 16.6 (1971), pp. 596–602. ISSN: 0018-9286. https://doi.org/10.1109/TAC.1971.1099826.
132. R. E. Kalman. "A New Approach to Linear Filtering and Prediction Problems". In: *Journal of Basic Engineering* 82.1 (1960), pp. 35–45. ISSN: 0021-9223. https://doi.org/10.1115/1.3662552.
133. Reiner Marchthaler and Sebastian Dingler. *Kalman-Filter*. Wiesbaden: Springer Fachmedien Wiesbaden, 2017. ISBN: 978-3-658-16727-1. https://doi.org/10.1007/978-3-658-16728-8.
134. Gerald L. Smith, Stanley F. Schmidt, and McGee Leonard A. "Application of statistical filter theory to the optimal estimation of position and velocity on board a circumlunar vehicle". In: *Technical Report TR R-135* (1962).
135. M. Athans, R. Wishner, and A. Bertolini. "Suboptimal state estimation for continuous-time nonlinear systems from discrete noisy measurements". In: *IEEE Transactions on Automatic Control* 13.5 (1968), pp. 504–514. ISSN: 0018-9286. https://doi.org/10.1109/TAC.1968.1098986.
136. Fredrik Gustafsson and Gustaf Hendeby. "Some Relations Between Extended and Unscented Kalman Filters". In: *IEEE Transactions on Signal Processing* 60.2 (2012), pp. 545–555. ISSN: 1053-587X. https://doi.org/10.1109/TSP.2011.2172431.
137. Norbert Henze. *Stochastik für Einsteiger: Eine Einführung in die faszinierende Welt des Zufalls*. 9., erweiterte Auflage 2012. Für Einsteiger. Wiesbaden: Vieweg+Teubner Verlag / GWV Fachverlage GmbH Wiesbaden, 2011. ISBN: 978-3-8348-1845-4. https://doi.org/10.1007/978-3-8348-8649-1.
138. Mohamed E. El-Hawary. *Advances in Electric Power and Energy*. Wiley, 2020. ISBN: 9781119480464. https://doi.org/10.1002/9781119480402.
139. Mohamed Esreraig and Joydeep Mitra. "An observer-based protection system for microgrids". In: *2011 IEEE Power and Energy Society General Meeting*. IEEE, 2011, pp. 1–7. ISBN: 978-1-4577-1000-1. https://doi.org/10.1109/PES.2011.6039818.
140. Mohamed Esreraig and Joydeep Mitra. "Microgrid protection using system observer and minimum measurement set". In: *International Transactions on Electrical Energy Systems* 25.4 (2015), pp. 607–622. ISSN: 20507038. https://doi.org/10.1002/etep.1849.
141. Rainer Cortain and Hans-Jürgen Koglin. "Demonstrationsmodell für Netzund Kraftwerkstechnik". In: *Sonderdruck aus der Siemens-Zeitschrift* 43.3 (1969), pp. 173–180.
142. OMICRON. "DANEO 400: Hybrid Signal Analyzer for Power Utility Automation Systems". In: (2022).
143. Florian Mahr, Stefan Henninger, Martin Biller, and Johann Jäger. *Elektrische Energiesysteme*. Wiesbaden: Springer Fachmedien Wiesbaden, 2021. ISBN: 978-3-658-34907-3. https://doi.org/10.1007/978-3-658-34908-0.
144. Markus Schroeder, Florian Mahr, Johann Jaeger, and Stefan Haensel. "Energy balancing in the modular multilevel converter under unbalanced grid conditions". In: *2017 19th European Conference on Power Electronics and Applications (EPE'17 ECCE Europe)*. IEEE, 2017, P.1–P.10. ISBN: 978-90-75815-27-6. https://doi.org/10.23919/EPE17ECCEEurope.2017.8099247.
145. Florian Mahr, Markus Schroeder, and Johann Jaeger. "Model-based Predictive Control of a Modular Multilevel Converter". In: *Center of Knowledge Interchange: Friedrich-Alexander University Erlangen-Nuremberg with Siemens AG (Presentation)* (2016).

146. Florian Mahr, Johann Jaeger, Stefan Henninger, and Hubert Rubenbauer. "Hybrid Energy Storage System for MVDC-Grids". In: *2020 22nd European Conference on Power Electronics and Applications (EPE'20 ECCE Europe)*. IEEE, 2020, P.1–P.10. ISBN: 978-9-0758-1536-8. https://doi.org/10.23919/EPE20ECCEEurope43536.2020.9215622.
147. Florian Mahr, Jakob Schindler, Edmond Petrossian, and Johann Jäger. "Energieversorgung mit Energiepaketen: Universelles Netzanschlusskonzept für Stromkunden". In: *ew-Magazin für die Energiewirtschaft* 12 (2018), pp. 42–47.
148. Florian Mahr, Johann Jaeger, and Edmond Petrossian. "Integrated Electrical and Chemical Energy Infrastructure via Energy Packages". In: *NEIS Conference on Sustainable Energy Supply and Energy Storage Systems* (2019).
149. Florian Mahr and Johann Jaeger. "Integrated Energy Management Concept for Residential Areas using Energy Packages". In: *NEIS Conference on Sustainable Energy Supply and Energy Storage Systems* (2022).
150. Florian Mahr. "Elektrische Energieversorgung mit Energiepaketen—flexibel, effizient und ausfallsicher". In: *EW Fachtagung Blackout: Netzprobleme—Auswirkungen und Lösungen (Presentation)* (2019).
151. Florian Mahr and Johann Jaeger. "Advanced Grid-Forming Control of HVDC Systems for Reliable Grid Restoration". In: *2018 IEEE Power & Energy Society General Meeting (PESGM)*. IEEE, 2018, pp. 1–5. ISBN: 978-1-5386-7703-2. https://doi.org/10.1109/PESGM.2018.8586200.
152. Florian Mahr and Johann Jaeger. "Frequency-Adaptive MPC of Grid-Forming VSC-HVDC Systems with Optimal Voltage Reference Tracking During Grid Restoration". In: *2018 20th European Conference on Power Electronics and Applications (EPE'18 ECCE Europe)* (2018).
153. Florian Mahr and Johann Jäger. "Operating Principles of VSCs in the Zero Sequence and Options for Neutral Point Treatment". In: *VDE ETG journal 01/2023* (2023).
154. Florian Mahr and Johann Jäger. "Aktivierung eines Reserveleiters für einen redundanten Betrieb nach einem einpoligen Fehler mittels Umrichter". In: *VDE ETG/FNN-Tutorial 2020 Schutz- und Leittechnik (Presentation)* (2020).
155. Florian Mahr and Johann Jaeger. "Ground Fault Current Control via Voltage Source Converters using Virtual Impedances in the Zero Sequence". In: *2023 25th European Conference on Power Electronics and Applications (EPE'23 ECCE Europe)*. IEEE, 2023, P.1–P.7. ISBN: 978-9-0758-1541-2. https://doi.org/10.23919/EPE23ECCEEurope58414.2023.10264249.
156. Florian Mahr and Johann Jaeger. "Regelung von Nullsystemgrößen mittels Stromrichtern". In: *VDE ETG/FNN-Tutorial 2022 Schutz- und Leittechnik (Presentation)* (2022).
157. Florian Mahr. "Erdschlusskompensation mittels Stromrichtern nach dem Prinzip der virtuellen Impedanz". In: *7. ETG Fachtagung Sternpunktbehandlung in Netzen bis 110 kV (D-A-CH) STE 2022 (Presentation)* (2022).
158. Florian Mahr and Johann Jaeger. "Advanced Fault Ride Through Operation of Grid-Forming Voltage Source Converters Using a Model Predictive Controller". In: *2021 IEEE 12th Energy Conversion Congress & Exposition—Asia (ECCE-Asia)*. IEEE, 2021, pp. 2248–2255. ISBN: 978-1-7281-6344-4. https://doi.org/10.1109/ECCE-Asia49820.2021.9479335.
159. Florian Mahr and Johann Jäger. "Modellbasierter Schutzalgorithmus für 100% stromrichterdominierte Netze". In: *VDE ETG/FNN-Tutorial 2024 Schutz- und Leittechnik (Presentation)* (2024).

The manufacturer's authorised representative in the EU is Springer Nature Customer Service Centre GmbH, Europaplatz 3, 69115 Heidelberg, Germany. If you have any concerns regarding our products, please contact ProductSafety@springernature.com

Printed and bound by CPI Group (UK) Ltd, Croydon, CR0 4YY
26/03/2026
02078969-0002